LINEAR MIXED MODELS

A Practical Guide Using Statistical Software

SECOND EDITION

LINEAR MIXED MODELS

A Practical Guide Using Statistical Software

SECOND EDITION

Brady T. West
Kathleen B. Welch
Andrzej T. Gałecki

University of Michigan
Ann Arbor, USA

with contributions from Brenda W. Gillespie

CRC Press
Taylor & Francis Group
Boca Raton London New York

CRC Press is an imprint of the
Taylor & Francis Group, an **informa** business

A CHAPMAN & HALL BOOK

First edition published in 2006.

CRC Press
Taylor & Francis Group
6000 Broken Sound Parkway NW, Suite 300
Boca Raton, FL 33487-2742

© 2015 by Taylor & Francis Group, LLC
CRC Press is an imprint of Taylor & Francis Group, an Informa business

No claim to original U.S. Government works

Printed on acid-free paper
Version Date: 20140326

International Standard Book Number-13: 978-1-4665-6099-4 (Hardback)

Library of Congress Cataloging-in-Publication Data

West, Brady T.
   Linear mixed models : a practical guide using statistical software / Brady T. West, Kathleen B. Welch, Andrzej T. Galecki ; with contributions from Brenda W. Gillespie. -- Second edition.
       pages cm
   "A Chapman & Hall Book."
   Includes bibliographical references and index.
   ISBN 978-1-4665-6099-4 (alk. paper)
   1. Linear models (Statistics)--Data processing. I. Welch, Kathleen B. II. Galecki, Andrzej T. III. Gillespie, Brenda W., 1950- IV. Title.

QA279.W47 2014
519.5'35--dc23
                                                     2014010347

**Visit the Taylor & Francis Web site at
http://www.taylorandfrancis.com**

**and the CRC Press Web site at
http://www.crcpress.com**

To Laura and Carter
To all of my mentors, advisors, and teachers, especially my parents and grandparents
—B.T.W.

To Jim, my children, and grandchildren
To the memory of Fremont and June
—K.B.W.

To Viola, my children, and grandchildren
To my teachers and mentors
In memory of my parents
—A.T.G.

Contents

List of Tables

List of Figures

Preface to the Second Edition

Books attempting to serve as practical guides on the use of statistical software are always at risk of becoming outdated as the software continues to develop, especially in an area of statistics and data analysis that has received as much research attention as linear mixed models. In fact, much has changed since the first publication of this book in early 2007, and while we tried to keep pace with these changes on the web site for this book, the demand for a second edition quickly became clear. There were also a number of topics that were only briefly referenced in the first edition, and we wanted to provide more comprehensive discussions of those topics in a new edition. This second edition of *Linear Mixed Models: A Practical Guide Using Statistical Software* aims to update the case studies presented in the first edition using the newest versions of the various software procedures, provide coverage of additional topics in the application of linear mixed models that we believe valuable for data analysts from all fields, and also provide up-to-date information on the options and features of the sofware procedures currently available for fitting linear mixed models in SAS, SPSS, Stata, R/S-plus, and HLM.

Based on feedback from readers of the first edition, we have included coverage of the following topics in this second edition:

- Models with *crossed* random effects, and software procedures capable of fitting these models (see Chapter 8 for a new case study);

- Power analysis methods for longitudinal and clustered study designs, including software options for power analyses and suggested approaches to writing simulations;

- Use of the `lmer()` function in the `lme4` package in R;

- Fitting linear mixed models to complex sample survey data;

- Bayesian approaches to making inferences based on linear mixed models; and

- Updated graphical procedures in the various software packages.

We hope that readers will find the updated coverage of these topics helpful for their research activities.

We have substantially revised the subject index for the book to enable more efficient reading and easier location of material on selected topics or software options. We have also added more practical recommendations based on our experiences using the software throughout each of the chapters presenting analysis examples. New sections discussing overall recommendations can be found at the end of each of these chapters. Finally, we have created an R package named `WWGbook` that contains all of the data sets used in the example chapters.

We will once again strive to keep readers updated on the web site for the book, and also continue to provide working, up-to-date versions of the software code used for all of the analysis examples on the web site. Readers can find the web site at the following address: `http://www.umich.edu/~bwest/almmussp.html`.

Preface to First Edition

The development of software for fitting linear mixed models was propelled by advances in statistical methodology and computing power in the late twentieth century. These developments, while providing applied researchers with new tools, have produced a sometimes confusing array of software choices. At the same time, parallel development of the methodology in different fields has resulted in different names for these models, including mixed models, multilevel models, and hierarchical linear models. This book provides a reference on the use of procedures for fitting linear mixed models available in five popular statistical software packages (SAS, SPSS, Stata, R/S-plus, and HLM). The intended audience includes applied statisticians and researchers who want a basic introduction to the topic and an easy-to-navigate software reference.

Several existing texts provide excellent theoretical treatment of linear mixed models and the analysis of variance components (e.g., McCulloch & Searle, 2001; Searle, Casella, & McCulloch, 1992; Verbeke & Molenberghs, 2000); this book is not intended to be one of them. Rather, we present the primary concepts and notation, and then focus on the software implementation and model interpretation. This book is intended to be a reference for practicing statisticians and applied researchers, and could be used in an advanced undergraduate or introductory graduate course on linear models.

Given the ongoing development and rapid improvements in software for fitting linear mixed models, the specific syntax and available options will likely change as newer versions of the software are released. The most up-to-date versions of selected portions of the syntax associated with the examples in this book, in addition to many of the data sets used in the examples, are available at the following web site: `http://www.umich.edu/~bwest/` `almmussp.html`.

The Authors

Brady T. West is a research assistant professor in the Survey Methodology Program, located within the Survey Research Center at the Institute for Social Research (ISR) on the University of Michigan–Ann Arbor campus. He also serves as a statistical consultant at the Center for Statistical Consultation and Research (CSCAR) on the University of Michigan–Ann Arbor campus. He earned his PhD from the Michigan program in survey methodology in 2011. Before that, he received an MA in applied statistics from the University of Michigan Statistics Department in 2002, being recognized as an Outstanding First-Year Applied Masters student, and a BS in statistics with Highest Honors and Highest Distinction from the University of Michigan Statistics Department in 2001. His current research interests include applications of interviewer observations in survey methodology, the implications of measurement error in auxiliary variables and survey paradata for survey estimation, survey nonresponse, interviewer variance, and multilevel regression models for clustered and longitudinal data. He has developed short courses on statistical analysis using SPSS, R, and Stata, and regularly consults on the use of procedures in SAS, SPSS, R, Stata, and HLM for the analysis of longitudinal and clustered data. He is also a coauthor of a book entitled *Applied Survey Data Analysis* (with Steven Heeringa and Patricia Berglund), which was published by Chapman Hall in April 2010. He lives in Dexter, Michigan with his wife Laura, his son Carter, and his American Cocker Spaniel Bailey.

Kathy Welch is a senior statistician and statistical software consultant at the Center for Statistical Consultation and Research (CSCAR) at the University of Michigan–Ann Arbor. She received a BA in sociology (1969), an MPH in epidemiology and health education (1975), and an MS in biostatistics (1984) from the University of Michigan (UM). She regularly consults on the use of SAS, SPSS, Stata, and HLM for analysis of clustered and longitudinal data, teaches a course on statistical software packages in the University of Michigan Department of Biostatistics, and teaches short courses on SAS software. She has also codeveloped and cotaught short courses on the analysis of linear mixed models and generalized linear models using SAS.

Andrzej Gałecki is a research professor in the Division of Geriatric Medicine, Department of Internal Medicine, and Institute of Gerontology at the University of Michigan Medical School, and has a joint appointment in the Department of Biostatistics at the University of Michigan School of Public Health. He received a MSc in applied mathematics (1977) from the Technical University of Warsaw, Poland, and an MD (1981) from the Medical Academy of Warsaw. In 1985 he earned a PhD in epidemiology from the Institute of Mother and Child Care in Warsaw (Poland). Since 1990, Dr. Galecki has collaborated with researchers in gerontology and geriatrics. His research interests lie in the development and application of statistical methods for analyzing correlated and overdispersed data. He developed the SAS macro NLMEM for nonlinear mixed-effects models, specified as a solution of ordinary differential equations. His research (Galecki, 1994) on a general class of covariance structures for two or more within-subject factors is considered to be one of the very first approaches to the joint modeling of multiple outcomes. Examples of these structures have been implemented in SAS `proc mixed`. He is also a co-author of more than 90 publications.

Brenda Gillespie is the associate director of the Center for Statistical Consultation and Research (CSCAR) at the University of Michigan in Ann Arbor. She received an AB in mathematics (1972) from Earlham College in Richmond, Indiana, an MS in statistics (1975) from The Ohio State University, and earned a PhD in statistics (1989) from Temple University in Philadelphia, Pennsylvania. Dr. Gillespie has collaborated extensively with researchers in health-related fields, and has worked with mixed models as the primary statistician on the Collaborative Initial Glaucoma Treatment Study (CIGTS), the Dialysis Outcomes Practice Pattern Study (DOPPS), the Scientific Registry of Transplant Recipients (SRTR), the University of Michigan Dioxin Study, and at the Complementary and Alternative Medicine Research Center at the University of Michigan.

Acknowledgments

First and foremost, we wish to thank Brenda Gillespie for her vision and the many hours she spent on making the first edition of this book a reality. Her contributions were invaluable.

We sincerely wish to thank Caroline Beunckens at the Universiteit Hasselt in Belgium, who patiently and consistently reviewed our chapters, providing her guidance and insight. We also wish to acknowledge, with sincere appreciation, the careful reading of our text and invaluable suggestions for its improvement provided by Tomasz Burzykowski at the Universiteit Hasselt in Belgium; Oliver Schabenberger at the SAS Institute; Douglas Bates and José Pinheiro, codevelopers of the `lme()` and `gls()` functions in R; Sophia Rabe-Hesketh, developer of the `gllamm` procedure in Stata; Chun-Yi Wu, Shu Chen, and Carrie Disney at the University of Michigan–Ann Arbor; and John Gillespie at the University of Michigan–Dearborn.

We would also like to thank the technical support staff at SAS and SPSS for promptly responding to our inquiries about the mixed modeling procedures in those software packages. We also thank the anonymous reviewers provided by Chapman & Hall/CRC Press for their constructive suggestions on our early draft chapters. The Chapman & Hall/CRC Press staff has consistently provided helpful and speedy feedback in response to our many questions, and we are indebted to Kirsty Stroud for her support of this project in its early stages. We especially thank Rob Calver at Chapman & Hall /CRC Press for his support and enthusiasm for this project, and his deft and thoughtful guidance throughout.

We thank our colleagues from the University of Michigan, especially Myra Kim and Julian Faraway (now at the University of Bath), for their perceptive comments and useful discussions. Our colleagues at the University of Michigan Center for Statistical Consultation and Research (CSCAR) have been wonderful, particularly Ed Rothman, who has provided encouragement and advice. We are very grateful to our clients who have allowed us to use their data sets as examples.

We are also thankful to individuals who have participated in our statistics.com course on mixed-effects modeling over the years, and provided us with feedback on the first edition of this book. In particular, we acknowledge Rickie Domangue from James Madison University, Robert E. Larzelere from the University of Nebraska, and Thomas Trojian from the University of Connecticut. We also gratefully acknowledge support from the Claude Pepper Center Grants AG08808 and AG024824 from the National Institute of Aging.

The transformation of the first edition of this book from Microsoft Word to LaTeX was not an easy one. This would not have been possible without the careful work and attention to detail provided by Alexandra Birg, who is currently a graduate student at Ludwig Maximilians University in Munich, Germany. We are extremely grateful to Alexandra for her extraordinary LaTeX skills and all of her hard work. We would also like to acknowledge the CRC Press / Chapman and Hall typesetting staff for their hard work and careful review of the new edition.

As was the case with the first edition of this book, we are once again especially indebted to our families and loved ones for their unconditional patience and support. It has been a long and sometimes arduous process that has been filled with hours of discussions and many late nights. The time we have spent writing this book has been a period of great learning and has developed a fruitful exchange of ideas that we have all enjoyed.

Brady, Kathy, and Andrzej

1

Introduction

1.1 What Are Linear Mixed Models (LMMs)?

LMMs are statistical models for continuous outcome variables in which the residuals are normally distributed but may not be independent or have constant variance. Study designs leading to data sets that may be appropriately analyzed using LMMs include (1) studies with clustered data, such as students in classrooms, or experimental designs with random blocks, such as batches of raw material for an industrial process, and (2) longitudinal or repeated-measures studies, in which subjects are measured repeatedly over time or under different conditions. These designs arise in a variety of settings throughout the medical, biological, physical, and social sciences. LMMs provide researchers with powerful and flexible analytic tools for these types of data.

Although software capable of fitting LMMs has become widely available in the past three decades, different approaches to model specification across software packages may be confusing for statistical practitioners. The available procedures in the general-purpose statistical software packages SAS, SPSS, R, and Stata take a similar approach to model specification, which we describe as the "general" specification of an LMM. The hierarchical linear model (HLM) software takes a hierarchical approach (Raudenbush & Bryk, 2002), in which an LMM is specified explicitly in multiple levels, corresponding to the levels of a clustered or longitudinal data set. In this book, we illustrate how the same models can be fitted using either of these approaches. We also discuss model specification in detail in Chapter 2 and present explicit specifications of the models fitted in each of our six example chapters (Chapters 3 through 8).

The name *linear mixed models* comes from the fact that these models are linear in the parameters, and that the covariates, or independent variables, may involve a mix of fixed and random effects. **Fixed effects** may be associated with continuous covariates, such as weight, baseline test score, or socioeconomic status, which take on values from a continuous (or sometimes a multivalued ordinal) range, or with factors, such as gender or treatment group, which are categorical. Fixed effects are unknown constant parameters associated with either continuous covariates or the levels of categorical factors in an LMM. Estimation of these parameters in LMMs is generally of intrinsic interest, because they indicate the relationships of the covariates with the continuous outcome variable. Readers familiar with linear regression models but not LMMs specifically may know fixed effects as *regression coefficients*.

When the levels of a categorical factor can be thought of as having been sampled from a sample space, such that each particular level is not of intrinsic interest (e.g., classrooms or clinics that are randomly sampled from a larger population of classrooms or clinics), the effects associated with the levels of those factors can be modeled as **random effects** in an LMM. In contrast to fixed effects, which are represented by constant parameters in an LMM, random effects are represented by (unobserved) random variables, which are usually assumed to follow a normal distribution. We discuss the distinction between fixed and random effects in more detail and give examples of each in Chapter 2.

With this book, we illustrate (1) a heuristic development of LMMs based on both general and hierarchical model specifications, (2) the step-by-step development of the model-building process, and (3) the estimation, testing, and interpretation of both fixed-effect parameters and covariance parameters associated with random effects. We work through examples of analyses of real data sets, using procedures designed specifically for the fitting of LMMs in SAS, SPSS, R, Stata, and HLM. We compare output from fitted models across the software procedures, address the similarities and differences, and give an overview of the options and features available in each procedure.

1.1.1 Models with Random Effects for Clustered Data

Clustered data arise when observations are made on subjects within the same randomly selected group. For example, data might be collected from students within the same classroom, patients in the same clinic, or rat pups in the same litter. These designs involve units of analysis *nested* within clusters. If the clusters can be considered to have been sampled from a larger population of clusters, their effects can be modeled as random effects in an LMM. In a designed experiment with blocking, such as a randomized block design, the blocks are *crossed* with treatments, meaning that each treatment occurs once in each block. Block effects are usually considered to be random. We could also think of blocks as clusters, where treatment is a factor with levels that vary within clusters.

LMMs allow for the inclusion of both individual-level covariates (such as age and sex) and cluster-level covariates (such as cluster size), while adjusting for the random effects associated with each cluster. Although individual cluster-specific coefficients are not explicitly estimated, most LMM software produces cluster-specific "predictions" (EBLUPs, or empirical best linear unbiased predictors) of the random cluster-specific effects. Estimates of the variability of the random effects associated with clusters can then be obtained, and inferences about the variability of these random effects in a greater population of clusters can be made.

We note that traditional approaches to analysis of variance (ANOVA) models with both fixed and random effects used expected mean squares to determine the appropriate denominator for each F-test. Readers who learned mixed models under the *expected mean squares* system will begin the study of LMMs with valuable intuition about model building, although expected mean squares per se are now rarely mentioned.

We examine a **two-level model** with random cluster-specific intercepts for a two-level clustered data set in Chapter 3 (the Rat Pup data). We then consider a **three-level model** for data from a study with students nested within classrooms and classrooms nested within schools in Chapter 4 (the Classroom data).

1.1.2 Models for Longitudinal or Repeated-Measures Data

Longitudinal data arise when multiple observations are made on the same subject or unit of analysis over time. Repeated-measures data may involve measurements made on the same unit over time, or under changing experimental or observational conditions. Measurements made on the same variable for the same subject are likely to be correlated (e.g., measurements of body weight for a given subject will tend to be similar over time). Models fitted to longitudinal or repeated-measures data involve the estimation of covariance parameters to capture this correlation.

The software procedures (e.g., the GLM, or General Linear Model, procedures in SAS and SPSS) that were available for fitting models to longitudinal and repeated-measures

Goodness of fit
nested models
Simplest model / least assumed

data prior to the advent of software for fitting LMMs accommodated only a limited range of models. These traditional repeated-measures ANOVA models assumed a multivariate normal (MVN) distribution of the repeated measures and required either estimation of all covariance parameters of the MVN distribution or an assumption of "sphericity" of the covariance matrix (with corrections such as those proposed by Geisser & Greenhouse (1958) or Huynh & Feldt (1976) to provide approximate adjustments to the test statistics to correct for violations of this assumption). In contrast, LMM software, although assuming the MVN distribution of the repeated measures, allows users to fit models with a broad selection of parsimonious covariance structures, offering greater efficiency than estimating the full variance-covariance structure of the MVN model, and more flexibility than models assuming sphericity. Some of these covariance structures may satisfy sphericity (e.g., independence or compound symmetry), and other structures may not (e.g., autoregressive or various types of heterogeneous covariance structures). The LMM software procedures considered in this book allow varying degrees of flexibility in fitting and testing covariance structures for repeated-measures or longitudinal data.

Software for LMMs has other advantages over software procedures capable of fitting traditional repeated-measures ANOVA models. First, LMM software procedures allow subjects to have missing time points. In contrast, software for traditional repeated-measures ANOVA drops an entire subject from the analysis if the subject has missing data for a single time point, known as *complete-case analysis* (Little & Rubin, 2002). Second, LMMs allow for the inclusion of time-varying covariates in the model (in addition to a covariate representing time), whereas software for traditional repeated-measures ANOVA does not. Finally, LMMs provide tools for the situation in which the trajectory of the outcome varies over time from one subject to another. Examples of such models include **growth curve models**, which can be used to make inference about the variability of growth curves in the larger population of subjects. Growth curve models are examples of **random coefficient models** (or Laird–Ware models), which will be discussed when considering the longitudinal data in Chapter 6 (the Autism data).

In Chapter 5, we consider LMMs for a small repeated-measures data set with two within-subject factors (the Rat Brain data). We consider models for a data set with features of both clustered and longitudinal data in Chapter 7 (the Dental Veneer data). Finally, we consider a unique educational data set with repeated measures on both students and teachers over time in Chapter 8 (the SAT score data), to illustrate the fitting of models with *crossed random effects*.

1.1.3 The Purpose of This Book

This book is designed to help applied researchers and statisticians use LMMs appropriately for their data analysis problems, employing procedures available in the SAS, SPSS, Stata, R, and HLM software packages. It has been our experience that examples are the best teachers when learning about LMMs. By illustrating analyses of real data sets using the different software procedures, we demonstrate the practice of fitting LMMs and highlight the similarities and differences in the software procedures.

We present a heuristic treatment of the basic concepts underlying LMMs in Chapter 2. We believe that a clear understanding of these concepts is fundamental to formulating an appropriate analysis strategy. We assume that readers have a general familiarity with ordinary linear regression and ANOVA models, both of which fall under the heading of general (or standard) linear models. We also assume that readers have a basic working knowledge of matrix algebra, particularly for the presentation in Chapter 2.

Nonlinear mixed models and generalized LMMs (in which the dependent variable may be a binary, ordinal, or count variable) are beyond the scope of this book. For a discussion of nonlinear mixed models, see Davidian & Giltinan (1995), and for references on generalized LMMs, see Diggle et al. (2002) or Molenberghs & Verbeke (2005). We also do not consider spatial correlation structures; for more information on spatial data analysis, see Gregoire et al. (1997). A general overview of current research and practice in multilevel modeling for all types of dependent variables can be found in the recently published (2013) edited volume entitled *The Sage Handbook of Multilevel Modeling*.

This book should not be substituted for the manuals of any of the software packages discussed. Although we present aspects of the LMM procedures available in each of the five software packages, we do not present an exhaustive coverage of all available options.

1.1.4 Outline of Book Contents

Chapter 2 presents the notation and basic concepts behind LMMs and is strongly recommended for readers whose aim is to understand these models. The remaining chapters are dedicated to case studies, illustrating some of the more common types of LMM analyses with real data sets, most of which we have encountered in our work as statistical consultants. Each chapter presenting a case study describes how to perform the analysis using each software procedure, highlighting features in one of the statistical software packages in particular.

In Chapter 3, we begin with an illustration of fitting an LMM to a simple two-level clustered data set and emphasize the SAS software. Chapter 3 presents the most detailed coverage of setting up the analyses in each software procedure; subsequent chapters do not provide as much detail when discussing the syntax and options for each procedure. Chapter 4 introduces models for three-level data sets and illustrates the estimation of variance components associated with nested random effects. We focus on the HLM software in Chapter 4. Chapter 5 illustrates an LMM for repeated-measures data arising from a randomized block design, focusing on the SPSS software. Examples in the second edition of this book were constructed using IBM SPSS Statistics Version 21, and all SPSS syntax presented should work in earlier versions of SPSS.

Chapter 6 illustrates the fitting of a random coefficient model (specifically, a growth curve model), and emphasizes the R software. Regarding the R software, the examples have been constructed using the `lme()` and `lmer()` functions, which are available in the **nlme** and **lme4** packages, respectively. Relative to the `lme()` function, the `lmer()` function offers improved estimation of LMMs with crossed random effects. More generally, each of these functions has particular advantages depending on the data structure and the model being fitted, and we consider these differences in our example chapters. Chapter 7 highlights the Stata software and combines many of the concepts introduced in the earlier chapters by introducing a model for clustered longitudinal data, which includes both random effects and correlated residuals. Finally, Chapter 8 discusses a case study involving crossed random effects, and highlights the use of the `lmer()` function in R.

The analyses of examples in Chapters 3, 5, and 7 all consider alternative, heterogeneous covariance structures for the residuals, which is a very important feature of LMMs that makes them much more flexible than alternative linear modeling tools. At the end of each chapter presenting a case study, we consider the similarities and differences in the results generated by the software procedures. We discuss reasons for any discrepancies, and make recommendations for use of the various procedures in different settings.

Appendix A presents several statistical software resources. Information on the background and availability of the statistical software packages SAS (Version 9.3), IBM SPSS Statistics (Version 21), Stata (Release 13), R (Version 3.0.2), and HLM (Version 7) is provided in addition to links to other useful mixed modeling resources, including web sites for important materials from this book. Appendix B revisits the Rat Brain analysis from Chapter 5 to illustrate the calculation of the marginal variance-covariance matrix implied by one of the LMMs considered in that chapter. This appendix is designed to provide readers with a detailed idea of how one models the covariance of dependent observations in clustered or longitudinal data sets. Finally, Appendix C presents some commonly used abbreviations and acronyms associated with LMMs.

1.2 A Brief History of LMMs

Some historical perspective on this topic is useful. At the very least, while LMMs might seem difficult to grasp at first, it is comforting to know that scores of people have spent over a hundred years sorting it all out. The following subsections highlight many (but not nearly all) of the important historical developments that have led to the widespread use of LMMs today. We divide the key historical developments into two categories: theory and software. Some of the terms and concepts introduced in this timeline will be discussed in more detail later in the book. For additional historical perspective, we refer readers to Brown & Prescott (2006).

1.2.1 Key Theoretical Developments

The following timeline presents the evolution of the theoretical basis of LMMs:

1861: The first known formulation of a one-way random-effects model (an LMM with one random factor and no fixed factors) is that by Airy, which was further clarified by Scheffé in 1956. Airy made several telescopic observations on the same night (clustered data) for several different nights and analyzed the data separating the variance of the random night effects from the random within-night residuals.

1863: Chauvenet calculated variances of random effects in a simple random-effects model.

1925: Fisher's book *Statistical Methods for Research Workers* outlined the general method for estimating variance components, or partitioning random variation into components from different sources, for balanced data.

1927: Yule assumed explicit dependence of the current residual on a limited number of the preceding residuals in building pure serial correlation models.

1931: Tippett extended Fisher's work into the linear model framework, modeling quantities as a linear function of random variations due to multiple random factors. He also clarified an ANOVA method of estimating the variances of random effects.

1935: Neyman, Iwaszkiewicz, and Kolodziejczyk examined the comparative efficiency of randomized blocks and Latin squares designs and made extensive use of LMMs in their work.

1938: The seventh edition of Fisher's 1925 work discusses estimation of the intraclass correlation coefficient (ICC).

1939: Jackson assumed normality for random effects and residuals in his description of an LMM with one random factor and one fixed factor. This work introduced the term *effect* in the context of LMMs. Cochran presented a one-way random-effects model for unbalanced data.

1940: Winsor and Clarke, and also Yates, focused on estimating variances of random effects in the case of unbalanced data. Wald considered confidence intervals for ratios of variance components. At this point, estimates of variance components were still not unique.

1941: Ganguli applied ANOVA estimation of variance components associated with random effects to nested mixed models.

1946: Crump applied ANOVA estimation to mixed models with interactions. Ganguli and Crump were the first to mention the problem that ANOVA estimation can produce negative estimates of variance components associated with random effects. Satterthwaite worked with approximate sampling distributions of variance component estimates and defined a procedure for calculating approximate degrees of freedom for approximate F-statistics in mixed models.

1947: Eisenhart introduced the "mixed model" terminology and formally distinguished between fixed- and random-effects models.

1950: Henderson provided the equations to which the BLUPs of random effects and fixed effects were the solutions, known as the *mixed model equations* (MMEs).

1952: Anderson and Bancroft published *Statistical Theory in Research,* a book providing a thorough coverage of the estimation of variance components from balanced data and introducing the analysis of unbalanced data in nested random-effects models.

1953: Henderson produced the seminal paper "Estimation of Variance and Covariance Components" in *Biometrics*, focusing on the use of one of three sums of squares methods in the estimation of variance components from unbalanced data in mixed models (the Type III method is frequently used, being based on a linear model, but all types are available in statistical software packages). Various other papers in the late 1950s and 1960s built on these three methods for different mixed models.

1965: Rao was responsible for the systematic development of the growth curve model, a model with a common linear time trend for all units and unit-specific random intercepts and random slopes.

1967: Hartley and Rao showed that unique estimates of variance components could be obtained using maximum likelihood methods, using the equations resulting from the matrix representation of a mixed model (Searle et al., 1992). However, the estimates of the variance components were biased downward because this method assumes that fixed effects are known and not estimated from data.

1968: Townsend was the first to look at finding minimum variance quadratic unbiased estimators of variance components.

1971: Restricted maximum likelihood (REML) estimation was introduced by Patterson & Thompson (1971) as a method of estimating variance components (without assuming that fixed effects are known) in a general linear model with unbalanced data. Likelihood-based methods developed slowly because they were computationally intensive. Searle described confidence intervals for estimated variance components in an LMM with one random factor.

1972: Gabriel developed the terminology of *ante-dependence of order p* to describe a model in which the conditional distribution of the current residual, given its predecessors, depends only on its p predecessors. This leads to the development of the first-order autoregressive AR(1) process (appropriate for equally spaced measurements on an individual over time), in which the current residual depends stochastically on the previous residual. Rao completed work on minimum-norm quadratic unbiased equation (MINQUE) estimators, which demand no distributional form for the random effects or residual terms (Rao, 1972). Lindley and Smith introduced HLMs.

1976: Albert showed that without any distributional assumptions at all, ANOVA estimators are the best quadratic unbiased estimators of variance components in LMMs, and the best unbiased estimators under an assumption of normality.

Mid-1970s onward: LMMs are frequently applied in agricultural settings, specifically split-plot designs (Brown & Prescott, 2006).

1982: Laird and Ware described the theory for fitting a random coefficient model in a single stage (Laird & Ware, 1982). Random coefficient models were previously handled in two stages: estimating time slopes and then performing an analysis of time slopes for individuals.

1985: Khuri and Sahai provided a comprehensive survey of work on confidence intervals for estimated variance components.

1986: Jennrich and Schluchter described the use of different covariance pattern models for analyzing repeated-measures data and how to choose between them (Jennrich & Schluchter, 1986). Smith and Murray formulated variance components as covariances and estimated them from balanced data using the ANOVA procedure based on quadratic forms. Green would complete this formulation for unbalanced data. Goldstein introduced iteratively reweighted generalized least squares.

1987: Results from Self & Liang (1987) and later from Stram & Lee (1994) made testing the significance of variance components feasible.

1990: Verbyla and Cullis applied REML in a longitudinal data setting.

1994: Diggle, Liang, and Zeger distinguished between three types of random variance components: random effects and random coefficients, serial correlation (residuals close to each other in time are more similar than residuals farther apart), and random measurement error (Diggle et al., 2002).

1990s onward: LMMs become increasingly popular in medicine (Brown & Prescott, 2006) and in the social sciences (Raudenbush & Bryk, 2002), where they are also known as *multilevel models* or *hierarchical linear models (HLMs)*.

1.2.2 Key Software Developments

Some important landmarks are highlighted here:

1982: Bryk and Raudenbush first published the HLM computer program.

1988: Schluchter and Jennrich first introduced the BMDP5-V software routine for unbalanced repeated-measures models.

1992: SAS introduced `proc mixed` as a part of the SAS/STAT analysis package.

1995: StataCorp released Stata Release 5, which offered the `xtreg` procedure for fitting models with random effects associated with a *single* random factor, and the `xtgee` procedure for fitting models to panel data using the Generalized Estimation Equations (GEE) methodology.

1998: Bates and Pinheiro introduced the generic linear mixed-effects modeling function `lme()` for the R software package.

2001: Rabe-Hesketh et al. collaborated to write the Stata command `gllamm` for fitting LMMs (among other types of models). SPSS released the first version of the `MIXED` procedure as part of SPSS version 11.0.

2005: Stata made the general LMM command `xtmixed` available as a part of Stata Release 9, and this would later become the `mixed` command in Stata Release 13. Bates introduced the `lmer()` function for the R software package.

2

Linear Mixed Models: An Overview

2.1 Introduction

A **linear mixed model** (LMM) is a parametric linear model for clustered, longitudinal, or repeated-measures data that quantifies the relationships between a continuous dependent variable and various predictor variables. An LMM may include both **fixed-effect parameters** associated with one or more continuous or categorical covariates and **random effects** associated with one or more random factors. The mix of fixed and random effects gives the *linear mixed model* its name. Whereas fixed-effect parameters describe the relationships of the covariates to the dependent variable for an entire population, random effects are specific to clusters or subjects within a population. Consequently, random effects are directly used in modeling the random variation in the dependent variable at different levels of the data.

In this chapter, we present a heuristic overview of selected concepts important for an understanding of the application of LMMs. In Subsection 2.1.1, we describe the types and structures of data that we analyze in the example chapters (Chapters 3 through 8). In Subsection 2.1.2, we present basic definitions and concepts related to fixed and random factors and their corresponding effects in an LMM. In Sections 2.2 through 2.4, we specify LMMs in the context of longitudinal data, and discuss parameter estimation methods. In Sections 2.5 through 2.10, we present other aspects of LMMs that are important when fitting and evaluating models.

We assume that readers have a basic understanding of standard linear models, including ordinary least-squares regression, analysis of variance (ANOVA), and analysis of covariance (ANCOVA) models. For those interested in a more advanced presentation of the theory and concepts behind LMMs, we recommend Verbeke & Molenberghs (2000).

2.1.1 Types and Structures of Data Sets

2.1.1.1 Clustered Data vs. Repeated-Measures and Longitudinal Data

In the example chapters of this book, we illustrate fitting linear mixed models to clustered, repeated-measures, and longitudinal data. Because different definitions exist for these types of data, we provide our definitions for the reader's reference.

We define **clustered data** as data sets in which the dependent variable is measured once for each subject (the unit of analysis), and the units of analysis are grouped into, or nested within, clusters of units. For example, in Chapter 3 we analyze the birth weights of rat pups (the units of analysis) nested within litters (clusters of units). We describe the Rat Pup data as a **two-level** clustered data set. In Chapter 4 we analyze the math scores of students (the units of analysis) nested within classrooms (clusters of units), which are in turn nested within schools (clusters of clusters). We describe the Classroom data as a **three-level** clustered data set.

We define **repeated-measures data** quite generally as data sets in which the dependent variable is measured more than once on the same unit of analysis across levels of a repeated-measures factor (or factors). The repeated-measures factors, which may be time or other experimental or observational conditions, are often referred to as *within-subject factors*. For example, in the Rat Brain example in Chapter 5, we analyze the activation of a chemical measured in response to two treatments across three brain regions within each rat (the unit of analysis). Both brain region and treatment are repeated-measures factors. Dropout of subjects is not usually a concern in the analysis of repeated-measures data, although there may be missing data because of an instrument malfunction or due to other unanticipated reasons.

By **longitudinal data**, we mean data sets in which the dependent variable is measured at several points in time for each unit of analysis. We usually conceptualize longitudinal data as involving at least two repeated measurements made over a relatively long period of time. For example, in the Autism example in Chapter 6, we analyze the socialization scores of a sample of autistic children (the subjects or units of analysis), who are each measured at up to five time points (ages 2, 3, 5, 9, and 13 years). In contrast to repeated-measures data, dropout of subjects is often a concern in the analysis of longitudinal data.

In some cases, when the dependent variable is measured over time, it may be difficult to classify data sets as either longitudinal or repeated-measures data. In the context of analyzing data using LMMs, this distinction is not critical. The important feature of both of these types of data is that the dependent variable is measured more than once for each unit of analysis, with the repeated measures likely to be correlated.

Clustered longitudinal data sets combine features of both clustered and longitudinal data. More specifically, the units of analysis are nested within clusters, and each unit is measured more than once. In Chapter 7 we analyze the Dental Veneer data, in which teeth

TABLE 2.1: Hierarchical Structures of the Example Data Sets Considered in Chapters 3 through 7

Data Type	Clustered Data		Repeated-Measures/ Longitudinal Data		
	Two-Level	Three-Level	Repeated-Measures	Longitudinal	Clustered Longitudinal
Data set (Chap.)	Rat Pup (Chap. 3)	Classroom (Chap. 4)	Rat Brain (Chap. 5)	Autism (Chap. 6)	Dental Veneer (Chap. 7)
Repeated/ longitudinal measures (t)			Spanned by brain region and treatment	Age in years	Time in months
Subject/unit of analysis (i)	*Rat Pup*	*Student*	*Rat*	*Child*	*Tooth*
Cluster of units (j)	Litter	Classroom			Patient
Cluster of clusters (k)		School			

Note: Italicized terms in boxes indicate the unit of analysis for each study; (t, i, j, k) indices shown here are used in the model notation presented later in this book.

(the units of analysis) are nested within a patient (a cluster of units), and each tooth is measured at multiple time points (i.e., at 3 months and 6 months post-treatment).

We refer to clustered, repeated-measures, and longitudinal data as **hierarchical** data sets, because the observations can be placed into levels of a hierarchy in the data. In Table 2.1, we present the hierarchical structures of the example data sets. The distinction between repeated-measures/longitudinal data and clustered data is reflected in the presence or absence of a blank cell in the row of Table 2.1 labeled "Repeated/Longitudinal Measures."

In Table 2.1 we also introduce the index notation used in the remainder of the book. In particular, we use the index t to denote repeated/longitudinal measurements, the index i to denote subjects or units of analysis, and the index j to denote clusters. The index k is used in models for three-level clustered data to denote "clusters of clusters."

We note that Table 2.1 does not include the example data set from Chapter 8, which features *crossed* random factors. In these cases, there is not an explicit hierarchy present in the data. We discuss crossed random effects in more detail in Subsection 2.1.2.5.

2.1.1.2 Levels of Data

We can also think of clustered, repeated-measures, and longitudinal data sets as multilevel data sets, as shown in Table 2.2. The concept of "levels" of data is based on ideas from the hierarchical linear modeling (HLM) literature (Raudenbush & Bryk, 2002). All data sets appropriate for an analysis using LMMs have at least two levels of data. We describe the example data sets that we analyze as **two-level** or **three-level data sets**, depending on how many levels of data are present. One notable exception is data sets with crossed random factors (Chapter 8), which do not have an explicit hierarchy due to the fact that levels of one random factor are not nested within levels of other random factors (see Subsection 2.1.2.5). We consider data with at most three levels (denoted as **Level 1**, **Level 2**, or **Level 3**) in the examples illustrated in this book, although data sets with additional levels may be encountered in practice:

Level 1 denotes observations at the most detailed level of the data. In a clustered data set, Level 1 represents the units of analysis (or subjects) in the study. In a repeated-measures or longitudinal data set, Level 1 represents the repeated measures made on the same unit of analysis. The continuous dependent variable is always measured at Level 1 of the data.

Level 2 represents the next level of the hierarchy. In clustered data sets, Level 2 observations represent clusters of units. In repeated-measures and longitudinal data sets, Level 2 represents the units of analysis.

Level 3 represents the next level of the hierarchy, and generally refers to clusters of units in clustered longitudinal data sets, or clusters of Level 2 units (clusters of clusters) in three-level clustered data sets.

We measure continuous and categorical variables at different levels of the data, and we refer to the variables as **Level 1**, **Level 2**, or **Level 3** variables (with the exception of models with crossed random effects, as in Chapter 8).

The idea of levels of data is explicit when using the HLM software, but it is implicit when using the other four software packages. We have emphasized this concept because we find it helpful to think about LMMs in terms of simple models defined at each level of the data hierarchy (the approach to specifying LMMs in the HLM software package), instead of only one model combining sources of variation from all levels (the approach to LMMs used in the other software procedures). However, when using the paradigm of *levels*

TABLE 2.2: Multiple Levels of the Hierarchical Data Sets Considered in Each Chapter

Data Type	Clustered Data		Repeated-Measures/ Longitudinal Data		
	Two-Level	Three-Level	Repeated-Measures	Longitu-dinal	Clustered Longitudinal
Data set (Chap.)	Rat Pup (Chap. 3)	Classroom (Chap. 4)	Rat Brain (Chap. 5)	Autism (Chap. 6)	Dental Veneer (Chap. 7)
Level 1	*Rat Pup*	*Student*	Repeated measures (spanned by brain region and treatment)	Longitudinal measures (age in years)	Longitudinal measures (time in months)
Level 2	Litter	Classroom	*Rat*	*Child*	*Tooth*
Level 3		School			Patient

Note: Italicized terms in boxes indicate the units of analysis for each study.

of data, the distinction between clustered vs. repeated-measures/longitudinal data becomes less obvious, as illustrated in Table 2.2.

2.1.2 Types of Factors and Their Related Effects in an LMM

The distinction between fixed and random factors and their related effects on a dependent variable is critical in the context of LMMs. We therefore devote separate subsections to these topics.

2.1.2.1 Fixed Factors

The concept of a fixed factor is most commonly used in the setting of a standard ANOVA or ANCOVA model. We define a **fixed factor** as a categorical or classification variable, for which the investigator has included **all levels** (or conditions) that are of interest in the study. Fixed factors might include qualitative covariates, such as gender; classification variables implied by a survey sampling design, such as region or stratum, or by a study design, such as the treatment method in a randomized clinical trial; or ordinal classification variables in an observational study, such as age group. Levels of a fixed factor are chosen so that they represent specific conditions, and they can be used to define contrasts (or sets of contrasts) of interest in the research study.

2.1.2.2 Random Factors

A **random factor** is a classification variable with levels that can be thought of as being **randomly sampled** from a population of levels being studied. All possible levels of the random factor are not present in the data set, but it is the researcher's intention to make inferences about the entire population of levels. The classification variables that identify the

Level 2 and Level 3 units in both clustered and repeated-measures/longitudinal data sets are often considered to be random factors. Random factors are considered in an analysis so that variation in the dependent variable across levels of the random factors can be assessed, and the results of the data analysis can be generalized to a greater population of levels of the random factor.

2.1.2.3 Fixed Factors vs. Random Factors

In contrast to the levels of fixed factors, the levels of random factors do not represent conditions chosen specifically to meet the objectives of the study. However, depending on the goals of the study, the same factor may be considered either as a fixed factor or a random factor, as we note in the following paragraph.

In the Dental Veneer data analyzed in Chapter 7, the dependent variable (Gingival Crevicular Fluid, or GCF) is measured repeatedly on selected teeth within a given patient, and the teeth are numbered according to their location in the mouth. In our analysis, we assume that the teeth measured within a given patient represent a random sample of all teeth within the patient, which allows us to generalize the results of the analysis to the larger hypothetical "population" of "teeth within patients." In other words, we consider "tooth within patient" to be a random factor. If the research had been focused on the specific differences between the selected teeth considered in the study, we might have treated "tooth within patient" as a fixed factor. In this latter case, inferences would have only been possible for the selected teeth in the study, and not for all teeth within each patient.

2.1.2.4 Fixed Effects vs. Random Effects

Fixed effects, called regression coefficients or fixed-effect parameters, describe the relationships between the dependent variable and predictor variables (i.e., fixed factors or continuous covariates) for an entire population of units of analysis, or for a relatively small number of subpopulations defined by levels of a fixed factor. Fixed effects may describe contrasts or differences between levels of a fixed factor (e.g., between males and females) in terms of mean responses for the continuous dependent variable, or they may describe the relationship of a continuous covariate with the dependent variable. Fixed effects are assumed to be unknown *fixed* quantities in an LMM, and we estimate them based on our analysis of the data collected in a given research study.

Random effects are random values associated with the levels of a random factor (or factors) in an LMM. These values, which are specific to a given level of a random factor, usually represent random deviations from the relationships described by fixed effects. For example, random effects associated with the levels of a random factor can enter an LMM as **random intercepts** (representing random deviations for a given subject or cluster from the overall fixed intercept), or as **random coefficients** (representing random deviations for a given subject or cluster from the overall fixed effects) in the model. In contrast to fixed effects, random effects are represented as random variables in an LMM.

In Table 2.3, we provide examples of the interpretation of fixed and random effects in an LMM, based on the analysis of the Autism data (a longitudinal study of socialization among autistic children) presented in Chapter 6. There are two covariates under consideration in this example: the continuous covariate AGE, which represents a child's age in years at which the dependent variable was measured, and the fixed factor SICDEGP, which identifies groups of children based on their expressive language score at baseline (age 2). The fixed effects associated with these covariates apply to the entire population of children. The classification variable CHILDID is a unique identifier for each child, and is considered to be a random factor in the analysis. The random effects associated with the levels of CHILDID apply to specific children.

TABLE 2.3: Examples of the Interpretation of Fixed and Random Effects in an LMM Based on the Autism Data Analyzed in Chapter 6

Effect Type	Predictor Variables Associated with Each Effect	Effect Applies to	Possible Interpretation of Effects
Fixed	Variable corresponding to the intercept (i.e., equal to 1 for all observations)	Entire population	Mean of the dependent variable when all covariates are equal to zero
	AGE	Entire population	Fixed slope for AGE (i.e., expected change in the dependent variable for a 1-year increase in AGE)
	SICDEGP1, SICDEGP2 (indicators for baseline expressive language groups; reference level is SICDEGP3)	Entire population within each subgroup of SICDEGP	Contrasts for different levels of SICDEGP (i.e., mean differences in the dependent variable for children in Level 1 and Level 2 of SICDEGP, relative to Level 3)
Random	Variable corresponding to the intercept	CHILDID (individual child)	Child-specific random deviation from the fixed intercept
	AGE	CHILDID (individual child)	Child-specific random deviation from the fixed slope for AGE

2.1.2.5 Nested vs. Crossed Factors and Their Corresponding Effects

When a particular level of a factor (random or fixed) can only be measured within a single level of another factor and not across multiple levels, the levels of the first factor are said to be **nested** within levels of the second factor. The effects of the nested factor on the response are known as **nested effects**. For example, in the Classroom data set analyzed in Chapter 4, both schools and classrooms within schools were randomly sampled. Levels of classroom (one random factor) are nested within levels of school (another random factor), because each classroom can appear within only one school.

When a given level of a factor (random or fixed) can be measured *across* multiple levels of another factor, one factor is said to be **crossed** with another, and the effects of these factors on the dependent variable are known as **crossed effects**. For example, in the analysis of the Rat Pup data in Chapter 3, we consider two crossed fixed factors: TREATMENT and SEX. Specifically, levels of TREATMENT are *crossed* with the levels of SEX, because both male and female rat pups are studied for each level of treatment.

We consider crossed *random* factors and their associated random effects in Chapter 8 of this book. In this chapter, we analyze a data set from an educational study in which there are multiple measures collected over time on a random sample of students within a school, and multiple students are instructed by the same randomly sampled teacher at a given point in time. A reasonable LMM for these data would include random student effects *and* random teacher effects, and the levels of the random student and teacher factors are *crossed* with each other.

Software Note: Estimation of the parameters in LMMs with crossed random effects is more computationally intensive than for LMMs with nested random effects, primarily due to the fact that the design matrices associated with the crossed random effects are no longer block-diagonal; see Chapter 15 of Galecki & Burzykowski (2013) for more discussion of this point. The `lmer()` function in R, which is available in the `lme4` package, was designed to optimize the estimation of parameters in LMMs with crossed random effects via the use of *sparse matrices* (see `http://pages.cs.wisc.edu/~bates/reports/MixedEffects.pdf`, or Fellner (1987)), and we recommend its use for such problems. SAS `proc hpmixed` also uses sparse matrices when fitting these models. Each of these procedures in SAS and R can increase the efficiency of model-fitting algorithms for larger data sets with crossed random factors; models with crossed random effects can also be fitted using SAS `proc mixed`. We present examples of fitting models with crossed random effects in the various software packages in Chapter 8.

Crossed and nested effects also apply to interactions of continuous covariates and categorical factors. For example, in the analysis of the Autism data in Chapter 6, we discuss the crossed effects of the continuous covariate, AGE, and the categorical factor, SICDEGP (expressive language group), on children's socialization scores.

2.2   Specification of LMMs

The general specification of an LMM presented in this section refers to a model for a longitudinal two-level data set, with the first index, t, being used to indicate a time point, and the second index, i, being used for subjects. We use a similar indexing convention (index t for Level 1 units, and index i for Level 2 units) in Chapters 5 through 7, which illustrate analyses involving repeated-measures and longitudinal data.

In Chapters 3 and 4, in which we consider analyses of clustered data, we specify the models in a similar way but follow a modified indexing convention. More specifically, we use the first index, i, for Level 1 units, the second index, j, for Level 2 units (in both chapters), and the third index, k, for Level 3 units (in Chapter 4 only). Refer to Table 2.1 for more details.

In both of these conventions, the unit of analysis is indexed by i. We define the index notation in Table 2.1 and in each of the chapters presenting example analyses.

2.2.1   General Specification for an Individual Observation

We begin with a simple and general formula that indicates how most of the components of an LMM can be written at the level of an individual observation in the context of a longitudinal two-level data set. The specification of the remaining components of the LMM, which in general requires matrix notation, is deferred to Subsection 2.2.2. In the example chapters we proceed in a similar manner; that is, we specify the models at the level of an individual observation for ease of understanding, followed by elements of matrix notation.

For the sake of simplicity, we specify an LMM in (2.1) for a hypothetical two-level longitudinal data set. In this specification, Y_{ti} represents the measure of the continuous response variable Y taken on the t-th occasion for the i-th subject.

$$
\begin{aligned}
Y_{ti} =& \beta_1 \times X_{ti}^{(1)} + \beta_2 \times X_{ti}^{(2)} + \beta_3 \times X_{ti}^{(3)} + \cdots + \beta_p \times X_{ti}^{(p)} && \textbf{(fixed)} \\
& + u_{1i} \times Z_{ti}^{(1)} + \cdots + u_{qi} \times Z_{ti}^{(q)} + \varepsilon_{ti} && \textbf{(random)} \quad (2.1)
\end{aligned}
$$

The value of t ($t = 1, \ldots, n_i$), indexes the n_i longitudinal observations of the dependent variable for a given subject, and i ($i = 1, \ldots, m$) indicates the i-th subject (unit of analysis). We assume that the model involves two sets of covariates, namely the X and Z covariates. The first set contains p covariates, $X^{(1)}, \ldots, X^{(p)}$, associated with the fixed effects $\beta_1, \ldots, \beta_p$. The second set contains q covariates, $Z^{(1)}, \ldots, Z^{(q)}$, associated with the random effects $u_{1i}, \ldots, u_{qi}$ that are specific to subject i. The X and/or Z covariates may be continuous or indicator variables. The indices for the X and Z covariates are denoted by superscripts so that they do not interfere with the subscript indices, t and i, for the elements in the design matrices, $\boldsymbol{X}_i$ and $\boldsymbol{Z}_i$, presented in Subsection 2.2.2.[1] For each X covariate, $X^{(1)}, \ldots, X^{(p)}$, the elements $X_{ti}^{(1)}, \ldots, X_{ti}^{(p)}$ represent the t-th observed value of the corresponding covariate for the i-th subject. We assume that the p covariates may be either **time-invariant** characteristics of the individual subject (e.g., gender) or **time-varying** for each measurement (e.g., time of measurement, or weight at each time point).

Each β parameter represents the fixed effect of a one-unit change in the corresponding X covariate on the mean value of the dependent variable, Y, assuming that the other covariates remain constant at some value. These β parameters are **fixed effects** that we wish to estimate, and their linear combination with the X covariates defines the fixed portion of the model.

The effects of the Z covariates on the response variable are represented in the random portion of the model by the q **random effects**, $u_{1i}, \ldots, u_{qi}$, associated with the i-th subject. In addition, ε_{ti} represents the residual associated with the t-th observation on the i-th subject. The random effects and residuals in (2.1) are random variables, with values drawn from distributions that are defined in (2.3) and (2.4) in the next section using matrix notation. We assume that for a given subject, the residuals are independent of the random effects.

The individual observations for the i-th subject in (2.1) can be combined into vectors and matrices, and the LMM can be specified more efficiently using matrix notation as shown in the next section. Specifying an LMM in matrix notation also simplifies the presentation of estimation and hypothesis tests in the context of LMMs.

2.2.2   General Matrix Specification

We now consider the general matrix specification of an LMM for a given subject i, by stacking the formulas specified in Subsection 2.2.1 for individual observations indexed by t into vectors and matrices.

---

[1] In Chapters 3 through 7, in which we analyze real data sets, our superscript notation for the covariates in (2.1) is replaced by actual variable names (e.g., for the Autism data in Chapter 6, $X_{ti}^{(1)}$ might be replaced by AGE_{ti}, the t-th age at which child i is measured).

$$\mathbf{Y}_i = \underbrace{\boldsymbol{X}_i\boldsymbol{\beta}}_{\text{fixed}} + \underbrace{\boldsymbol{Z}_i\mathbf{u}_i + \boldsymbol{\varepsilon}_i}_{\text{random}}$$

(2.2)

$$\mathbf{u}_i \sim \mathcal{N}(\mathbf{0}, \boldsymbol{D})$$
$$\boldsymbol{\varepsilon}_i \sim \mathcal{N}(\mathbf{0}, \boldsymbol{R}_i)$$

In (2.2), $\mathbf{Y}_i$ represents a vector of continuous responses for the i-th subject. We present elements of the $\mathbf{Y}_i$ vector as follows, drawing on the notation used for an individual observation in (2.1):

$$\mathbf{Y}_i = \begin{pmatrix} Y_{1i} \\ Y_{2i} \\ \vdots \\ Y_{n_i i} \end{pmatrix}$$

Note that the number of elements, n_i, in the vector $\mathbf{Y}_i$ may vary from one subject to another.

The $\boldsymbol{X}_i$ in (2.2) is an $n_i \times p$ design matrix, which represents the known values of the p covariates, $X^{(1)}, \ldots, X^{(p)}$, for each of the n_i observations collected on the i-th subject:

$$\boldsymbol{X}_i = \begin{pmatrix} X_{1i}^{(1)} & X_{1i}^{(2)} & \cdots & X_{1i}^{(p)} \\ X_{2i}^{(1)} & X_{2i}^{(2)} & \cdots & X_{2i}^{(p)} \\ \vdots & \vdots & \ddots & \vdots \\ X_{n_i i}^{(1)} & X_{n_i i}^{(2)} & \cdots & X_{n_i i}^{(p)} \end{pmatrix}$$

In a model including an intercept term, the first column would simply be equal to 1 for all observations. Note that all elements in a column of the $\boldsymbol{X}_i$ matrix corresponding to a time-invariant (or subject-specific) covariate will be the same. For ease of presentation, we assume that the $\boldsymbol{X}_i$ matrices are of full rank; that is, none of the columns (or rows) is a linear combination of the remaining ones. In general, $\boldsymbol{X}_i$ matrices may not be of full rank, and this may lead to an aliasing (or parameter identifiability) problem for the fixed effects stored in the vector $\boldsymbol{\beta}$ (see Subsection 2.9.3).

The $\boldsymbol{\beta}$ in (2.2) is a vector of p unknown regression coefficients (or fixed-effect parameters) associated with the p covariates used in constructing the $\boldsymbol{X}_i$ matrix:

$$\boldsymbol{\beta} = \begin{pmatrix} \beta_1 \\ \beta_2 \\ \vdots \\ \beta_p \end{pmatrix}$$

The $n_i \times q$ matrix $\boldsymbol{Z}_i$ in (2.2) is a design matrix that represents the known values of the q covariates, $Z^{(1)}, \ldots, Z^{(q)}$, for the i-th subject. This matrix is very much like the $\boldsymbol{X}_i$ matrix in that it represents the observed values of covariates; however, it usually has fewer columns than the $\boldsymbol{X}_i$ matrix:

$$\boldsymbol{Z}_i = \begin{pmatrix} Z_{1i}^{(1)} & Z_{1i}^{(2)} & \cdots & Z_{1i}^{(q)} \\ Z_{2i}^{(1)} & Z_{2i}^{(2)} & \cdots & Z_{2i}^{(q)} \\ \vdots & \vdots & \ddots & \vdots \\ Z_{n_i i}^{(1)} & Z_{n_i i}^{(2)} & \cdots & Z_{n_i i}^{(q)} \end{pmatrix}$$

The columns in the Z_i matrix represent observed values for the q predictor variables for the i-th subject, which have effects on the continuous response variable that vary randomly across subjects. In many cases, predictors with effects that vary randomly across subjects are represented in both the X_i matrix and the Z_i matrix. In an LMM in which only the intercepts are assumed to vary randomly from subject to subject, the Z_i matrix would simply be a column of 1's.

The $\mathbf{u}_i$ vector for the i-th subject in (2.2) represents a vector of q **random effects** (defined in Subsection 2.1.2.4) associated with the q covariates in the Z_i matrix:

$$\mathbf{u}_i = \begin{pmatrix} u_{1i} \\ u_{2i} \\ \vdots \\ u_{qi} \end{pmatrix}$$

Recall that by definition, random effects are random variables. We assume that the q random effects in the $\mathbf{u}_i$ vector follow a multivariate normal distribution, with mean vector $\mathbf{0}$ and a variance-covariance matrix denoted by D:

$$\mathbf{u}_i \sim \mathcal{N}(\mathbf{0}, D) \tag{2.3}$$

Elements along the main diagonal of the D matrix represent the **variances** of each random effect in $\mathbf{u}_i$, and the off-diagonal elements represent the **covariances** between two corresponding random effects. Because there are q random effects in the model associated with the i-th subject, D is a $q \times q$ matrix that is symmetric and positive-definite.[2] Elements of this matrix are shown as follows:

$$D = Var(\mathbf{u}_i) = \begin{pmatrix} Var(u_{1i}) & cov(u_{1i}, u_{2i}) & \cdots & cov(u_{1i}, u_{qi}) \\ cov(u_{1i}, u_{2i}) & Var(u_{2i}) & \cdots & cov(u_{2i}, u_{qi}) \\ \vdots & \vdots & \ddots & \vdots \\ cov(u_{1i}, u_{qi}) & cov(u_{2i}, u_{qi}) & \cdots & Var(u_{qi}) \end{pmatrix}$$

The elements (variances and covariances) of the D matrix are defined as functions of a (usually) small set of covariance parameters stored in a vector denoted by $\boldsymbol{\theta}_D$. Note that the vector $\boldsymbol{\theta}_D$ imposes structure (or constraints) on the elements of the D matrix. We discuss different structures for the D matrix in Subsection 2.2.2.1.

Finally, the $\boldsymbol{\varepsilon}_i$ vector in (2.2) is a vector of n_i **residuals**, with each element in $\boldsymbol{\varepsilon}_i$ denoting the residual associated with an observed response at occasion t for the i-th subject. Because some subjects might have more observations collected than others (e.g., if data for one or more time points are not available when a subject drops out), the $\boldsymbol{\varepsilon}_i$ vectors may have a different number of elements.

$$\boldsymbol{\varepsilon}_i = \begin{pmatrix} \varepsilon_{1i} \\ \varepsilon_{2i} \\ \vdots \\ \varepsilon_{n_i i} \end{pmatrix}$$

In contrast to the standard linear model, the residuals associated with repeated observations on the same subject in an LMM can be correlated. We assume that the n_i residuals in the $\boldsymbol{\varepsilon}_i$ vector for a given subject, i, are random variables that follow a multivariate normal

---

[2]For more details on positive-definite matrices, interested readers can visit `http://en.wikipedia.org/wiki/Positive-definite_matrix`.

distribution with a mean vector $\mathbf{0}$ and a positive-definite symmetric variance-covariance matrix $\boldsymbol{R}_i$:

$$\varepsilon_i \sim \mathcal{N}(\mathbf{0}, \boldsymbol{R}_i) \tag{2.4}$$

We also assume that residuals associated with different subjects are independent of each other. Further, we assume that the vectors of residuals, $\varepsilon_1, \ldots, \varepsilon_m$, and random effects, $\mathbf{u}_1, \ldots, \mathbf{u}_m$, are independent of each other. We represent the general form of the $\boldsymbol{R}_i$ matrix as shown below:

$$\boldsymbol{R}_i = Var(\boldsymbol{\varepsilon}_i) = \begin{pmatrix} Var(\varepsilon_{1i}) & cov(\varepsilon_{1i}, \varepsilon_{2i}) & \cdots & cov(\varepsilon_{1i}, \varepsilon_{n_i i}) \\ cov(\varepsilon_{1i}, \varepsilon_{2i}) & Var(\varepsilon_{2i}) & \cdots & cov(\varepsilon_{2i}, \varepsilon_{n_i i}) \\ \vdots & \vdots & \ddots & \vdots \\ cov(\varepsilon_{1i}, \varepsilon_{n_i i}) & cov(\varepsilon_{2i}, \varepsilon_{n_i i}) & \cdots & Var(\varepsilon_{n_i i}) \end{pmatrix}$$

The elements (variances and covariances) of the $\boldsymbol{R}_i$ matrix are defined as functions of another (usually) small set of covariance parameters stored in a vector denoted by $\boldsymbol{\theta}_R$. Many different covariance structures are possible for the $\boldsymbol{R}_i$ matrix; we discuss some of these structures in Subsection 2.2.2.2.

To complete our notation for the LMM, we introduce the vector $\boldsymbol{\theta}$ used in subsequent sections, which combines all covariance parameters contained in the vectors $\boldsymbol{\theta}_D$ and $\boldsymbol{\theta}_R$.

2.2.2.1 Covariance Structures for the D Matrix

We consider different covariance structures for the $\boldsymbol{D}$ matrix in this subsection.

A $\boldsymbol{D}$ matrix with no additional constraints on the values of its elements (aside from positive-definiteness and symmetry) is referred to as an **unstructured** (or **general**) $\boldsymbol{D}$ matrix. This structure is often used for **random coefficient models** (discussed in Chapter 6). The symmetry in the $q \times q$ matrix $\boldsymbol{D}$ implies that the $\boldsymbol{\theta}_D$ vector has $q \times (q+1)/2$ parameters. The following matrix is an example of an unstructured $\boldsymbol{D}$ matrix, in the case of an LMM having two random effects associated with the i-th subject.

$$\boldsymbol{D} = Var(\mathbf{u}_i) = \begin{pmatrix} \sigma_{u1}^2 & \sigma_{u1,u2} \\ \sigma_{u1,u2} & \sigma_{u2}^2 \end{pmatrix}$$

In this case, the vector $\boldsymbol{\theta}_D$ contains three covariance parameters:

$$\boldsymbol{\theta}_D = \begin{pmatrix} \sigma_{u1}^2 \\ \sigma_{u1,u2} \\ \sigma_{u2}^2 \end{pmatrix}$$

We also define other more parsimonious structures for $\boldsymbol{D}$ by imposing certain constraints on the structure of $\boldsymbol{D}$. A very commonly used structure is the **variance components** (or diagonal) structure, in which each random effect in $\mathbf{u}_i$ has its own variance, and all covariances in $\boldsymbol{D}$ are defined to be zero. In general, the $\boldsymbol{\theta}_D$ vector for the variance components structure requires q covariance parameters, defining the variances on the diagonal of the $\boldsymbol{D}$ matrix. For example, in an LMM having two random effects associated with the i-th subject, a variance component $\boldsymbol{D}$ matrix has the following form:

$$\boldsymbol{D} = Var(\mathbf{u}_i) = \begin{pmatrix} \sigma_{u1}^2 & 0 \\ 0 & \sigma_{u2}^2 \end{pmatrix}$$

In this case, the vector $\boldsymbol{\theta}_D$ contains two parameters:

$$\boldsymbol{\theta}_D = \begin{pmatrix} \sigma_{u1}^2 \\ \sigma_{u2}^2 \end{pmatrix}$$

The unstructured $\boldsymbol{D}$ matrix and variance components structures for the matrix are the most commonly used in practice, although other structures are available in some software procedures. For example, the parameters representing the variances and covariances of the random effects in the vector $\boldsymbol{\theta}_D$ could be allowed to vary across different subgroups of cases (e.g., males and females in a longitudinal study), if greater between-subject variance in selected effects was to be expected in one subgroup compared to another (e.g., males have more variability in their intercepts); see Subsection 2.2.2.3. We discuss the structure of the $\boldsymbol{D}$ matrices for specific models in the example chapters.

2.2.2.2 Covariance Structures for the R_i Matrix

In this section, we discuss some of the more commonly used covariance structures for the $\boldsymbol{R}_i$ matrix.

The simplest covariance matrix for $\boldsymbol{R}_i$ is the **diagonal** structure, in which the residuals associated with observations on the same subject are assumed to be uncorrelated and to have equal variance. The diagonal $\boldsymbol{R}_i$ matrix for each subject i has the following structure:

$$\boldsymbol{R}_i = Var(\boldsymbol{\varepsilon}_i) = \sigma^2 \boldsymbol{I}_{n_i} = \begin{pmatrix} \sigma^2 & 0 & \cdots & 0 \\ 0 & \sigma^2 & \cdots & 0 \\ \vdots & \vdots & \ddots & \vdots \\ 0 & 0 & \cdots & \sigma^2 \end{pmatrix}$$

The diagonal structure requires one parameter in $\boldsymbol{\theta}_R$, which defines the constant variance at each time point:

$$\boldsymbol{\theta}_R = (\sigma^2)$$

All software procedures that we discuss use the diagonal structure as the default structure for the $\boldsymbol{R}_i$ matrix.

The **compound symmetry** structure is frequently used for the $\boldsymbol{R}_i$ matrix. The general form of this structure for each subject i is as follows:

$$\boldsymbol{R}_i = Var(\boldsymbol{\varepsilon}_i) = \begin{pmatrix} \sigma^2 + \sigma_1 & \sigma_1 & \cdots & \sigma_1 \\ \sigma_1 & \sigma^2 + \sigma_1 & \cdots & \sigma_1 \\ \vdots & \vdots & \ddots & \vdots \\ \sigma_1 & \sigma_1 & \cdots & \sigma^2 + \sigma_1 \end{pmatrix}$$

In the compound symmetry covariance structure, there are two parameters in the $\boldsymbol{\theta}_R$ vector that define the variances and covariances in the $\boldsymbol{R}_i$ matrix:

$$\boldsymbol{\theta}_R = \begin{pmatrix} \sigma^2 \\ \sigma_1 \end{pmatrix}$$

Note that the n_i residuals associated with the observed response values for the i-th subject are assumed to have a constant covariance, σ_1, and a constant variance, $\sigma^2 + \sigma_1$, in the compound symmetry structure. This structure is often used when an assumption of equal correlation of residuals is plausible (e.g., repeated trials under the same condition in an experiment).

The **first-order autoregressive** structure, denoted by **AR(1)**, is another commonly used covariance structure for $\boldsymbol{R}_i$ the matrix. The general form of the $\boldsymbol{R}_i$ matrix for this covariance structure is as follows:

$$\boldsymbol{R}_i = Var(\boldsymbol{\varepsilon}_i) = \begin{pmatrix} \sigma^2 & \sigma^2\rho & \cdots & \sigma^2\rho^{n_i-1} \\ \sigma^2\rho & \sigma^2 & \cdots & \sigma^2\rho^{n_i-2} \\ \vdots & \vdots & \ddots & \vdots \\ \sigma^2\rho^{n_i-1} & \sigma^2\rho^{n_i-2} & \cdots & \sigma^2 \end{pmatrix}$$

The AR(1) structure has only two parameters in the $\boldsymbol{\theta}_R$ vector that define all the variances and covariances in the $\boldsymbol{R}_i$ matrix: a variance parameter, σ^2, and a correlation parameter, ρ.

$$\boldsymbol{\theta}_R = \begin{pmatrix} \sigma^2 \\ \rho \end{pmatrix}$$

Note that σ^2 must be positive, whereas ρ can range from -1 to 1. In the AR(1) covariance structure, the variance of the residuals, σ^2, is assumed to be constant, and the covariance of residuals of observations that are w units apart is assumed to be equal to $\sigma^2\rho^w$. This means that all adjacent residuals (i.e., the residuals associated with observations next to each other in a sequence of longitudinal observations for a given subject) have a covariance of $\sigma^2\rho$, and residuals associated with observations two units apart in the sequence have a covariance of $\sigma^2\rho^2$, and so on.

The AR(1) structure is often used to fit models to data sets with equally spaced longitudinal observations on the same units of analysis. This structure implies that observations closer to each other in time exhibit higher correlation than observations farther apart in time.

Other covariance structures, such as the **Toeplitz** structure, allow more flexibility in the correlations, but at the expense of using more covariance parameters in the $\boldsymbol{\theta}_R$ vector. In any given analysis, we try to determine the structure for the $\boldsymbol{R}_i$ matrix that seems most appropriate and parsimonious, given the observed data and knowledge about the relationships between observations on an individual subject.

2.2.2.3 Group-Specific Covariance Parameter Values for the $\boldsymbol{D}$ and $\boldsymbol{R}_i$ Matrices

The $\boldsymbol{D}$ and $\boldsymbol{R}_i$ covariance matrices can also be specified to allow **heterogeneous variances** for different groups of subjects (e.g., males and females). Specifically, we might assume the same *structures* for the matrices in different groups, but with different *values* for the covariance parameters in the $\boldsymbol{\theta}_D$ and $\boldsymbol{\theta}_R$ vectors. Examples of heterogeneous $\boldsymbol{R}_i$ matrices defined for different groups of subjects and observations are given in Chapter 3, Chapter 5, and Chapter 7. We do not consider examples of heterogeneity in the $\boldsymbol{D}$ matrix. For a recently published example of this type of heterogeneity (many exist in the literature), interested readers can refer to West & Elliott (Forthcoming in 2014).

2.2.3 Alternative Matrix Specification for All Subjects

In (2.2), we presented a general matrix specification of the LMM for a given subject i. An alternative specification, based on all subjects under study, is presented in (2.5):

$$\mathbf{Y} = \underbrace{\boldsymbol{X\beta}}_{\text{fixed}} + \underbrace{\boldsymbol{Z}\mathbf{u}}_{\text{random}} + \boldsymbol{\varepsilon} \tag{2.5}$$

$$\mathbf{u} \sim \mathcal{N}(\mathbf{0}, \boldsymbol{G})$$

$$\boldsymbol{\varepsilon} \sim \mathcal{N}(\mathbf{0}, \boldsymbol{R})$$

In (2.5), the $n \times 1$ vector $\mathbf{Y}$, where $n = \sum_i n_i$, is the result of "stacking" the $\mathbf{Y}_i$ vectors for all subjects vertically. The $n \times p$ design matrix $\boldsymbol{X}$ is obtained by stacking all $\boldsymbol{X}_i$ matrices vertically as well. In two-level models or models with nested random effects, the $\boldsymbol{Z}$ matrix is a block-diagonal matrix, with blocks on the diagonal defined by the $\boldsymbol{Z}_i$ matrices. The $\mathbf{u}$ vector stacks all $\mathbf{u}_i$ vectors vertically, and the vector $\boldsymbol{\varepsilon}$ stacks all $\boldsymbol{\varepsilon}_i$ vectors vertically. The $\boldsymbol{G}$ matrix is a block-diagonal matrix representing the variance-covariance matrix for all random effects (not just those associated with a single subject i), with blocks on the diagonal defined by the $\boldsymbol{D}$ matrix. The $n \times n$ matrix $\boldsymbol{R}$ is a block-diagonal matrix representing the variance-covariance matrix for all residuals, with blocks on the diagonal defined by the $\boldsymbol{R}_i$ matrices.

This "all subjects" specification is used in the documentation for SAS `proc mixed` and the `MIXED` command in SPSS, but we primarily refer to the $\boldsymbol{D}$ and $\boldsymbol{R}_i$ matrices for a single subject (or cluster) throughout the book.

2.2.4   Hierarchical Linear Model (HLM) Specification of the LMM

It is often convenient to specify an LMM in terms of an explicitly defined hierarchy of simpler models, which correspond to the levels of a clustered or longitudinal data set. When LMMs are specified in such a way, they are often referred to as **hierarchical linear models (HLMs)**, or **multilevel models (MLMs)**. The HLM software is the only program discussed in this book that requires LMMs to be specified in a hierarchical manner.

The HLM specification of an LMM is equivalent to the general LMM specification introduced in Subsection 2.2.2, and may be implemented for any LMM. We do not present a general form for the HLM specification of LMMs here, but rather introduce examples of the HLM specification in Chapters 3 through 8. The levels of the example data sets considered in the HLM specification of models for these data sets are displayed in Table 2.2.

2.3   The Marginal Linear Model

In Section 2.2, we specified the general LMM. In this section, we specify a closely related marginal linear model. The key difference between the two models lies in the presence or absence of random effects. Specifically, random effects are explicitly used in LMMs to explain the between-subject or between-cluster variation, but they are not used in the specification of marginal models. This difference implies that the LMM allows for subject-specific inference, whereas the marginal model does not. For the same reason, LMMs are often referred to as *subject-specific* models, and marginal models are called *population-averaged* models. In Subsection 2.3.1, we specify the marginal model in general, and in Subsection 2.3.2, we present the marginal model implied by an LMM.

2.3.1 Specification of the Marginal Model

The general matrix specification of the marginal model for subject i is

$$\mathbf{Y}_i = \boldsymbol{X}_i \boldsymbol{\beta} + \boldsymbol{\varepsilon}_i^* \qquad (2.6)$$

where

$$\boldsymbol{\varepsilon}_i^* \sim \mathcal{N}(0, \boldsymbol{V}_i^*)$$

In (2.6), the $n_i \times p$ design matrix $\boldsymbol{X}_i$ is constructed the same way as in an LMM. Similarly, $\boldsymbol{\beta}$ is a vector of fixed effects. The vector $\boldsymbol{\varepsilon}_i^*$ represents a vector of **marginal residual errors**. Elements in the $n_i \times n_i$ marginal variance-covariance matrix $\boldsymbol{V}_i^*$ are usually defined by a small set of covariance parameters, which we denote as $\boldsymbol{\theta}^*$. All structures used for the $\boldsymbol{R}_i$ matrix in LMMs (described in Subsection 2.2.2.2) can be used to specify a structure for $\boldsymbol{V}_i^*$. Other structures for $\boldsymbol{V}_i^*$, such as those shown in Subsection 2.3.2, are also allowed.

Note that the entire random part of the marginal model is described in terms of the marginal residuals $\boldsymbol{\varepsilon}_i^*$ only. In contrast to the LMM, the marginal model does not involve the random effects, $\mathbf{u}_i$, so inferences cannot be made about them and consequently this model is not a *mixed* model.

Software Note: Several software procedures designed for fitting LMMs, including the procedures in SAS, SPSS, R, and Stata, also allow users to specify a marginal model directly. The most natural way to specify selected marginal models in these procedures is to make sure that random effects are not included in the model, and then specify an appropriate covariance structure for the $\boldsymbol{R}_i$ matrix, which in the context of the marginal model will be used for $\boldsymbol{V}_i^*$. A marginal model of this form is **not** an LMM, because no random effects are included in the model. This type of model cannot be specified using the HLM software, because HLM generally requires the specification of at least one set of random effects (e.g., a random intercept). Examples of fitting a marginal model by omitting random effects and using an appropriate $\boldsymbol{R}_i$ matrix are given in alternative analyses of the Rat Brain data at the end of Chapter 5, and the Autism data at the end of Chapter 6.

2.3.2 The Marginal Model Implied by an LMM

The LMM introduced in (2.2) implies the following marginal linear model:

$$\mathbf{Y}_i = \boldsymbol{X}_i \boldsymbol{\beta} + \boldsymbol{\varepsilon}_i^* \qquad (2.7)$$

where

$$\boldsymbol{\varepsilon}_i^* \sim \mathcal{N}(\mathbf{0}, \boldsymbol{V}_i)$$

and the variance-covariance matrix, $\boldsymbol{V}_i$, is defined as

$$\boldsymbol{V}_i = \boldsymbol{Z}_i \boldsymbol{D} \boldsymbol{Z}_i' + \boldsymbol{R}_i$$

A few observations are in order. First, the implied marginal model is an example of the marginal model defined in Subsection 2.3.1. Second, the LMM in (2.2) and the corresponding implied marginal model in (2.7) involve the same set of covariance parameters $\boldsymbol{\theta}$ (i.e., the $\boldsymbol{\theta}_D$ and $\boldsymbol{\theta}_R$ vectors combined). The important difference is that there are more restrictions imposed on the covariance parameter space in the LMM than in the implied marginal model. In general, the $\boldsymbol{D}$ and $\boldsymbol{R}_i$ matrices in LMMs have to be positive-definite, whereas the only requirement in the implied marginal model is that the $\boldsymbol{V}_i$ matrix be positive-definite. Third, interpretation of the covariance parameters in a marginal model is different from that in an LMM, because inferences about random effects are no longer valid.

The concept of the implied marginal model is important for at least two reasons. First, estimation of fixed-effect and covariance parameters in the LMM (see Subsection 2.4.1.2) is carried out in the framework of the implied marginal model. Second, in the case in which a software procedure produces a nonpositive-definite (i.e., invalid) estimate of the $\boldsymbol{D}$ matrix in an LMM, we may be able to fit the implied marginal model, which has fewer restrictions. Consequently, we may be able to diagnose problems with nonpositive-definiteness of the $\boldsymbol{D}$ matrix or, even better, we may be able to answer some relevant research questions in the context of the implied marginal model.

The implied marginal model defines the marginal distribution of the $\mathbf{Y}_i$ vector:

$$\mathbf{Y}_i \sim \mathcal{N}(\boldsymbol{X}_i\boldsymbol{\beta}, \boldsymbol{Z}_i\boldsymbol{D}\boldsymbol{Z}_i' + \boldsymbol{R}_i) \tag{2.8}$$

The marginal mean (or expected value) and the marginal variance-covariance matrix of the vector $\mathbf{Y}_i$ are equal to

$$E(\mathbf{Y}_i) = \boldsymbol{X}_i\boldsymbol{\beta} \tag{2.9}$$

and

$$Var(\mathbf{Y}_i) = \boldsymbol{V}_i = \boldsymbol{Z}_i\boldsymbol{D}\boldsymbol{Z}_i' + \boldsymbol{R}_i$$

The off-diagonal elements in the $n_i \times n_i$ matrix $\boldsymbol{V}_i$ represent the marginal covariances of the $\mathbf{Y}_i$ vector. These covariances are in general different from zero, which means that in the case of a longitudinal data set, repeated observations on a given individual i are correlated. We present an example of calculating the $\boldsymbol{V}_i$ matrix for the marginal model implied by an LMM fitted to the Rat Brain data (Chapter 5) in Appendix B. The marginal distribution specified in (2.8), with mean and variance defined in (2.9), is a focal point of the likelihood estimation in LMMs outlined in the next section.

Software Note: The software discussed in this book is primarily designed to fit LMMs. In some cases, we may be interested in fitting the marginal model implied by a given LMM using this software:

1. For some fairly simple LMMs, it is possible to specify the implied marginal model directly using the software procedures in SAS, SPSS, R, and Stata, as described in Subsection 2.3.1. As an example, consider an LMM with random intercepts and constant residual variance. The $\boldsymbol{V}_i$ matrix for the marginal model implied by this LMM has a compound symmetry structure (see Appendix B), which can be specified by omitting the random intercepts from the model and choosing a compound symmetry structure for the $\boldsymbol{R}_i$ matrix.

2. Another very general method available in the LMM software procedures is to "emulate" fitting the implied marginal model by fitting the LMM itself. By emulation, we mean using the same syntax as for an LMM, i.e., including specification of random effects, but interpreting estimates and other results as if they were obtained for the marginal model. In this approach, we simply take advantage of the fact that estimation of the LMM and of the implied marginal model are performed using the same algorithm (see Section 2.4).

3. Note that the general emulation approach outlined in item 2 has some limitations related to less restrictive constraints in the implied marginal model compared to LMMs. In most software procedures that fit LMMs, it is difficult to relax the positive-definiteness constraints on the $\boldsymbol{D}$ and $\boldsymbol{R}_i$ matrices as required by the implied marginal model. The `nobound` option in SAS `proc mixed` is the only exception among the software procedures discussed in this book that allows users to remove the positive-definiteness constraints on the $\boldsymbol{D}$ and $\boldsymbol{R}_i$ matrices and allows user-defined constraints to be imposed on the covariance parameters in the $\boldsymbol{\theta}_D$ and $\boldsymbol{\theta}_R$ vectors. An example of using the `nobound` option to relax the constraints on covariance parameters applicable to the fitted linear mixed model is given in Subsection 6.4.1.

2.4 Estimation in LMMs

In the LMM, we estimate the fixed-effect parameters, $\boldsymbol{\beta}$, and the covariance parameters, $\boldsymbol{\theta}$ (i.e., $\boldsymbol{\theta}_D$ and $\boldsymbol{\theta}_R$ for the $\boldsymbol{D}$ and $\boldsymbol{R}_i$ matrices, respectively). In this section, we discuss maximum likelihood (ML) and restricted maximum likelihood (REML) estimation, which are methods commonly used to estimate these parameters.

2.4.1 Maximum Likelihood (ML) Estimation

In general, **maximum likelihood (ML) estimation** is a method of obtaining estimates of unknown parameters by optimizing a **likelihood function**. To apply ML estimation, we first construct the likelihood as a function of the parameters in the specified model, based on distributional assumptions. The **maximum likelihood estimates (MLEs)** of the parameters are the values of the arguments that maximize the likelihood function (i.e., the values of the parameters that make the observed values of the dependent variable most likely, given the distributional assumptions). See Casella & Berger (2002) for an in-depth discussion of ML estimation.

In the context of the LMM, we construct the likelihood function of $\boldsymbol{\beta}$ and $\boldsymbol{\theta}$ by referring to the marginal distribution of the dependent variable $\mathbf{Y}_i$ defined in (2.8). The corresponding **multivariate normal probability density function**, $f(\mathbf{Y}_i|\boldsymbol{\beta}, \boldsymbol{\theta})$, is:

$$f(\mathbf{Y}_i|\boldsymbol{\beta}, \boldsymbol{\theta}) = (2\pi)^{\frac{-n_i}{2}} \det(\boldsymbol{V}_i)^{\frac{-1}{2}} \exp(-0.5 \times (\mathbf{Y}_i - \boldsymbol{X}_i\boldsymbol{\beta})'\boldsymbol{V}_i^{-1}(\mathbf{Y}_i - \boldsymbol{X}_i\boldsymbol{\beta})) \qquad (2.10)$$

where det refers to the determinant. Recall that the elements of the $\boldsymbol{V}_i$ matrix are functions of the covariance parameters in $\boldsymbol{\theta}$.

Based on the probability density function (pdf) defined in (2.10), and given the observed data $\mathbf{Y}_i = \mathbf{y}_i$, the likelihood function contribution for the i-th subject is defined as follows:

$$L_i(\boldsymbol{\beta}, \boldsymbol{\theta}; \mathbf{y}_i) = (2\pi)^{\frac{-n_i}{2}} \det(\boldsymbol{V}_i)^{\frac{-1}{2}} \exp(-0.5 \times (\mathbf{y}_i - \boldsymbol{X}_i\boldsymbol{\beta})'\boldsymbol{V}_i^{-1}(\mathbf{y}_i - \boldsymbol{X}_i\boldsymbol{\beta})) \qquad (2.11)$$

We write the **likelihood function**, $L(\boldsymbol{\beta}, \boldsymbol{\theta})$ as the product of the m independent contributions defined in (2.11) for the individuals ($i = 1, \ldots, m$):

$$L(\boldsymbol{\beta},\boldsymbol{\theta}) = \prod_i L_i(\boldsymbol{\beta},\boldsymbol{\theta}) = \prod_i (2\pi)^{\frac{-n_i}{2}} \det(\boldsymbol{V}_i)^{\frac{-1}{2}} \exp(-0.5 \times (\mathbf{y}_i - \boldsymbol{X}_i\boldsymbol{\beta})' \boldsymbol{V}_i^{-1}(\mathbf{y}_i - \boldsymbol{X}_i\boldsymbol{\beta}))$$
$$(2.12)$$

The corresponding **log-likelihood function**, $\ell(\boldsymbol{\beta},\boldsymbol{\theta})$, is defined using (2.12) as

$$\ell(\boldsymbol{\beta},\boldsymbol{\theta}) = \ln L(\boldsymbol{\beta},\boldsymbol{\theta}) \quad = \quad -0.5n \times \ln(2\pi) - 0.5 \times \sum_i \ln(\det(\boldsymbol{V}_i))$$

$$-0.5 \times \sum_i (\mathbf{y}_i - \boldsymbol{X}_i\boldsymbol{\beta})' \boldsymbol{V}_i^{-1}(\mathbf{y}_i - \boldsymbol{X}_i\boldsymbol{\beta}) \quad\quad (2.13)$$

where n $(= \sum n_i)$ is the number of observations (rows) in the data set, and "ln" refers to the natural logarithm.

Although it is often possible to find estimates of $\boldsymbol{\beta}$ and $\boldsymbol{\theta}$ simultaneously, by optimization of $\ell(\boldsymbol{\beta},\boldsymbol{\theta})$ with respect to both $\boldsymbol{\beta}$ and $\boldsymbol{\theta}$, many computational algorithms simplify the optimization by **profiling out** the $\boldsymbol{\beta}$ parameters from $\ell(\boldsymbol{\beta},\boldsymbol{\theta})$, as shown in Subsections 2.4.1.1 and 2.4.1.2.

2.4.1.1 Special Case: Assume θ Is Known

In this section, we consider a special case of ML estimation for LMMs, in which we assume that $\boldsymbol{\theta}$, and as a result the matrix $\boldsymbol{V}_i$, are known. Although this situation does not occur in practice, it has important computational implications, so we present it separately.

Because we assume that $\boldsymbol{\theta}$ is known, the only parameters that we estimate are the fixed effects, $\boldsymbol{\beta}$. The log-likelihood function, $\ell(\boldsymbol{\beta},\theta)$, thus becomes a function of $\boldsymbol{\beta}$ only, and its optimization is equivalent to finding a minimum of an objective function q($\boldsymbol{\beta}$), defined by the last term in (2.13):

$$\mathrm{q}(\boldsymbol{\beta}) = 0.5 \times \sum_i (\mathbf{y}_i - \boldsymbol{X}_i\boldsymbol{\beta})' \boldsymbol{V}_i^{-1}(\mathbf{y}_i - \boldsymbol{X}_i\boldsymbol{\beta}) \quad\quad (2.14)$$

The function in (2.14) looks very much like the matrix formula for the sum of squared errors that is minimized in the standard linear model, but with the addition of the nondiagonal "weighting" matrix $\boldsymbol{V}_i^{-1}$.

Note that optimization of q($\boldsymbol{\beta}$) with respect to $\boldsymbol{\beta}$ can be carried out by applying the method of **generalized least squares** (GLS). The optimal value of $\boldsymbol{\beta}$ can be obtained analytically:

$$\widehat{\boldsymbol{\beta}} = \left(\sum_i \boldsymbol{X}_i' \widehat{\boldsymbol{V}}_i^{-1} \boldsymbol{X}_i \right)^{-1} \sum_i \boldsymbol{X}_i' \widehat{\boldsymbol{V}}_i^{-1} \mathbf{y}_i \quad\quad (2.15)$$

The estimate $\widehat{\boldsymbol{\beta}}$ has the desirable statistical property of being the best linear unbiased estimator (BLUE) of $\boldsymbol{\beta}$.

The closed-form formula in (2.15) also defines a functional relationship between the covariance parameters, $\boldsymbol{\theta}$, and the value of $\boldsymbol{\beta}$ that maximizes $\ell(\boldsymbol{\beta},\boldsymbol{\theta})$. We use this relationship in the next section to **profile out** the fixed-effect parameters, $\boldsymbol{\beta}$, from the log-likelihood, and make it strictly a function of $\boldsymbol{\theta}$.

2.4.1.2 General Case: Assume θ Is Unknown

In this section, we consider ML estimation of the covariance parameters, $\boldsymbol{\theta}$, and the fixed effects, $\boldsymbol{\beta}$, assuming $\boldsymbol{\theta}$ is unknown.

First, to obtain estimates for the covariance parameters in $\boldsymbol{\theta}$, we construct a **profile log-likelihood function** $\ell_{ML}(\boldsymbol{\theta})$. The function $\ell_{ML}(\boldsymbol{\theta})$ is derived from $\ell(\boldsymbol{\beta}, \theta)$ by replacing the $\boldsymbol{\beta}$ parameters with the expression defining $\widehat{\boldsymbol{\beta}}$ in (2.15). The resulting function is

$$\ell_{ML}(\boldsymbol{\theta}) = -0.5n \times \ln(2\pi) - 0.5 \times \sum_i \ln(\det(\boldsymbol{V}_i)) - 0.5 \times \sum_i \mathbf{r}_i' \boldsymbol{V}_i^{-1} \mathbf{r}_i \qquad (2.16)$$

where

$$\mathbf{r}_i = \mathbf{y}_i - \boldsymbol{X}_i \left(\left(\sum_i \boldsymbol{X}_i' \boldsymbol{V}_i^{-1} \boldsymbol{X}_i \right)^{-1} \sum_i \boldsymbol{X}_i' \boldsymbol{V}_i^{-1} \mathbf{y}_i \right) \qquad (2.17)$$

In general, maximization of $\ell_{ML}(\boldsymbol{\theta})$, as shown in (2.16), with respect to $\boldsymbol{\theta}$ is an example of a nonlinear optimization, with inequality constraints imposed on $\boldsymbol{\theta}$ so that positive-definiteness requirements on the $\boldsymbol{D}$ and $\boldsymbol{R}_i$ matrices are satisfied. There is no closed-form solution for the optimal $\boldsymbol{\theta}$, so the estimate of $\boldsymbol{\theta}$ is obtained by performing computational iterations until convergence (see Subsection 2.5.1).

After the ML estimates of the covariance parameters in $\boldsymbol{\theta}$ (and consequently, estimates of the variances and covariances in $\boldsymbol{D}$ and $\boldsymbol{R}_i$) are obtained through an iterative computational process, we are ready to calculate $\widehat{\boldsymbol{\beta}}$. This can be done without an iterative process, using (2.18) and (2.19). First, we replace the $\boldsymbol{D}$ and $\boldsymbol{R}_i$ matrices in (2.9) by their ML estimates, $\widehat{\boldsymbol{D}}$ and $\widehat{\boldsymbol{R}}_i$, to calculate $\widehat{\boldsymbol{V}}_i$, an estimate of $\boldsymbol{V}_i$:

$$\widehat{\boldsymbol{V}}_i = \boldsymbol{Z}_i \widehat{\boldsymbol{D}} Z_i' + \widehat{\boldsymbol{R}}_i \qquad (2.18)$$

Then, we use the generalized least-squares formula, (2.15), for $\widehat{\boldsymbol{\beta}}$, with $\boldsymbol{V}_i$ replaced by its estimate defined in (2.18) to obtain $\widehat{\boldsymbol{\beta}}$:

$$\widehat{\boldsymbol{\beta}} = \left(\sum_i \boldsymbol{X}_i' \widehat{\boldsymbol{V}}_i^{-1} \boldsymbol{X}_i \right)^{-1} \sum_i \boldsymbol{X}_i' \widehat{\boldsymbol{V}}_i^{-1} \mathbf{y}_i \qquad (2.19)$$

Because we replaced $\boldsymbol{V}_i$ by its estimate, $\widehat{\boldsymbol{V}}_i$, we say that $\widehat{\boldsymbol{\beta}}$ is the empirical best linear unbiased estimator (EBLUE) of $\boldsymbol{\beta}$.

The variance of $\widehat{\boldsymbol{\beta}}$, $var(\widehat{\boldsymbol{\beta}})$, is a $p \times p$ variance-covariance matrix calculated as follows:

$$var(\widehat{\boldsymbol{\beta}}) = \left(\sum_i \boldsymbol{X}_i' \widehat{\boldsymbol{V}}_i^{-1} \boldsymbol{X}_i \right)^{-1} \qquad (2.20)$$

We discuss issues related to the estimates of $var(\widehat{\boldsymbol{\beta}})$ in Subsection 2.4.3, because they apply to both ML and REML estimation.

The ML estimates of $\boldsymbol{\theta}$ are biased because they do not take into account the loss of degrees of freedom that results from estimating the fixed-effect parameters in $\boldsymbol{\beta}$ (see Verbeke & Molenberghs (2000) for a discussion of the bias in ML estimates of $\boldsymbol{\theta}$ in the context of LMMs). An alternative form of the maximum likelihood method known as REML estimation is frequently used to eliminate the bias in the ML estimates of the covariance parameters. We discuss REML estimation in Subsection 2.4.2.

2.4.2 REML Estimation

REML estimation is an alternative way of estimating the covariance parameters in $\boldsymbol{\theta}$. REML estimation (sometimes called residual maximum likelihood estimation) was introduced in the early 1970s by Patterson & Thompson (1971) as a method of estimating variance components in the context of unbalanced incomplete block designs. Alternative and more general derivations of REML are given by Harville (1977), Cooper & Thompson (1977), and Verbyla (1990).

REML is often preferred to ML estimation, because it produces unbiased estimates of covariance parameters by taking into account the loss of degrees of freedom that results from estimating the fixed effects in $\boldsymbol{\beta}$.

The REML estimates of $\boldsymbol{\theta}$ are based on optimization of the following **REML log-likelihood function**:

$$
\begin{aligned}
\ell_{REML}(\boldsymbol{\theta}) \;=\; & -0.5 \times (n - p) \times \ln(2\pi) - 0.5 \times \sum_i \ln(\det(\boldsymbol{V}_i)) \\
& -0.5 \times \sum_i \mathbf{r}_i' \boldsymbol{V}_i^{-1} \mathbf{r}_i - 0.5 \times \sum_i \ln(\det(\boldsymbol{X}_i' \boldsymbol{V}_i^{-1} \boldsymbol{X}_i))
\end{aligned}
\tag{2.21}
$$

In the function shown in (2.21), $\mathbf{r}_i$ is defined as in (2.17). Once an estimate, $\widehat{\boldsymbol{V}}_i$, of the $\boldsymbol{V}_i$ matrix has been obtained, REML-based estimates of the fixed-effect parameters, $\widehat{\boldsymbol{\beta}}$, and $var(\widehat{\boldsymbol{\beta}})$ can be computed. In contrast to ML estimation, the REML method does not provide a formula for the estimates. Instead, we use (2.18) and (2.19) from ML estimation to estimate the fixed-effect parameters and their standard errors.

Although we use the same formulas in (2.18) and (2.19) for REML and ML estimation of the fixed-effect parameters, it is important to note that the resulting $\widehat{\boldsymbol{\beta}}$ and corresponding $var(\widehat{\boldsymbol{\beta}})$ from REML and ML estimation are different, because the $\widehat{\boldsymbol{V}}_i$ matrix is different in each case.

2.4.3   REML vs. ML Estimation

In general terms, we use maximum likelihood methods (either REML or ML estimation) to obtain estimates of the covariance parameters in $\boldsymbol{\theta}$ in an LMM. We then obtain estimates of the fixed-effect parameters in $\boldsymbol{\beta}$ using results from generalized least squares. However, ML estimates of the covariance parameters are biased, whereas REML estimates are not.

When used to estimate the covariance parameters in $\boldsymbol{\theta}$, ML and REML estimation are computationally intensive; both involve the optimization of some objective function, which generally requires starting values for the parameter estimates and several subsequent iterations to find the values of the parameters that maximize the likelihood function (iterative methods for optimizing the likelihood function are discussed in Subsection 2.5.1). Statistical software procedures capable of fitting LMMs often provide a choice of either REML or ML as an estimation method, with the default usually being REML. Table 2.4 provides information on the estimation methods available in the software procedures discussed in this book.

Note that the variances of the estimated fixed effects, i.e., the diagonal elements in $var(\widehat{\boldsymbol{\beta}})$ as presented in (2.20), are biased downward in both ML and REML estimation, because they do not take into account the uncertainty introduced by replacing $\boldsymbol{V}_i$ with $\widehat{\boldsymbol{V}}_i$ in (2.15). Consequently, the standard errors of the estimated fixed effects, $se(\widehat{\boldsymbol{\beta}})$, are also biased downward. In the case of ML estimation, this bias is compounded by the bias in the estimation of $\boldsymbol{\theta}$ and hence in the elements of $\boldsymbol{V}_i$. To take this bias into account, approximate degrees of freedom are estimated for the t-tests or F-tests that are used for hypothesis tests about the fixed-effect parameters (see Subsection 2.6.3.1). Kenward & Roger (1997) proposed an adjustment to account for the extra variability in using $\widehat{\boldsymbol{V}}_i$ as an estimator of $\boldsymbol{V}_i$, which has been implemented in SAS `proc mixed`.

The estimated variances of the estimated fixed-effect parameters contained in $var(\widehat{\boldsymbol{\beta}})$ depend on how close $\widehat{\boldsymbol{V}}_i$ is to the "true" value of $\boldsymbol{V}_i$. To get the best possible estimate of $\boldsymbol{V}_i$ in practice, we often use REML estimation to fit LMMs with different structures for the $\boldsymbol{D}$ and $\boldsymbol{R}_i$ matrices and use model selection tools (discussed in Section 2.6) to find the best estimate for $\boldsymbol{V}_i$. We illustrate the selection of appropriate structures for the $\boldsymbol{D}$ and $\boldsymbol{R}_i$ variance-covariance matrices in detail for the LMMs that we fit in the example chapters.

Although we dealt with estimation in the LMM in this section, a very similar algorithm can be applied to the estimation of fixed effects and covariance parameters in the marginal model specified in Section 2.3.

2.5 Computational Issues

2.5.1 Algorithms for Likelihood Function Optimization

Having defined the ML and REML estimation methods, we briefly introduce the computational algorithms used to carry out the estimation for an LMM.

The key computational difficulty in the analysis of LMMs is estimation of the covariance parameters, using iterative **numerical optimization** of the log-likelihood functions introduced in Subsection 2.4.1.2 for ML estimation and in Subsection 2.4.2 for REML estimation, subject to constraints imposed on the parameters to ensure positive-definiteness of the $\boldsymbol{D}$ and $\boldsymbol{R}_i$ matrices. The most common iterative algorithms used for this optimization problem in the context of LMMs are the **expectation-maximization (EM) algorithm**, the **Newton–Raphson (N–R) algorithm** (the preferred method), and the **Fisher scoring algorithm**.

The EM algorithm is often used to maximize complicated likelihood functions or to find good starting values of the parameters to be used in other algorithms (this latter approach is currently used by the procedures in R, Stata, and HLM, as shown in Table 2.4). General descriptions of the EM algorithm, which alternates between expectation (E) and maximization (M) steps, can be found in Dempster et al. (1977) and Laird et al. (1987). For "incomplete" data sets arising from studies with unbalanced designs, the E-step involves, at least conceptually, creation of a "complete" data set based on a hypothetical scenario, in which we assume that data have been obtained from a balanced design and there are no missing observations for the dependent variable. In the context of the LMM, the complete data set is obtained by augmenting observed values of the dependent variable with expected values of the sum of squares and sum of products of the unobserved random effects and residuals. The complete data are obtained using the information available at the current iteration of the algorithm, i.e., the current values of the covariance parameter estimates and the observed values of the dependent variable. Based on the complete data, an objective function called the **complete data log-likelihood function** is constructed and maximized in the M-step, so that the vector of estimated $\boldsymbol{\theta}$ parameters is updated at each iteration. The underlying assumption behind the EM algorithm is that optimization of the complete data log-likelihood function is simpler than optimization of the likelihood based on the observed data.

The main drawback of the EM algorithm is its slow rate of convergence. In addition, the precision of estimators derived from the EM algorithm is overly optimistic, because the estimators are based on the likelihood from the last maximization step, which uses complete data instead of observed data. Although some solutions have been proposed to overcome these shortcomings, the EM algorithm is rarely used to fit LMMs, except to provide starting values for other algorithms.

The N–R algorithm and its variations are the most commonly used algorithms in ML and REML estimation of LMMs. The N–R algorithm minimizes an objective function defined as –2 times the log-likelihood function for the covariance parameters specified in Subsection 2.4.1.2 for ML estimation or in Subsection 2.4.2 for REML estimation. At every iteration, the N–R algorithm requires calculation of the vector of partial derivatives (the **gradient**), and the second derivative matrix with respect to the covariance parameters (the **observed Hessian matrix**). Analytical formulas for these matrices are given in Jennrich & Schluchter (1986) and Lindstrom & Bates (1988). Owing to Hessian matrix calculations, N–R iterations are more time consuming, but convergence is usually achieved in fewer iterations than when using the EM algorithm. Another advantage of using the N–R algorithm is that the Hessian matrix from the last iteration can be used to obtain an asymptotic variance-covariance matrix for the estimated covariance parameters in $\boldsymbol{\theta}$, allowing for calculation of standard errors of $\widehat{\boldsymbol{\theta}}$.

The Fisher scoring algorithm can be considered as a modification of the N–R algorithm. The primary difference is that Fisher scoring uses the **expected Hessian matrix** rather than the observed one. Although Fisher scoring is often more stable numerically, more likely to converge, and calculations performed at each iteration are simplified compared to the N–R algorithm, Fisher scoring is not recommended to obtain final estimates. The primary disadvantage of the Fisher scoring algorithm, as pointed out by Little & Rubin (2002), is that it may be difficult to determine the expected value of the Hessian matrix because of difficulties with identifying the appropriate sampling distribution. To avoid problems with determining the expected Hessian matrix, use of the N–R algorithm instead of the Fisher scoring algorithm is recommended.

To initiate optimization of the N–R algorithm, a sensible choice of starting values for the covariance parameters is needed. One method for choosing starting values is to use a noniterative method based on method-of-moment estimators (Rao, 1972). Alternatively, a small number of EM iterations can be performed to obtain starting values. In other cases, initial values may be assigned explicitly by the analyst.

The optimization algorithms used to implement ML and REML estimation need to ensure that the estimates of the $\boldsymbol{D}$ and $\boldsymbol{R}_i$ matrices are positive-definite. In general, it is preferable to ensure that estimates of the covariance parameters in $\boldsymbol{\theta}$, updated from one iteration of an optimization algorithm to the next, imply positive-definiteness of $\boldsymbol{D}$ and $\boldsymbol{R}_i$ at every step of the estimation process. Unfortunately, it is difficult to meet these requirements, so software procedures set much simpler conditions that are necessary, but not sufficient, to meet positive-definiteness constraints. Specifically, it is much simpler to ensure that elements on the diagonal of the estimated $\boldsymbol{D}$ and $\boldsymbol{R}_i$ matrices are greater than zero during the entire iteration process, and this method is often used by software procedures in practice. At the last iteration, estimates of the $\boldsymbol{D}$ and $\boldsymbol{R}_i$ matrices are checked for being positive-definite, and a warning message is issued if the positive-definiteness constraints are not satisfied. See Subsection 6.4.1 for a discussion of a nonpositive-definite $\boldsymbol{D}$ matrix (called the $\boldsymbol{G}$ matrix in SAS), in the analysis of the Autism data using `proc mixed` in SAS.

An alternative way to address positive-definiteness constraints is to apply a log-Cholesky decomposition (or other transformations) to the $\boldsymbol{D}$ and/or $\boldsymbol{R}_i$ matrices, which results in substantial simplification of the optimization problem. This method changes the problem from a constrained to an unconstrained one and ensures that the $\boldsymbol{D}, \boldsymbol{R}_i$, or both matrices are positive-definite during the entire estimation process (see Pinheiro & Bates, 1996, for more details on the log-Cholesky decomposition method). Table 2.4 details the computational algorithms used to implement both ML and REML estimation by the LMM procedures in the five software packages presented in this book.

TABLE 2.4: Computational Algorithms Used by the Software Procedures for Estimation of the Covariance Parameters in an LMM

Software Procedures	Available Estimation Methods, *Default Method*	Computational Algorithms
SAS `proc mixed`	ML, *REML*	Ridge-stabilized N–R,[a] Fisher scoring
SPSS `MIXED`	ML, *REML*	N–R, Fisher scoring
R: `lme()` function	ML, *REML*	EM[b] algorithm,[c] N–R
R: `lmer()` function	ML, *REML*	EM[b] algorithm,[c] N–R
Stata: `mixed` command	*ML*, REML	EM algorithm, N–R (default)
HLM: HLM2 (Chapters 3, 5, 6)	ML, *REML*	EM algorithm, Fisher scoring
HLM: HLM3 (Chapter 4)	*ML*	EM algorithm, Fisher scoring
HLM: HMLM2 (Chapter 7)	*ML*	EM algorithm, Fisher scoring
HLM: HCM2 (Chapter 8)	*ML*	EM algorithm, Fisher scoring

[a]N–R denotes the Newton–Raphson algorithm (see Subsection 2.5.1).
[b]EM denotes the Expectation-Maximization algorithm (see Subsection 2.5.1).
[c]The `lme()` function in R actually use the ECME (expectation conditional maximization either) algorithm, which is a modification of the EM algorithm. For details, see Liu and Rubin (1994).

2.5.2 Computational Problems with Estimation of Covariance Parameters

The random effects in the $\mathbf{u}_i$ vector in an LMM are assumed to arise from a multivariate normal distribution with variances and covariances described by the positive-definite variance-covariance matrix $\boldsymbol{D}$. Occasionally, when one is using a software procedure to fit an LMM, depending on (1) the nature of a clustered or longitudinal data set, (2) the degree of similarity of observations within a given level of a random factor, or (3) model misspecification, the iterative estimation routines converge to a value for the estimate of a covariance parameter in $\boldsymbol{\theta}_D$ that lies very close to or outside the boundary of the parameter space. Consequently, the estimate of the $\boldsymbol{D}$ matrix may not be positive-definite.

Note that in the context of estimation of the $\boldsymbol{D}$ matrix, we consider positive-definiteness in a numerical, rather than mathematical, sense. By numerical, we mean that we take into account the finite numeric precision of a computer.

Each software procedure produces different error messages or notes when computational problems are encountered in estimating the $\boldsymbol{D}$ matrix. In some cases, some software procedures (e.g., `proc mixed` in SAS, or `MIXED` in SPSS) stop the estimation process, assume that an estimated variance in the $\boldsymbol{D}$ matrix lies on a boundary of the parameter space, and report that the estimated D matrix is not positive-definite (in a numerical sense). In other cases, computational algorithms elude the positive-definiteness criteria and converge to an estimate of the D matrix that is outside the allowed parameter space (a nonpositive-definite matrix). We encounter this type of problem when fitting Model 6.1 in Chapter 6 (see Subsection 6.4.1).

In general, when fitting an LMM, analysts should be aware of warning messages indicating that the estimated D matrix is not positive-definite and interpret parameter estimates with extreme caution when these types of messages are produced by a software procedure.

We list some alternative approaches for fitting the model when problems arise with estimation of the covariance parameters:

1. **Choose alternative starting values for covariance parameter estimates:** If a computational algorithm does not converge or converges to possibly suboptimal values for the covariance parameter estimates, the problem may lie in the choice of starting values for covariance parameter estimates. To remedy this problem, we may choose alternative starting values or initiate computations using a more stable algorithm, such as the EM algorithm (see Subsection 2.5.1).

2. **Rescale the covariates:** In some cases, covariance parameters are very different in magnitude and may even be several orders of magnitude apart. Joint estimation of covariance parameters may cause one of the parameters to become extremely small, approaching the boundary of the parameter space, and the D matrix may become nonpositive-definite (within the numerical tolerance of the computer being used). If this occurs, one could consider rescaling the covariates associated with the small covariance parameters. For example, if a covariate measures time in minutes and a study is designed to last several days, the values on the covariate could become very large and the associated variance component could be small (because the incremental effects of time associated with different subjects will be relatively small). Dividing the time covariate by a large number (e.g., 60, so that time would be measured in hours instead of minutes) may enable the corresponding random effects and their variances to be on a scale more similar to that of the other covariance parameters. Such rescaling may improve numerical stability of the optimization algorithm and may circumvent convergence problems. We do not consider this alternative in any of the examples that we discuss.

3. **Based on the design of the study, simplify the model by removing random effects that may not be necessary:** In general, we recommend removing higher-order terms (e.g., higher-level interactions and higher-level polynomials) from a model first for both random and fixed effects. This method helps to ensure that the reduced model remains well formulated (Morrell et al., 1997).

   However, in some cases, it may be appropriate to remove lower-order random effects first, while retaining higher-order random effects in a model; such an approach requires thorough justification. For instance, in the analysis of the longitudinal data for the Autism example in Chapter 6, we remove the random effects associated with the intercept (which contribute to variation at all time points for a given subject) first, while retaining random effects associated with the linear and quadratic effects of age. By doing this, we assume that all variation between measurements of the dependent variable at the initial time point is attributable to residual variation (i.e., we assume that none of the overall variation at the first time point is attributable to between-subject variation). To implement this in an LMM, we define additional random effects (i.e., the random linear and quadratic effects associated with age) in such a way that they do not contribute to the variation at the initial time point, and consequently, all variation at this time point is due to residual error. Another implication of this choice is that between-subject variation is described using random linear and quadratic effects of age only.

4. **Fit the implied marginal model:** As mentioned in Section 2.3, one can sometimes fit the marginal model implied by a given LMM. The important difference when fitting the implied marginal model is that there are fewer restrictions on the covariance parameters being estimated. We present two examples of this approach:

   (a) If one is fitting an LMM with random intercepts only and a homogeneous residual covariance structure, one can directly fit the marginal model implied by this LMM by fitting a model with random effects omitted, and with a compound symmetry covariance structure for the residuals. We present an example of this approach in the analysis of the Dental Veneer data in Subsection 7.11.1.

   (b) Another approach is to "emulate" the fit of an implied marginal model by fitting an LMM and, if needed, removing the positive-definiteness constraints on the D and the R_i matrices. The option of relaxing constraints on the D and R_i matrices is currently only available in SAS `proc mixed`, via use of the `nobound` option. We consider this approach in the analysis of the Autism data in Subsection 6.4.1.

5. **Fit the marginal model with an unstructured covariance matrix:** In some cases, software procedures are not capable of fitting an implied marginal model, which involves less restrictive constraints imposed on the covariance parameters. If measurements are taken at a relatively small number of prespecified time points for all subjects, one can instead fit a marginal model (without any random effects specified) with an unstructured covariance matrix for the residuals. We consider this alternative approach in the analysis of the Autism data in Chapter 6.

Note that none of these alternative methods guarantees convergence to the optimal and properly constrained values of covariance parameter estimates. The methods that involve fitting a marginal model (items 4 and 5 in the preceding text) shift a more restrictive requirement for the D and R_i matrices to be positive-definite to a less restrictive requirement for the matrix V_i (or V_i^*) to be positive-definite, but they still do not guarantee convergence. In addition, methods involving marginal models do not allow for inferences about random effects and their variances.

2.6 Tools for Model Selection

When analyzing clustered and repeated-measures/longitudinal data sets using LMMs, researchers are faced with several competing models for a given data set. These competing models describe sources of variation in the dependent variable and at the same time allow researchers to test hypotheses of interest. It is an important task to select the "best" model, i.e., a model that is parsimonious in terms of the number of parameters used, and at the same time is best at predicting (or explaining variation in) the dependent variable.

In selecting the best model for a given data set, we take into account research objectives, sampling and study design, previous knowledge about important predictors, and important subject matter considerations. We also use analytic tools, such as the hypothesis tests and the information criteria discussed in this section. Before we discuss specific hypothesis tests and information criteria in detail, we introduce the basic concepts of nested models and hypothesis specification and testing in the context of LMMs. For readers interested in additional details on current practice and research in model selection techniques for LMMs, we suggest Steele (2013).

2.6.1 Basic Concepts in Model Selection

2.6.1.1 Nested Models

An important concept in the context of model selection is to establish whether, for any given pair of models, there is a "nesting" relationship between them. Assume that we have two competing models: Model A and Model B. We define Model A to be **nested** in Model B if Model A is a "special case" of Model B. By special case, we mean that the parameter space for the nested Model A is a subspace of that for the more general Model B. Less formally, we can say that the parameters in the nested model can be obtained by imposing certain constraints on the parameters in the more general model. In the context of LMMs, a model is nested within another model if a set of fixed effects and/or covariance parameters in a nested model can be obtained by imposing constraints on parameters in a more general model (e.g., constraining certain parameters to be equal to zero or equal to each other).

2.6.1.2 Hypotheses: Specification and Testing

Hypotheses about parameters in an LMM are specified by providing null (H_0) and alternative (H_A) hypotheses about the parameters in question. Hypotheses can also be formulated in the context of two models that have a nesting relationship. A more general model encompasses both the null and alternative hypotheses, and we refer to it as a **reference model**. A second simpler model satisfies the null hypothesis, and we refer to this model as a **nested (null hypothesis) model**. Briefly speaking, the only difference between these two models is that the reference model contains the parameters being tested, but the nested (null) model does not.

Hypothesis tests are useful tools for making decisions about which model (nested vs. reference) to choose. The likelihood ratio tests presented in Subsection 2.6.2 require analysts to fit both the reference and nested models. In contrast, the alternative tests presented in Subsection 2.6.3 require fitting only the reference model.

We refer to nested and reference models explicitly in the example chapters when testing various hypotheses. We also include a diagram in each of the example chapters (e.g., Figure 3.3) that indicates the nesting of models, and the choice of preferred models based on results of formal hypothesis tests or other considerations.

2.6.2 Likelihood Ratio Tests (LRTs)

LRTs are a class of tests that are based on comparing the values of likelihood functions for two models (i.e., the nested and reference models) defining a hypothesis being tested. LRTs can be employed to test hypotheses about covariance parameters or fixed-effect parameters in the context of LMMs. In general, LRTs require that both the nested (null hypothesis) model and reference model corresponding to a specified hypothesis are fitted to the *same* subset of the data. The LRT statistic is calculated by subtracting –2 times the log-likelihood for the reference model from that for the nested model, as shown in the following equation:

$$-2\ln\left(\frac{L_{nested}}{L_{reference}}\right) = -2\ln(L_{nested}) - (-2\ln(L_{reference})) \sim \chi^2_{df} \qquad (2.22)$$

In (2.22), L_{nested} refers to the value of the likelihood function evaluated at the ML or REML estimates of the parameters in the nested model, and $L_{reference}$ refers to the value of the likelihood function in the reference model. Likelihood theory states that under mild regularity conditions the LRT statistic asymptotically follows a χ^2 distribution, in which the number of degrees of freedom, *df*, is obtained by subtracting the number of parameters in the nested model from the number of parameters in the reference model.

Using the result in (2.22), hypotheses about the parameters in LMMs can be tested. The significance of the likelihood ratio test statistic can be determined by referring it to a χ^2 distribution with the appropriate degrees of freedom. If the LRT statistic is sufficiently large, there is evidence against the null hypothesis model and in favor of the reference model. If the likelihood values of the two models are very close, and the resulting LRT statistic is small, we have evidence in favor of the nested (null hypothesis) model.

2.6.2.1 Likelihood Ratio Tests for Fixed-Effect Parameters

The likelihood ratio tests that we use to test linear hypotheses about fixed-effect parameters in an LMM are based on ML estimation; using REML estimation is not appropriate in this context (Morrell, 1998; Pinheiro & Bates, 2000; Verbeke & Molenberghs, 2000). For LRTs of fixed effects, the nested and reference models have the same set of covariance parameters but different sets of fixed-effect parameters. The test statistic is calculated by subtracting the –2 ML log-likelihood for the reference model from that for the nested model. The asymptotic null distribution of the test statistic is a χ^2 with degrees of freedom equal to the difference in the number of fixed-effect parameters between the two models.

2.6.2.2 Likelihood Ratio Tests for Covariance Parameters

When testing hypotheses about covariance parameters in an LMM, REML estimation should be used for both the reference and nested models, especially in the context of small sample size. REML estimation has been shown to reduce the bias inherent in ML estimates of covariance parameters (Morrell, 1998). We assume that the nested and reference models have the same set of fixed-effect parameters, but different sets of covariance parameters.

To carry out a REML-based likelihood ratio test for covariance parameters, the –2 REML log-likelihood value for the reference model is subtracted from that for the nested model. The null distribution of the test statistic depends on whether the null hypothesis values for the covariance parameters lie on the boundary of the parameter space for the covariance parameters or not.

Case 1: The covariance parameters satisfying the null hypothesis do not lie on the boundary of the parameter space.

When carrying out a REML-based likelihood ratio test for covariance parameters in which the null hypothesis does not involve testing whether any parameters lie on the boundary of the parameter space (e.g., testing a model with heterogeneous residual variance vs. a model with constant residual variance, or testing whether a covariance between two random effects is equal to zero), the test statistic is asymptotically distributed as a χ^2 with degrees of freedom calculated by subtracting the number of covariance parameters in the nested model from that in the reference model. An example of such a test is given in Subsection 5.5.2, in which we test a heterogeneous residual variance model vs. a model with constant residual variance (Hypothesis 5.2 in the Rat Brain example).

Case 2: The covariance parameters satisfying the null hypothesis lie on the boundary of the parameter space.

Tests of null hypotheses in which covariance parameters have values that lie on the boundary of the parameter space often arise in the context of testing whether a given random effect should be kept in a model or not. We do not directly test hypotheses about the random effects themselves. Instead, we test whether the corresponding variances and covariances of the random effects are equal to zero.

In the case in which we have a single random effect in a model, we might wish to test the null hypothesis that the random effect can be omitted. Self & Liang (1987), Stram & Lee (1994), and Verbeke & Molenberghs (2000) have shown that the test statistic in this case has an asymptotic null distribution that is a mixture of χ_0^2 and χ_1^2 distributions, with each having an equal weight of 0.5. Note that the χ_0^2 distribution is concentrated entirely at zero, so calculations of p-values can be simplified and effectively are based on the χ_1^2 distribution only. An example of this type of test is given in the analysis of the Rat Pup data, in which we test whether the variance of the random intercepts associated with litters is equal to zero in Subsection 3.5.1 (Hypothesis 3.1).

In the case in which we have two random effects in a model and we wish to test whether one of them can be omitted, we need to test whether the variance for the given random effect that we wish to test and the associated covariance of the two random effects are both equal to zero. The asymptotic null distribution of the test statistic in this case is a mixture of χ_1^2 and χ_2^2 distributions, with each having an equal weight of 0.5 (Verbeke & Molenberghs, 2000). An example of this type of likelihood ratio test is shown in the analysis of the Autism data in Chapter 6, in which we test whether the variance associated with the random quadratic age effects and the associated covariance of these random effects with the random linear age effects are both equal to zero in Subsection 6.5.1 (Hypothesis 6.1).

Because most statistical software procedures capable of fitting LMMs provide the option of using either ML estimation or REML estimation for a given model, one can choose to use REML estimation to fit the reference and nested models when testing hypotheses about covariance parameters, and ML estimation when testing hypotheses about fixed effects.

Finally, we note that Crainiceanu & Ruppert (2004) have defined the exact null distribution of the likelihood ratio test statistic for a single variance component under more general conditions (including small samples), and software is available in R implementing exact likelihood ratio tests based on simulations from this distribution (the `exactLRT()` function in the `RLRsim` package). Galecki & Burzykowski (2013) present examples of the use of this function. Appropriate null distributions of likelihood ratio test statistics for multiple covariance parameters have not been derived to date, and classical likelihood ratio tests comparing nested models with multiple variance components constrained to be 0 in the reduced model should be considered conservative. The `mixed` command in Stata, for example, makes explicit note of this when users fit models with multiple random effects (for example, see Section 5.4.4).

2.6.3   Alternative Tests

In this section we present alternatives to likelihood ratio tests of hypotheses about the parameters in a given LMM. Unlike the likelihood ratio tests discussed in Subsection 2.6.2, these tests require fitting only a reference model.

2.6.3.1   Alternative Tests for Fixed-Effect Parameters

A *t*-test is often used for testing hypotheses about a single fixed-effect parameter (e.g., $H_0 : \beta = 0$ vs. $H_A : \beta \neq 0$) in an LMM. The corresponding t-statistic is calculated as follows:

$$t = \frac{\widehat{\beta}}{\text{se}(\widehat{\beta})} \tag{2.23}$$

In the context of an LMM, the null distribution of the t-statistic in (2.23) does not in general follow an exact t distribution. Unlike the case of the standard linear model, the number of degrees of freedom for the null distribution of the test statistic is not equal to $n - p$ (where p is the total number of fixed-effect parameters estimated). Instead, we use approximate methods to estimate the degrees of freedom. The approximate methods for degrees of freedom for both t-tests and F-tests are discussed later in this section.

Software Note: The `mixed` command in Stata calculates z-statistics for tests of single fixed-effect parameters in an LMM using the same formula as specified for the t-test in (2.23). These z-statistics assume large sample sizes and refer to the standard normal distribution, and therefore do not require the calculation of degrees of freedom to derive a p-value.

An **F-test** can be used to test linear hypotheses about multiple fixed effects in an LMM. For example, we may wish to test whether any of the parameters associated with the levels of a fixed factor are different from zero. In general, when testing a linear hypothesis of the form

$$\text{H}_0\colon \boldsymbol{L\beta} = \boldsymbol{0} \quad \text{vs.} \quad \text{H}_A\colon \boldsymbol{L\beta} \neq \boldsymbol{0}$$

where $\boldsymbol{L}$ is a known matrix, the F-statistic defined by

$$F = \frac{\widehat{\boldsymbol{\beta}}' \boldsymbol{L}' \left(\boldsymbol{L} \left(\sum_i \boldsymbol{X}_i' \boldsymbol{V}_i^{-1} \boldsymbol{X}_i \right)^{-1} \boldsymbol{L}' \right)^{-1} \boldsymbol{L}\widehat{\boldsymbol{\beta}}}{\text{rank}(\boldsymbol{L})} \tag{2.24}$$

follows an approximate F distribution, with numerator degrees of freedom equal to the rank of the matrix $\boldsymbol{L}$ (recall that the rank of a matrix is the number of linearly independent rows or columns), and an approximate denominator degrees of freedom that can be estimated using various methods (Verbeke & Molenberghs, 2000).

Similar to the case of the t-test, the F-statistic in general does not follow an exact F distribution, with known numerator and denominator degrees of freedom. Instead, the denominator degrees of freedom are approximated. The approximate methods that apply to both t-tests and F-tests take into account the presence of random effects and correlated residuals in an LMM. Several of these approximate methods (e.g., the Satterthwaite method, or the "between-within" method) involve different choices for the degrees of freedom used in the approximate t-tests and F-tests. The Kenward–Roger method goes a step further. In addition to adjusting the degrees of freedom using the Satterthwaite method, this method also modifies the estimated covariance matrix to reflect uncertainty in using $\widehat{\boldsymbol{V}}_i$ as a substitute for $\boldsymbol{V}_i$ in (2.19) and (2.20). We discuss these approximate methods in more detail in Subsection 3.11.6.

Different types of F-tests are often used in practice. We focus on **Type I F-tests** and **Type III F-tests**. Briefly, Type III F-tests are conditional on the effects of all other terms in a given model, whereas Type I (sequential) F-tests are conditional on just the fixed effects listed in the model prior to the effects being tested. Type I and Type III F-tests are therefore equivalent only for the term entered last in the model (except for certain models for balanced data). We compare these two types of F-tests in more detail in the example chapters.

An **omnibus Wald test** can also be used to test linear hypotheses of the form H_0: $\boldsymbol{L\beta} = \boldsymbol{0}$ vs. $H_A : \boldsymbol{L\beta} \neq \boldsymbol{0}$. The test statistic for a Wald test is the numerator in (2.24), and it asymptotically follows a χ^2 distribution with degrees of freedom equal to the rank of the $\boldsymbol{L}$ matrix. We consider Wald tests for fixed effects using the Stata and HLM software in the example chapters.

2.6.3.2 Alternative Tests for Covariance Parameters

A simple test for covariance parameters is the Wald z-test. In this test, a z-statistic is computed by dividing an estimated covariance parameter by its estimated standard error. The p-value for the test is calculated by referring the test statistic to a standard normal distribution. The Wald z-test is asymptotic, and requires that the random factor with which the random effects are associated has a large number of levels. This test statistic also has unfavorable properties when a hypothesis test about a covariance parameter involves values on the boundary of its parameter space. Because of these drawbacks, we do not recommend using Wald z-tests for covariance parameters, and instead recommend the use of likelihood ratio tests, with p-values calculated using appropriate χ^2 distributions or mixtures of χ^2 distributions.

The procedures in the HLM software package by default generate alternative chi-square tests for covariance parameters in an LMM (see Subsection 4.7.2 for an example). These tests are described in detail in Raudenbush & Bryk (2002).

2.6.4 Information Criteria

Another set of tools useful in model selection are referred to as **information criteria**. The information criteria (sometimes referred to as *fit* criteria) provide a way to assess the fit of a model based on its optimum log-likelihood value, after applying a penalty for the number of parameters that are estimated in fitting the model. A key feature of the information criteria discussed in this section is that they provide a way to compare any two models fitted to the same set of observations; i.e., the models do not need to be nested. We use the "smaller is better" form for the information criteria discussed in this section; that is, a smaller value of the criterion indicates a "better" fit.

The **Akaike information criterion (AIC)** may be calculated based on the (ML or REML) log-likelihood, $\ell(\boldsymbol{\beta}, \boldsymbol{\theta})$, of a fitted model as follows (Akaike, 1973):

$$AIC = -2 \times \ell(\widehat{\boldsymbol{\beta}}, \widehat{\boldsymbol{\theta}}) + 2p \qquad (2.25)$$

In (2.25), p represents the total number of parameters being estimated in the model for both the fixed and random effects. Note that the AIC in effect "penalizes" the fit of a model for the number of parameters being estimated by adding $2p$ to the -2 log-likelihood. Some software procedures calculate the AIC using slightly different formulas, depending on whether ML or REML estimation is being used (see Subsection 3.6.1 for a discussion of the calculation formulas used for the AIC in the different software procedures).

The **Bayes information criterion (BIC)** is also commonly used and may be calculated as follows:

$$BIC = -2 \times \ell(\widehat{\boldsymbol{\beta}}, \widehat{\boldsymbol{\theta}}) + p \times \ln(n) \qquad (2.26)$$

The BIC in (2.26) applies a greater penalty for models with more parameters than does the AIC, because we multiply the number of parameters being estimated by the natural logarithm of n, where n is the total number of observations used in estimation of the model.

Recent work (Steele, 2013; Gurka, 2006) suggests that no one information criterion stands apart as the best criterion to be used when selecting LMMs, and that more work still needs to be done in understanding the role that information criteria play in the selection of LMMs. Consistent with Steele (2013), we recommend that analysts compute a variety of information criteria when choosing among competing models, and identify models favored by multiple criteria.

2.7   Model-Building Strategies

A primary goal of model selection is to choose the simplest model that provides the best fit to the observed data. There may be several choices concerning which fixed and random effects should be included in an LMM. There are also many possible choices of covariance structures for the D and R_i matrices. All of these considerations have an impact on both the estimated marginal mean ($X_i\beta$) and the estimated marginal variance-covariance matrix $V_i(= Z_i D Z_i' + R_i)$ for the observed responses in Y_i based on the specified model.

The process of building an LMM for a given set of longitudinal or clustered data is an iterative one that requires a series of model-fitting steps and investigations, and selection of appropriate mean and covariance structures for the observed data. Model building typically involves a balance of statistical and subject matter considerations; there is no single strategy that applies to every application.

2.7.1   The Top-Down Strategy

The following broadly defined steps are suggested by Verbeke & Molenberghs (2000) for building an LMM for a given data set. We refer to these steps as a *top-down* strategy for model building, because they involve starting with a model that includes the maximum number of fixed effects that we wish to consider in a model.

1. **Start with a well-specified mean structure for the model:** This step typically involves adding the fixed effects of as many covariates (and interactions between the covariates) as possible to the model to make sure that the systematic variation in the responses is well explained before investigating various covariance structures to describe random variation in the data. In the example chapters we refer to this as a model with a *loaded* mean structure.

2. **Select a structure for the random effects in the model:** This step involves the selection of a set of random effects to include in the model. The need for including the selected random effects can be tested by performing REML-based likelihood ratio tests for the associated covariance parameters (see Subsection 2.6.2.2 for a discussion of likelihood ratio tests for covariance parameters).

3. **Select a covariance structure for the residuals in the model:** Once fixed effects and random effects have been added to the model, the remaining variation in the observed responses is due to residual error, and an appropriate covariance structure for the residuals should be investigated.

4. **Reduce the model:** This step involves using appropriate statistical tests (see Subsections 2.6.2.1 and 2.6.3.1) to determine whether certain fixed-effect parameters are needed in the model.

In general, one would iterate between Steps 2 and 4 when building a model. We use this top-down approach to model building for the data sets that we analyze in Chapter 3, and Chapters 5 through 7.

2.7.2 The Step-Up Strategy

An alternative approach to model building, which we refer to as the *step-up* strategy, has been developed in the literature on HLMs. We use the step-up model-building strategy in the analysis of the Classroom data in Chapter 4. This approach is outlined in both Raudenbush & Bryk (2002) and Snijders & Bosker (1999), and is described in the following text:

1. **Start with an "unconditional" (or means-only) Level 1 model for the data:** This step involves fitting an initial Level 1 model having the fixed intercept as the only fixed-effect parameter. The model also includes random effects associated with the Level 2 units, and Level 3 units in the case of a three-level data set. This model allows one to assess the variation in the response values across the different levels of the clustered or longitudinal data set without adjusting for the effects of any covariates.

2. **Build the model by adding Level 1 covariates to the Level 1 model. In the Level 2 model, consider adding random effects to the equations for the coefficients of the Level 1 covariates:** In this step, Level 1 covariates and their associated fixed effects are added to the Level 1 model. These Level 1 covariates may help to explain variation in the residuals associated with the observations on the Level 1 units. The Level 2 model can also be modified by adding random effects to the equations for the coefficients of the Level 1 covariates. These random effects allow for random variation in the effects of the Level 1 covariates across Level 2 units.

3. **Build the model by adding Level 2 covariates to the Level 2 model. For three-level models, consider adding random effects to the Level 3 equations for the coefficients of the Level 2 covariates:** In this step, Level 2 covariates and their associated fixed effects can be added to the Level 2 model. These Level 2 covariates may explain some of the random variation in the effects of the Level 1 covariates that is captured by the random effects in the Level 2 models. In the case of a three-level data set, the effects of the Level 2 covariates in the Level 2 model might also be allowed to vary randomly across Level 3 units. After appropriate equations for the effects of the Level 1 covariates have been specified in the Level 2 model, one can assess assumptions about the random effects in the Level 2 model (e.g., normality and constant variance). This process is then repeated for the Level 3 model in the case of a three-level analysis (e.g., Chapter 4).

The model-building steps that we present in this section are meant to be guidelines and are not hard-and-fast rules for model selection. In the example chapters, we illustrate aspects of the top-down and step-up model-building strategies when fitting LMMs to real data sets. Our aim is to illustrate specific concepts in the analysis of longitudinal or clustered data, rather than to construct the best LMM for a given data set.

2.8 Checking Model Assumptions (Diagnostics)

After fitting an LMM, it is important to carry out **model diagnostics** to check whether distributional assumptions for the residuals are satisfied and whether the fit of the model is sensitive to unusual observations. The process of carrying out model diagnostics involves several informal and formal techniques.

Diagnostic methods for standard linear models are well established in the statistics literature. In contrast, diagnostics for LMMs are more difficult to perform and interpret, because the model itself is more complex, due to the presence of random effects and different covariance structures. In this section, we focus on the definitions of a selected set of terms related to residual and influence diagnostics in LMMs. We refer readers to Claeskens (2013) and Schabenberger (2004) for more detailed descriptions of existing diagnostic methods for LMMs.

In general, model diagnostics should be part of the model-building process throughout the analysis of a clustered or longitudinal data set. We consider diagnostics only for the final model fitted in each of the example chapters for simplicity of presentation.

2.8.1 Residual Diagnostics

Informal techniques are commonly used to check residual diagnostics; these techniques rely on the human mind and eye, and are used to decide whether or not a specific pattern exists in the residuals. In the context of the standard linear model, the simplest example is to decide whether a given set of residuals plotted against predicted values represents a random pattern or not. These residual vs. fitted plots are used to verify model assumptions and to detect outliers and potentially influential observations.

In general, residuals should be assessed for normality, constant variance, and outliers. In the context of LMMs, we consider conditional residuals and their "studentized" versions, as described in the following subsections.

2.8.1.1 Raw Residuals

A **conditional residual** is the difference between the observed value and the conditional predicted value of the dependent variable. For example, we write an equation for the vector of conditional residuals for a given individual i in a two-level longitudinal data set as follows (refer to Subsection 2.9.1 for the calculation of $\widehat{u}_i$):

$$\widehat{\varepsilon}_i = \mathbf{y}_i - \boldsymbol{X}_i\widehat{\boldsymbol{\beta}} - \boldsymbol{Z}_i\widehat{\mathbf{u}}_i \tag{2.27}$$

In general, conditional residuals in their basic form in (2.27) are not well suited for verifying model assumptions and detecting outliers. Even if the true model residuals are uncorrelated and have equal variance, conditional residuals will tend to be correlated and their variances may be different for different subgroups of individuals. The shortcomings of raw conditional residuals apply to models other than LMMs as well. We discuss alternative forms of the conditional residuals in Subsection 2.8.1.2.

In contrast to conditional residuals, **marginal residuals** are based on models that do not include explicit random effects:

$$\widehat{\varepsilon}_i^* = \mathbf{y}_i - \boldsymbol{X}_i\widehat{\boldsymbol{\beta}} \tag{2.28}$$

Although we consider diagnostic tools for conditional residuals in this section, a separate class of diagnostic tools exists for the marginal residuals defined in (2.28) (see Galecki &

Burzykowski, 2013, for more details). We consider examples of different covariance structures for marginal residuals in later chapters.

2.8.1.2   Standardized and Studentized Residuals

To alleviate problems with the interpretation of conditional residuals that may have unequal variances, we consider scaling (i.e., dividing) the residuals by their true or estimated standard deviations. Ideally, we would like to scale residuals by their true standard deviations to obtain **standardized residuals**. Unfortunately, the true standard deviations are rarely known in practice, so scaling is done using estimated standard deviations instead. Residuals obtained in this manner are called **studentized residuals**.

Another method of scaling residuals is to divide them by the estimated standard deviation of the dependent variable. The resulting residuals are called **Pearson residuals**. Pearson-type scaling is appropriate if we assume that variability of $\widehat{\boldsymbol{\beta}}$ can be ignored. Other scaling choices are also possible, although we do not consider them.

The calculation of a studentized residual may also depend on whether the observation corresponding to the residual in question is included in the estimation of the standard deviation or not. If the corresponding observation is included, we refer to it as **internal studentization**. If the observation is excluded, we refer to it as **external studentization**.

We discuss studentized residuals in the model diagnostics section in the analysis of the Rat Pup data in Chapter 3. Studentized residuals are directly available in SAS `proc mixed`, but are not readily available in the other software that we feature, and require additional calculation.

2.8.2   Influence Diagnostics

Likelihood-based estimation methods (both ML and REML) are sensitive to unusual observations. **Influence diagnostics** are formal techniques that allow one to identify observations that heavily influence estimates of the parameters in either $\boldsymbol{\beta}$ or $\boldsymbol{\theta}$.

Influence diagnostics for LMMs is an active area of research. The idea of influence diagnostics for a given observation (or subset of observations) is to quantify the effect of omission of those observations on the results of the analysis of the entire data set. Schabenberger discusses several influence diagnostics for LMMs in detail (Schabenberger, 2004), and a recent review of current practice and research in this area can be found in Claeskens (2013).

Influence diagnostics may be used to investigate various aspects of the model fit. Because LMMs are more complicated than standard linear models, the influence of observations on the model fit can manifest itself in more varied and complicated ways. It is generally recommended to follow a top-down approach when carrying out influence diagnostics in mixed models. First, check overall influence diagnostics. Assuming that there are influential sets of observations based on the overall influence diagnostics, proceed with other diagnostics to see what aspect of the model a given subset of observations affects: fixed effects, covariance parameters, the precision of the parameter estimates, or predicted values.

Influence diagnostics play an important role in the interpretation of the results. If a given subset of data has a strong influence on the estimates of covariance parameters, but limited impact on the fixed effects, then it is appropriate to interpret the model with respect to prediction. However, we need to keep in mind that changes in estimates of covariance parameters may affect the precision of tests for fixed effects and, consequently, confidence intervals.

We focus on a selected group of influence diagnostics, which are summarized in Table 2.5. Following Schabenberger's notation, we use the subscript (U) to denote quantities calculated based on the data having a subset, U, excluded from calculations. For instance, consider the overall influence calculations for an arbitrarily chosen vector of parameters, ψ (which can include parameters in $\boldsymbol{\beta}$ or $\boldsymbol{\theta}$). The vector $\widehat{\psi}_u$ used in the calculation formulas denotes an estimate of ψ computed based on the reduced "leave-U-out" data. These methods include, but are not limited to, overall influence, change in parameter estimates, change in precision of parameter estimates, and effect on predicted values.

All methods for influence diagnostics presented in Table 2.5 clearly depend on the subset, U, of observations that is being considered. The main difference between the Cook's distance statistic and the MDFFITS statistic shown in Table 2.5 is that the MDFFITS statistic uses "externalized" estimates of $var(\widehat{\boldsymbol{\beta}})$, which are based on recalculated covariance estimates using the reduced data, whereas Cook's distance does not recalculate the covariance parameter estimates in $var(\widehat{\boldsymbol{\beta}})$ (see (2.20)).

Calculations for influence statistics can be performed using either noniterative or iterative methods. Noniterative methods are based on explicit (closed-form) updated formulas (not shown in Table 2.5). The advantage of noniterative methods is that they are more time efficient than iterative methods. The disadvantage is that they require the rather strong assumption that all covariance parameters are known, and thus are not updated, with the exception of the profiled residual variance. Iterative influence diagnostics require refitting the model without the observations in question; consequently, the covariance parameters are updated at each iteration, and computational execution time is much longer.

Software Note: All the influence diagnostic methods presented in Table 2.5 are currently supported by `proc mixed` in SAS. A class of leverage-based methods is also available in `proc mixed`, but we do not discuss them in the example chapters. In Chapter 3, we present and interpret several influence diagnostics generated by `proc mixed` for the final model fitted to the Rat Pup data. Selected influence diagnostics can also be computed when using the `nlmeU` and `HLMdiag` packages in R (see Section 20.3 of Galecki & Burzykowski, 2013, or Loy & Hofmann, 2014, for computational details). To our knowledge, influence diagnostic methods are not currently available in the other software procedures. The theory behind these and other diagnostic methods is outlined in more detail by Claeskens (2013).

2.8.3 Diagnostics for Random Effects

The natural choice to diagnose random effects is to consider the empirical Bayes (EB) predictors defined in Subsection 2.9.1. EB predictors are also referred to as random-effects predictors or, due to their properties, empirical best linear unbiased predictors (EBLUPs). We recommend using standard diagnostic plots (e.g., histograms, quantile–quantile (Q–Q) plots, and scatter-plots) to investigate EBLUPs for potential outliers that may warrant further investigation. In general, checking EBLUPs for normality is of limited value, because their distribution does not necessarily reflect the true distribution of the random effects. We consider informal diagnostic plots for EBLUPs in the example chapters.

TABLE 2.5: Summary of Influence Diagnostics for LMMs

Group	Name	Par. of Interest	Formula	Description[a]
Overall influence	Likelihood distance / displacement	ψ	$LD_{(u)} = 2\{\ell(\widehat{\psi}) - \ell(\widehat{\psi}_{(u)})\}$	Change in ML log-likelihood for all data with ψ estimated for all data vs. reduced data
	Restricted likelihood distance / displacement	ψ	$RLD_{(u)} = 2\{\ell_R(\widehat{\psi}) - \ell_R(\widehat{\psi}_{(u)})\}$	Change in REML log-likelihood for all data with ψ estimated for all data vs. reduced data
Change in parameter estimates	Cook's D	β	$D(\beta) = (\widehat{\beta} - \widehat{\beta}_{(u)})'\widehat{var}[\widehat{\beta}]^{-1}(\widehat{\beta} - \widehat{\beta}_{(u)})/rank(X)$	Scaled change in entire estimated β vector
		θ	$D(\theta) = (\widehat{\theta} - \widehat{\theta}_{(u)})'\widehat{var}[\widehat{\theta}]^{-1}(\widehat{\theta} - \widehat{\theta}_{(u)})$	Scaled change in entire estimated θ vector
	Multivariate DFFITS statistic	β	$MDFFITS(\beta) = (\widehat{\beta} - \widehat{\beta}_{(u)})'\widehat{var}[\widehat{\beta}_{(u)}]^{-1}(\widehat{\beta} - \widehat{\beta}_{(u)})/rank(X)$	Scaled change in entire estimated β vector, using externalized estimates of var ($\widehat{\beta}$)
		θ	$MDFFITS(\theta) = (\widehat{\theta} - \widehat{\theta}_{(u)})'\widehat{var}[\widehat{\theta}_{(u)}]^{-1}(\widehat{\theta} - \widehat{\theta}_{(u)})$	Scaled change in entire estimated θ vector, using externalized estimates of var ($\widehat{\theta}$)

TABLE 2.5: (Continued)

Group	Name	Par. of Interest	Formula	Description[a]
Change in precision of parameter estimates	Trace of covariance matrix	β	$COVTRACE(\boldsymbol{\beta}) = $ $\left| trace(\widehat{var}[\widehat{\boldsymbol{\beta}}]^{-1}\widehat{var}[\widehat{\boldsymbol{\beta}}_{(u)}]) - rank(\mathbf{X}) \right|$	Change in precision of estimated $\boldsymbol{\beta}$ vector, based on trace of var ($\widehat{\boldsymbol{\beta}}$)
		θ	$COVTRACE(\boldsymbol{\theta}) = \left| trace(\widehat{var}[\widehat{\boldsymbol{\theta}}]^{-1}\widehat{var}[\widehat{\boldsymbol{\theta}}_{(u)}]) - q \right|$	Change in precision of estimated $\boldsymbol{\theta}$ vector, based on trace of var ($\widehat{\boldsymbol{\theta}}$)
	Covariance ratio[b]	β	$COVRATIO(\boldsymbol{\beta}) = \dfrac{\det_{ns}(\widehat{var}[\widehat{\boldsymbol{\beta}}_{(u)}])}{\det_{ns}(\widehat{var}[\widehat{\boldsymbol{\beta}}])}$	Change in precision of estimated $\boldsymbol{\beta}$ vector, based on determinant of var ($\widehat{\boldsymbol{\beta}}$)
		θ	$COVRATIO(\boldsymbol{\theta}) = \dfrac{\det_{ns}(\widehat{var}[\widehat{\boldsymbol{\theta}}_{(u)}])}{\det_{ns}(\widehat{var}[\widehat{\boldsymbol{\theta}}])}$	Change in precision of estimated $\boldsymbol{\theta}$ vector, based on determinant of var ($\widehat{\boldsymbol{\theta}}$)
Effect on predicted value	Sum of squared PRESS residuals	N/A	$PRESS_{(u)} = \sum\limits_{i \in u} (\mathbf{y}_i - x_i'\widehat{\boldsymbol{\beta}}_{(u)})$	Sum of PRESS residuals calculated by deleting observations in U

[a]The "change" in the parameters estimates for each influence statistic is calculated by using all data compared to the reduced "leave-U-out" data.

[b]$\det_{ns}$ means the determinant of the nonsingular part of the matrix.

2.9 Other Aspects of LMMs

In this section, we discuss additional aspects of fitting LMMs that may be considered when analyzing clustered or longitudinal data sets.

2.9.1 Predicting Random Effects: Best Linear Unbiased Predictors

One aspect of LMMs that is different from standard linear models is the prediction of the values in the random-effects vector, $\mathbf{u}_i$. The values in $\mathbf{u}_i$ are not fixed, unknown parameters that can be estimated, as is the case for the values of $\boldsymbol{\beta}$ in a linear model. Rather, they are random variables that are assumed to follow some multivariate normal distribution. As a result, we predict the values of these random effects, rather than estimate them (Carlin & Louis, 2009).

Thus far, we have discussed the variances and covariances of the random effects in the $\boldsymbol{D}$ matrix without being particularly interested in predicting the values that these random effects may take. However, in some research settings, it may be useful to predict the values of the random effects associated with specific levels of a random factor.

Unlike fixed effects, we are not interested in estimating the mean (i.e., the expected value) of a set of random effects, because we assume that the expected value of the multivariate normal distribution of random effects is a vector of zeroes. However, assuming that the expected value of a random effect is zero does not make any use of the observed data. In the context of an LMM, we take advantage of all the data collected for those observations sharing the same level of a particular random factor and use that information to predict the values of the random effects in the LMM. To do this, we look at the conditional expectations of the random effects, given the observed response values, $\mathbf{y}_i$, in $\mathbf{Y}_i$. The conditional expectation for $\mathbf{u}_i$ is

$$\widehat{\mathbf{u}}_i = E(\mathbf{u}_i | \mathbf{Y}_i = \mathbf{y}_i) = \widehat{\boldsymbol{D}} \boldsymbol{Z}_i' \widehat{\boldsymbol{V}}_i^{-1} (\mathbf{y}_i - \boldsymbol{X}_i \widehat{\boldsymbol{\beta}}) \tag{2.29}$$

The predicted values in (2.29) are the expected values of the random effects, $\mathbf{u}_i$, associated with the i-th level of a random factor, given the observed data in $\mathbf{y}_i$. These conditional expectations are known as **best linear unbiased predictors (BLUPs)** of the random effects. We refer to them as **EBLUPs** (or **empirical BLUPs**), because they are based on the estimates of the $\boldsymbol{\beta}$ and $\boldsymbol{\theta}$ parameters.

The variance-covariance matrix of the EBLUPs can be written as follows:

$$Var(\widehat{\mathbf{u}}_i) = \widehat{\boldsymbol{D}} \boldsymbol{Z}_i' (\widehat{\boldsymbol{V}}_i^{-1} - \widehat{\boldsymbol{V}}_i^{-1} \boldsymbol{X}_i (\sum_i \boldsymbol{X}_i \widehat{\boldsymbol{V}}_i^{-1} \boldsymbol{X}_i)^{-1} \boldsymbol{X}_i \widehat{\boldsymbol{V}}_i^{-1}) \boldsymbol{Z}_i \widehat{\boldsymbol{D}} \tag{2.30}$$

EBLUPs are "linear" in that they are linear functions of the observed data, $\mathbf{y}_i$. They are "unbiased" in that their expectation is equal to the expectation of the random effects for a single subject i. They are "best" in that they have minimum variance (see (2.30)) among all linear unbiased estimators (i.e., they are the most precise linear unbiased estimators; Robinson (1991)). And finally, they are "predictions" of the random effects based on the observed data.

EBLUPs are also known as **shrinkage estimators** because they tend to be closer to zero than the estimated effects would be if they were computed by treating a random factor as if it were fixed. We include a discussion of shrinkage estimators on the web page for the book (see Appendix A).

2.9.2 Intraclass Correlation Coefficients (ICCs)

In general, the **intraclass correlation coefficient (ICC)** is a measure describing the similarity (or homogeneity) of the responses on the dependent variable within a cluster (in a clustered data set) or a unit of analysis (in a repeated-measures or longitudinal data set). We consider different forms of the ICC in the analysis of a two-level clustered data set (the Rat Pup data) in Chapter 3, and the analysis of a three-level data set (the Classroom data) in Chapter 4.

2.9.3 Problems with Model Specification (Aliasing)

In this subsection we informally discuss **aliasing** (and related concepts) in general terms. We then illustrate these concepts with two hypothetical examples. In our explanation, we follow the work of Nelder (1977).

We can think of aliasing as an ambiguity that may occur in the specification of a parametric model (e.g., an LMM), in which multiple parameter sets (aliases) imply models that are indistinguishable from each other. There are two types of aliasing:

1. **Intrinsic aliasing:** Aliasing attributable to the model formula specification.

2. **Extrinsic aliasing:** Aliasing attributable to the particular characteristics of a given data set.

Nonidentifiability and **overparameterization** are other terms often used to refer to intrinsic aliasing. In this section we use the term *aliasing* to mean intrinsic aliasing; however, most of the remarks apply to both intrinsic and extrinsic aliasing.

Aliasing should be detected by the researcher at the time that a model is specified; otherwise, if unnoticed, it may lead to difficulties in the estimation of the model parameters and/or incorrect interpretation of the results.

Aliasing has important implications for parameter estimation. More specifically, aliasing implies that only certain linear combinations of parameters are estimable and other combinations of the parameters are not. "Nonestimability" due to aliasing is caused by the fact that there are infinitely many sets of parameters that lead to the same set of predicted values (i.e., imply the same model). Consequently, each value of the likelihood function (including the maximum value) can be obtained with infinitely many sets of parameters.

To resolve a problem with aliasing so that a unique solution in a given parameter space can be obtained, the common practice is to impose additional constraints on the parameters in a specified model. Although constraints can be chosen arbitrarily out of infinitely many, some choices are more natural than others. We choose constraints in such a way as to facilitate interpretation of parameters in the model. At the same time, it is worthwhile to point out that the choice of constraints does not affect the meaning (or interpretation) of the model itself. It should also be noted that constraints imposed on parameters should not be considered as part of the model specification. Rather, constraints are a convenient way to resolve the issue of nonestimability caused by aliasing.

In the case of aliasing of the β parameters in standard linear models (Example 1 following), many software packages by default impose constraints on the parameters to avoid aliasing, and it is the user's responsibility to determine what constraints are used. In Example 2, we consider aliasing of covariance parameters.

Example 1. A linear model with an intercept and a gender factor (Model E1).
Most commonly, intrinsic aliasing is encountered in linear models involving categorical fixed factors as covariates. Consider for instance a hypothetical linear model, **Model E1**, with an intercept and gender considered as a fixed factor. Suppose that this model involves three

corresponding fixed-effect parameters: μ (the intercept), μ_F for females, and μ_M for males. The $\boldsymbol{X}$ design matrix for this model has three columns: a column containing an indicator variable for the intercept (a column of ones), a column containing an indicator variable for females, and a column containing an indicator variable for males. Note that this design matrix is not of full rank.

Consider transformation $\mathbf{T}_1$ of the fixed-effect parameters μ, μ_F, and μ_M, such that a constant C is added to μ and the same constant is subtracted from both μ_F and from μ_M. Transformation $\mathbf{T}_1$ is artificially constructed in such a way that any transformed set of parameters $\mu' = \mu + C$, $\mu'_F = \mu_F - C$, and $\mu'_M = \mu_M - C$ generates predicted values that are the same as in Model E1. In other words, the model implied by any transformed set of parameters is indistinguishable from Model E1.

Note that the linear combinations $\mu + \mu_F$, $\mu + \mu_M$, or $C_1 \times \mu_F + C_2 \times \mu_M$, where $C_1 + C_2 = 0$, are not affected by transformation $\mathbf{T}_1$, because $\mu' + \mu'_F = \mu + \mu_F$, $\mu' + \mu'_M = \mu + \mu_M$, and $C_1 \times \mu'_F + C_2 \times \mu'_M = C_1 \times \mu_F + C_2 \times \mu_M$. All linear combinations of parameters unaffected by transformation $\mathbf{T}_1$ are estimable. In contrast, the individual parameters μ, μ_F, and μ_M are affected by transformation $\mathbf{T}_1$ and, consequently, are not estimable.

To resolve this issue of nonestimability, we impose constraints on μ, μ_F, and μ_M. Out of an infinite number of possibilities, we arbitrarily constrain μ to be zero. This constraint was selected so that it allows us to directly interpret μ_F and μ_M as the means of a dependent variable for females and males, respectively. In SAS `proc mixed`, for example, such a constraint can be accomplished by using the `noint` option in the `model` statement. By default, using the `solution` option in the `model` statement of `proc mixed` would constrain μ_M to be equal to zero, meaning that μ would be interpreted as the mean of the dependent variable for males, and μ_F would represent the difference in the mean for females compared to males.

Example 2. An LMM with aliased covariance parameters (Model E2). Consider an LMM (**Model E2**) with the only fixed effect being the intercept, one random effect associated with the intercept for each subject (resulting in a single covariance parameter, σ^2_{int}), and a compound symmetry covariance structure for the residuals associated with repeated observations on the same subject (resulting in two covariance parameters, σ^2 and σ_1; see the compound symmetry covariance structure for the $\boldsymbol{R}_i$ matrix in Subsection 2.2.2.2).

In the marginal $\boldsymbol{V}_i$ matrix for observations on subject i that is implied by this model, the diagonal elements (i.e., the marginal variances) are equal to $\sigma^2 + \sigma_1 + \sigma^2_{int}$, and the off-diagonal elements (i.e., the marginal covariances) are equal to $\sigma_1 + \sigma^2_{int}$.

Consider transformation $\mathbf{T}_2$, such that a constant C is added to σ^2_{int}, and the same constant C is subtracted from σ_1. We assume that the possible values of C in transformation $\mathbf{T}_2$ should be constrained to those for which the matrices $\boldsymbol{D}$ and $\boldsymbol{R}_i$ remain positive-definite. Transformation $\mathbf{T}_2$ is constructed in such a way that any transformed set of parameters $\sigma^2_{int} + C$ and $\sigma_1 - C$ implies the same marginal variance-covariance matrix, $\boldsymbol{V}_i$, and consequently, the marginal distribution of the dependent variable is the same as in Model E2, which means that all these models are indistinguishable. Moreover, after applying transformation $\mathbf{T}_2$, the matrix $\boldsymbol{R}_i$ remains compound symmetric, as needed.

The linear combinations of covariance parameters $\sigma^2 + \sigma_1 + \sigma^2_{int}$ and $\sigma_1 + \sigma^2_{int}$ (i.e., the elements in the $\boldsymbol{V}_i$ matrix) are not affected by transformation $\mathbf{T}_2$. In other words, these linear combinations are estimable. Due to aliasing, the individual parameters σ^2_{int} and σ_1 are not estimable.

To resolve this issue of nonestimability, we impose constraints on σ^2_{int} and σ_1. One possible constraint to consider, out of infinitely many, is $\sigma_1 = 0$, which is equivalent to assuming that the residuals are not correlated and have constant variance (σ^2). In other words, the $\boldsymbol{R}_i$ matrix no longer has a compound symmetry structure, but rather has a structure with constant variance on the diagonal and all covariances equal to zero.

If such a constraint is not defined by the user, then the corresponding likelihood function based on all parameters has an infinite number of ML solutions. Consequently, the algorithm used for optimization in software procedures may not converge to a solution at all, or it may impose arbitrary constraints on the parameters and converge. In such a case, software procedures will generally issue a warning, such as "Invalid likelihood" or "Hessian not positive-definite" or "Convergence not achieved" (among others). In all these instances, parameter estimates and their standard errors may be invalid and should be interpreted with caution. We discuss aliasing of covariance parameters and illustrate how each software procedure handles it in the analysis of the Dental Veneer data in Chapter 7.

2.9.4   Missing Data

In general, analyses using LMMs are carried out under the assumption that missing data in clustered or longitudinal data sets are **missing at random (MAR)** (see Little & Rubin, 2002, or Allison, 2001, for a more thorough discussion of missing data patterns and mechanisms). Under the assumption that missing data are MAR, inferences based on methods of ML estimation in LMMs are valid (Verbeke & Molenberghs, 2000).

The MAR pattern means that the probability of having missing data on a given variable may depend on other observed information, but does not depend on the data that would have been observed but were in fact missing. For example, if subjects in a study do not report their weight because the actual (unobserved or missing) weights are too large or too small, then the missing weight data are not MAR. Likewise, if a rat pup's birth weight is not collected because it is too small or too large for a measurement device to accurately detect it, the information is not MAR. However, if a subject's current weight is not related to the probability that he or she reports it, but rather the likelihood of failing to report it depends on other observed information (e.g., illness or previous weight), then the data can be considered MAR. In this case, an LMM for the outcome of current weight should consider the inclusion of covariates, such as previous weight and illness, which are related to the nonavailability of current weight.

Missing data are quite common in longitudinal studies, often due to dropout. Multivariate repeated-measures ANOVA models are often used in practice to analyze repeated-measures or longitudinal data sets, but LMMs offer two primary advantages over these multivariate approaches when there are missing data.

First, they allow subjects being followed over time to have unequal numbers of measurements (i.e., some subjects may have missing data at certain time points). If a subject does not have data for the response variable present at all time points in a longitudinal or repeated-measures study, the subject's entire set of data is omitted in a multivariate ANOVA (this is known as *listwise deletion*); the analysis therefore involves complete cases only. In an LMM analysis, all observations that are available for a given subject are used in the analysis.

Second, when analyzing longitudinal data with repeated-measures ANOVA techniques, time is considered to be a within-subject factor, where the levels of the time factor are assumed to be the same for all subjects. In contrast, LMMs allow the time points when measurements are collected to vary for different subjects.

Because of these key differences, LMMs are much more flexible analytic tools for longitudinal data than repeated-measures ANOVA models, under the assumption that any missing data are MAR. We advise readers to inspect longitudinal data sets thoroughly for problems with missing data. If the vast majority of subjects in a longitudinal study have data present at only a single time point, an LMM approach may not be warranted, because there may not be enough information present to estimate all of the desired covariance parameters in the model. In this situation, simpler regression models should probably be considered because issues of within-subject dependency in the data may no longer apply.

When analyzing clustered data sets (such as students nested within classrooms), clusters may be of unequal sizes, or there may be data within clusters that are MAR. These problems result in unbalanced data sets, in which an unequal number of observations are collected for each cluster. LMMs can be fitted to unbalanced clustered data sets, again under the assumption that any missing data are MAR. Quite similar to the analysis of longitudinal data sets, multivariate techniques or techniques requiring balanced data break down when attempting to analyze unbalanced clustered data. LMMs allow one to make valid inferences when modeling these types of clustered data, which arise frequently in practice.

2.9.5   Centering Covariates

Centering covariates at specific values (i.e., subtracting a specific value, such as the mean, from the observed values of a continuous covariate) has the effect of changing the intercept in the model, so that it represents the expected value of the dependent value at a specific value of the covariate (e.g., the mean), rather than the expected value when the covariate is equal to zero (which is often outside the range of the observed data). This type of centering is often known as **grand mean centering**. We consider grand mean centering of covariates in the analysis of the Autism data in Chapter 6.

An alternative centering procedure is to center continuous covariates at the mean values of the higher-level clusters or groups within which the units measured on the continuous covariates are nested. This type of centering procedure, sometimes referred to as **group mean centering**, changes the interpretation of both the intercept and the estimated fixed effects for the centered covariates, unlike grand mean centering. This type of centering requires an interpretation of the fixed effects associated with the centered covariates that is relative to other units within a higher-level group (or cluster). A fixed effect in this case now reflects the expected change in the dependent variable for a one-unit *within-cluster* increase in the centered covariate.

A thorough overview of these centering options in the context of LMMs can be found in Enders (2013). We revisit this issue in Chapter 6.

2.9.6   Fitting Linear Mixed Models to Complex Sample Survey Data

In many applied fields, scientific advances are driven by secondary analyses of large survey data sets. Survey data are often collected from large, representative **probability samples** of specified populations, where a formal sample design assigns a known probability of selection to all individual elements (people, households, establishments, etc.) in the target population. Many of the larger, nationally representative survey data sets that survey organizations make available for public consumption (e.g., the National Health and Nutrition Examination Survey, or NHANES, in the United States; see `http://www.cdc.gov/nchs/nhanes/about_nhanes.htm`) are collected from probability samples with so-called "complex" designs, giving rise to the term **complex sample survey data**. These complex sample designs generally have the following features (Heeringa et al., 2010):

- The sampling frame, or list of population elements (or clusters of population elements, e.g., schools where students will be sampled and surveyed) is **stratified** into mutually exclusive divisions of elements that are homogeneous within (in terms of the features of survey interest) and heterogeneous between. This stratification serves to increase the precision of survey estimates and ensure that the sample is adequately representative of the target population. Public-use survey data sets will sometimes include a variable containing unique codes for each stratum of the population.

- To reduce the costs associated with data collection, naturally-occurring clusters of elements (e.g., schools, hospitals, neighborhoods, Census blocks, etc.) are randomly sampled within strata, rather than randomly sampling individual elements. This is often done because sampling frames only contain clusters, rather than the individual elements of interest. This type of **cluster sampling** gives rise to a hierarchical structure in the data set, where sampled elements nested within the sampled clusters tend to have similar values on the survey features of interest (e.g., similar attitudes within a neighborhood, or similar test performance within a school). This **intracluster correlation** needs to be accounted for in analyses, so that standard errors of parameter estimates appropriately reflect this dependency among the survey respondents. Public-use survey data sets will sometimes include a variable containing unique codes for each sampled cluster.

- The known probabilities of selection assigned to each case in the target population need not be equal (as in the case of a simple random sample), and these probabilities are inverted to compute what is known as a **design weight** for each sampled case. These design weights may also be adjusted to account for differential nonresponse among different subgroups of the sample, and further calibrated to produce weighted sample distributions that match known population distributions (Valliant et al., 2013). Public-use survey data sets will include variables containing the final survey weight values for responding cases in a survey, and these weights enable analysts to compute unbiased estimates of population quantities.

Analysts of complex sample survey data are therefore faced with some decisions when working with data sets containing variables that reflect these sampling features. When fitting linear mixed models to complex sample survey data that recognize the hierarchical structures of these data sets (given the cluster sampling), how exactly should these complex design features be accounted for? Completely ignoring these design features when fitting linear mixed models can be problematic for inference purposes, especially if the design features are predictive of (or informative about) the dependent variable of interest. The existing literature generally suggests two approaches that could be used to fit linear mixed models to survey data sets, and we discuss implementation of these two approaches using existing software in the following subsections.

2.9.6.1 Purely Model-Based Approaches

The majority of the analyses discussed in this book can be thought of as "model-based" approaches to data analysis, where a probability model is specified for a dependent variable of interest, and one estimates the parameters defining that probability model. The probability models in this book are defined by fixed effects and covariance parameters associated with random effects, and the dependent variables considered have marginal normal distributions defined by these parameters. When analyzing survey data following this framework, the key complex sample design features outlined above are simply thought of as additional relevant information for the dependent variable that should be accounted for in the specification of a given probability model.

First, the sampling strata are generally thought of as categorical characteristics of the randomly sampled clusters (which may define Level 2 or Level 3 units, depending on the sample design), and the effects of the strata (or recoded versions of the original stratum variable, to reduce the number of categories) are considered as fixed effects. The reason that the effects of strata enter into linear mixed models as *fixed* effects is that all possible strata (or divisions of the population) will be included in each hypothetical replication of the sample; there is no sampling variability introduced by having different strata in different samples. The effects associated with the different strata are therefore considered as fixed. The fixed effects of strata can be incorporated into models by following the standard approaches for categorical fixed factors discussed in detail in this book.

Second, the intracluster correlations in the survey data introduced by the cluster sampling are generally accounted for by including *random* effects of the randomly sampled clusters in the specification of a given linear mixed model. In this sense, the data hierarchy of a given survey data set needs to be carefully considered. For example, an analyst may be interested in modeling between-school variance when fitting a linear mixed model, but the cluster variable in that survey data set may represent a higher-level unit of clustering (such as a county, where multiple schools may be included in the sample from a given county). In this example, the analyst may need to consider a three-level linear mixed model, where students are nested within schools, and schools are nested within clusters (or counties). Model-based approaches to fitting linear mixed models will be most powerful when the models are correctly specified, and correctly accounting for the different possible levels of clustering in a given data set therefore becomes quite important, so that important random effects are not omitted from a model.

Third, when following a purely model-based approach to the analysis of complex sample survey data using linear mixed models, the weights (or characteristics of sample units used to form the weights, e.g., ethnicity in a survey that over-samples minority ethnic groups relative to known population distributions) are sometimes considered as *covariates* associated with the sampled units. As discussed by Gelman (2007), model-based approaches to the analysis of survey data arising from complex samples need to account for covariates that affect sample selection or nonresponse (and of course also potentially predict the dependent variable of interest). Gelman (2007) proposed a model-based approach to estimating population means and differences in population means (which can be estimated using regression) based on the ideas of *post-stratification* weighting, which is a technique commonly used to adjust for differences between samples and populations in survey data. In this approach, implicit adjustment weights for survey respondents are based on regression models that condition on variables defining post-stratification cells, where corresponding population distributions across those cells are also known. Gelman (2007) called for more work to develop model-based approaches to fitting regression models (possibly with random cluster effects) that simultaneously adjust for these differences between samples and populations using the ideas of post-stratification.

Other purely model-based approaches to accounting for sampling weights in multilevel models for survey data are outlined by Korn & Graubard (1999) and Heeringa et al. (2010). In general, these model-based approaches to accounting for survey weights can be readily implemented using existing software capable of fitting multilevel models, given that the weights (or features of respondents used to define the weights) can be included as covariates in the specified models.

2.9.6.2   Hybrid Design- and Model-Based Approaches

In contrast to the purely model-based approaches discussed above, an alternative approach is to incorporate complex sampling weights into the estimation of the parameters in a linear mixed model. We refer to this as a "hybrid" approach, given that it includes features of model-based approaches (where the models include random effects of sampling clusters and fixed effects of sampling strata) and design-based approaches (incorporating sampling weights into the likelihood functions used for estimation, in an effort to compute design-unbiased estimates of model parameters). It is well known that if sampling weights informative about a dependent variable in a linear mixed model are not accounted for when estimating that model, estimates of the parameters may be biased (Carle, 2009). These "hybrid" approaches offer analysts an advantage over purely model-based approaches, in that the variables used to compute sampling weights may not always be well-known (or well-communicated by survey organizations in the documentation for their surveys) for inclusion in models. Incorporating the sampling weights into estimation of the parameters in a linear mixed model will generally lead to unbiased estimates of the parameters defining the larger population model that is the objective of the estimation, even if the model has been misspecified in some way. In this sense, unbiased estimates of the parameters in a poorly-specified model would be preferred over biased estimates of the parameters in a poorly-specified model.

What is critical for these "hybrid" approaches is that software be used that correctly implements the theory of weighted estimation for multilevel models. Initial theory for parameter estimation and variance estimation in this context was communicated by Pfeffermann et al. (1998), and later elaborated on by Rabe-Hesketh & Skrondal (2006), who have implemented the theory in the Stata software procedure `gllamm` (available at www.gllamm.org). This theory requires that *conditional* weights be used for estimation at lower levels of the data hierarchy; for example, in a two-level cross-sectional data set with people nested within clusters, the weights associated with individual people need to be *conditional* on their cluster having been sampled. Making this more concrete, suppose that the weight provided in a survey data set for a given individual is 100, meaning that (ignoring possible nonresponse and post-stratification adjustments for illustration purposes), their probability of selection into the sample was 1/100. This probability of selection was actually determined based on the probability that their cluster was selected into the sample (say, 1/20), and then the *conditional* probability that the individual was selected given that the cluster was sampled (say, 1/5). In this sense, the appropriate "Level-1 weight" for implementing this theory for that individual would be 5, or the inverse of their conditional selection probability. Furthermore, the "Level-2 weight" of 20 for the sampled cluster is also needed in the data set; this is because this theory requires these two components of the overall weight at different places in the likelihood function.

There are two important issues that arise in this case. First, survey agencies seldom release these conditional weights (or the information needed to compute them) in public-use data files, and cluster-level weights are also seldom released; in general, only weights for the ultimate Level-1 units accounting for the final probabilities of selection across all stages of sampling are released. Second, these weights need to be *scaled* for inclusion in the likelihood function used to estimate the model parameters. Scaling becomes most important when an analyst only has access to final weights (as opposed to conditional weights) for Level-1 and Level-2 units, and the final Level-1 weights (representing overall probabilities of selection) need to be appropriately re-scaled to remove their dependence on the Level-2 weights (for the clusters). Scaling can still be important even if the *conditional* Level-1 weights are in fact provided in a data set (Rabe-Hesketh & Skrondal, 2006). Carle (2009) presents the results of simulation studies examining the performance of the two most popular scaling methods that have emerged in this literature, concluding that they do not tend to differ in performance and still provide better inference than ignoring the weights entirely. Based on this work, we recommend that analysts following these "hybrid" approaches use existing software to examine the sensitivity of their inferences to the different scaling approaches (and differences in inference compared to unweighted analyses).

At the time of the writing for the second edition of this book, two of the software procedures emphasized in this book implement the appropriate theory for this hybrid type of analysis approach: the `mixed` command in Stata, and the various procedures in the HLM software package. When using the `mixed` command, the appropriate scaling approaches are only currently implemented for two-level models, and in general, conditional weights need to be computed at each level for higher-level models. The following syntax indicates the general setup of the `mixed` command, for a two-level cross-sectional survey data set with individuals nested within sampling clusters:

```
. mixed depvar indvar1 i.strata ... [pw = finwgt] || cluster:,
pweight(level2wgt) pwscale(size)
```

In this syntax, we assume that `finwgt` is the final weight associated with Level-1 respondents, and reflects the overall probability of selection (including the conditional probability of selection for the individuals and the cluster probability of selection). This is the type of weight that is commonly provided in survey data sets. The `pwscale(size)` option is then a tool for re-scaling these weights to remove their dependence on the cluster-level weights, which reflect the probabilities of selection for the clusters (`level2wgt`). If *conditional* Level-1 weights were available in the data set, the `pwscale(gk)` scaling option could also be used. In general, all three scaling options (including `pwscale(effective)`) should be considered in a given analysis, to examine the sensitivity of the results to the different scaling approaches. We note that random cluster intercepts are included in the model to reflect the within-cluster dependency in values on the dependent variable, and fixed effects of the categorical strata are included as well (using `i.strata`).

Obviously the use of these options requires the presence of cluster weights and possibly conditional Level-1 weights in the survey data set. In the absence of cluster weights, one would have to assume that the weights associated with each cluster were 1 (essentially, a simple random sample of clusters was selected). This is seldom the case in practice, and would likely lead to biased parameter estimates.

In the HLM software, the conditional lower-level weights and the higher-level cluster weights can be selected by clicking on the **Other Settings** menu, and then **Estimation Settings**. Next, click on the button for **Weighting**, and select the appropriate weight variables at each level. If weights are only available at Level 1 and are not conditional (the usual case), HLM will automatically normalize the weights so that they have a mean of 1. If Level-1 and Level-2 weights are specified, HLM will assume that the Level-1 weights are conditional weights, and normalize these weights so that they sum to the size of the sample within each Level-2 unit (or cluster). In this sense, HLM automatically performs scaling of the weights. See `http://www.ssicentral.com/hlm/example6-2.html` for more details for higher-level models.

Although the approaches above are described for two-level, cross-sectional, clustered survey data sets, there are also interesting applications of these hybrid modeling approaches for **longitudinal survey data**, where repeated measures are nested within individuals, and individuals may be nested within sampling clusters. The same basic theory with regard to weighting at each level still holds, but the longitudinal data introduce the possibility for unique error covariance structures at Level 1. Heeringa et al. (2010) present an example of fitting a model to data from the longitudinal Health and Retirement Study (HRS) using this type of hybrid approach (as implemented in the `gllamm` procedure of Stata), and review some of the literature that has advanced the use of these methods for longitudinal survey data. More recently, Veiga et al. (2014) described the theory and methods required to fit multilevel models with specific error covariance structures for the repeated measures at Level 1, and showed how accounting for the survey weights at each level can make a difference in terms of inferences related to trends over time. These authors also make software available that implements this hybrid approach for this specific class of multilevel models. The analysis of data from longitudinal surveys is a ripe area for future research, and we anticipate more developments in this area in the near future.

Other software packages that currently implement the theory for this hybrid approach include Mplus (see Asparouhov, 2006, 2008), the `gllamm` add-on command for Stata (Rabe-Hesketh & Skrondal, 2006; see `www.gllamm.org` for examples) and the MLwiN software package (`http://www.bristol.ac.uk/cmm/software/mlwin/`). We are hopeful that these approaches will become available in more of the general-purpose software packages in the near future, and we will provide any updates in this regard on the web page for this book.

2.9.7 Bayesian Analysis of Linear Mixed Models

In general, we present a **frequentist** approach to the analysis of linear mixed models in this book. These approaches involve the specification of a likelihood function corresponding to a given linear mixed model, and subsequent application of the various numerical optimization algorithms discussed earlier in this chapter to find the parameter estimates that maximize that likelihood function. In contrast, one could also take a **Bayesian** approach to the analysis of linear mixed models. Heuristically speaking, this approach makes use of Bayes' Theorem, and defines a **posterior distribution** for a parameter vector $\boldsymbol{\theta}$ given data denoted by $\mathbf{y}$, $f(\boldsymbol{\theta}|\mathbf{y})$, as a function of the product of a **prior distribution** for that parameter vector, $f(\boldsymbol{\theta})$, and the **likelihood function** defined by the model of interest and the observed data, $f(\mathbf{y}|\boldsymbol{\theta})$. The Bayesian approach treats the data $\mathbf{y}$ as fixed and the parameter vector $\boldsymbol{\theta}$ as random, allowing analysts to ascribe uncertainty to the parameters of interest based on previous investigations via the prior distributions. Another important distinction between the Bayesian and frequentist approaches is that uncertainty in estimates of variance components describing distributions of random coefficients is incorporated into inferential procedures, as described below.

Given a prior distribution for the vector of parameters $\boldsymbol{\theta}$ (specified by the analyst) and the likelihood function defined by the data and a given model, various computational methods (e.g., **Markov Chain Monte Carlo (MCMC)** methods) can be used to simulate draws from the resulting posterior distribution for the parameters of interest. The prior distribution is essentially "updated" by the likelihood defined by a given data set and model to form the posterior distribution. Inferences for the parameters are then based on many simulated draws from the posterior distribution; for example, a 95% **credible set** for a parameter of interest could be defined by the 0.025 and 0.975 percentiles of the simulated draws from the posterior distribution for that parameter. This approach to inference is considered by Bayesians to be much more "natural" than frequentist inferences based on confidence intervals (which would suggest, for example, that 95% of intervals computed a certain way will cover a true population parameters across hypothetical repeated samples). With the Bayesian approach, which essentially treats parameters of interest as random variables, one could suggest that the probability that a parameter takes a value between the limits defined by a 95% credible set is actually 0.95.

Bayesian approaches also provide a natural method of inference for data sets having a hierarchical structure, such as those analyzed in the case studies in this book, given that the parameters of interest are treated as random variables rather than fixed constants. For example, in a two-level cross-sectional clustered data set, observations on the dependent variable at Level 1 (the "data") may arise from a normal distribution governed by random coefficients specific to a higher-level cluster; the coefficients in this model define the likelihood function. The random cluster coefficients themselves may arise from a normal distribution governed by mean and variance parameters. Then, these parameters may be governed by some prior distribution, reflecting uncertainty associated with those parameters based on prior studies. Following a Bayesian approach, one can then simulate draws from the posterior distribution for all of the random coefficients and parameters that is defined by these conditional distributions, making inference about either the random coefficients or the parameters of interest.

There is a vast literature on Bayesian approaches to fitting linear mixed models; we only provide a summary of the approach in this section. General texts providing very accessible overviews of this topic with many worked examples using existing software include Gelman et al. (2004), Carlin & Louis (2009), and Jackman (2009). Gelman & Hill (2006) provide a very practical overview of using Bayesian approaches to fit multilevel models using the R software in combination with the BUGS software. We will aim to include additional examples using other software on the book's web page as the various software packages covered in this book progress in making Bayesian approaches more available and accessible. At the time of this writing, fully Bayesian analysis approaches are only available in `proc mixed` in SAS, and we will provide examples of these approaches on the book's web page.

2.10 Power Analysis for Linear Mixed Models

At the stage where researchers are designing studies that will produce longitudinal or clustered data, a **power analysis** is often necessary to ensure that sample sizes at each level of the data hierarchy will be large enough to permit detection of effects (or parameters) of scientific interest. In this section, we review methods that can be used to perform power analyses for linear mixed models. We focus on *a priori* power calculations, which would be performed at the stage of designing a given study. We first consider direct power computations based on known analytic results. Readers interested in more details with regard to study design considerations for LMMs should review van Breukelen & Moerbeek (2013).

2.10.1 Direct Power Computations

Methods and software for performing *a priori* power calculations for a study that will employ linear mixed models are readily available, and much of the available software can fortunately be obtained free of charge. Helms (1992) and Verbeke & Molenberghs (2000) describe power calculations for linear mixed models based on known analytic formulas, and Spybrook et al. (2011) provide a detailed account of power analysis methods for linear mixed models that do not require simulations.

In terms of existing software, Galecki & Burzykowski (2013) describe the use of available tools in the freely available R software for performing *a priori* power calculations for linear mixed models. Spybrook et al. (2011) have developed the freely available Optimal Design software package, which can be downloaded from `http://sitemaker.umich.edu/group-based/optimal_design_software`. The documentation for this software, which can also be downloaded free-of-charge from the same web site, provides several detailed examples of *a priori* power calculations using known analytic results for linear mixed models that are employed by this software. We find these tools to be extremely valuable for researchers designing studies that will collect longitudinal or clustered data sets, and we urge readers to consult these references when designing such studies.

For some study designs where closed-form results for power calculations are not readily available, simulations will likely need to be employed. In the next section, we discuss a general approach to the use of simulations to perform power analyses for studies that will employ linear mixed models.

2.10.2 Examining Power via Simulation

In general, there are three steps involved in using simulations to compute the power of a given study design to detect specified parameters of interest in a linear mixed model:

1. Specify the model of interest, using actual values for the parameters of interest which correspond to the values that the research team would like to be able to detect at a specified level of significance. For example, in a simple random intercept model for a continuous outcome variable in a two-level clustered data set with one binary covariate at Level 1 (e.g., treatment), there are four parameters to be estimated: the fixed intercept (or the mean for the control group), the fixed treatment effect, the variance of the random intercepts, and the variance of the errors at Level 1. A researcher may wish to detect a mean difference of 5 units between the treatment and control groups, where the control group is expected to have a mean of 50 on the continuous outcome. The researcher could also use estimated values of the variance of the random intercepts and the variance of the errors at Level 1 based on pilot studies or previous publications.

2. Randomly generate a predetermined number of observations based on the specified model, at each level of the data hierarchy. In the context of the two-level data set described above, the researcher may be able afford data collection in 30 clusters, with 20 observations per cluster. This would involve randomly selecting 30 random effects from the specified distribution, and then randomly selecting 20 draws of values for the outcome variable conditional on each selected random effect (based on the prespecified values of the fixed-effect parameters and 20 random draws of residuals from the specified distribution for the errors).

3. After generating a hypothetical data set, fit the linear mixed model of interest, and use appropriate methods to test the hypothesis of interest or generate inference regarding the target parameter(s). Record the result of the hypothesis test, and repeat these last two steps several hundred (or even thousands) of times, recording the proportion of simulated data sets where the effect of interest can be detected at a specified level of significance. This proportion is the power of the proposed design to detect the effect of interest.

We now present an example of a macro in SAS that illustrates this three-step simulation methodology, in the case of a fairly simple linear mixed model. We suppose that a researcher is interested in being able to detect a between-cluster variance component in a two-level cross-sectional design, where individual observations on a continuous dependent variable are nested within clusters. The research question is whether the between-cluster variance component is greater than zero. Based on previous studies, the researcher estimates that the dependent variable has a mean of 45, and a within-cluster variance (error variance) of 64. We do not consider any covariates in this example, meaning that 64 would be the "raw" (or unconditional) variance of the dependent variable within a given cluster.

The researcher would like to be able to detect a between-cluster variance component of 2 as being significantly greater than zero with at least 80 percent power when using a 0.05 level of significance. This variance component corresponds to a within-cluster correlation (or ICC) of 0.03. The 80 percent power means that if this ICC exists in the target population, the researcher would be able to detect it 80 times out of 100 when using a significance level of 0.05 and a likelihood ratio test based on a mixture of chi-square distributions (see Subsection 2.6.2.2). The researcher needs to know how many clusters to sample, and how many individuals to sample within each cluster.

We have prepared a SAS program on the book's web page (see Appendix A) for performing this simulation. We use `proc glimmix` in SAS to fit the linear mixed model to each simulated data set, which enables the appropriate likelihood ratio test of the null hypothesis that the between-cluster variance is equal to zero based on a mixture of chi-square distributions (via the `covtest glm` statement). We also take advantage of the `ods output` functionality in SAS to save the likelihood ratio test p-values in a SAS data file (`lrtresult`) that can be analyzed further. This program requests 100 simulated data sets, where 40 observations are sampled from each of 30 clusters. One run of this macro results in a computed power value of 0.91, suggesting that these choices for the number of clusters and the number of observations per cluster would result in sufficient power (and that the researcher may be able to sample fewer clusters or observations, given that a power value of 0.8 was desired).

This simple example introduces the idea of using simulation for power analysis when fitting linear mixed models and designing longitudinal studies or cluster samples. With a small amount of SAS code, these simulations are straightforward to implement, and similar macros can be written using other languages (SPSS, Stata, R, etc.). For additional reading on the use of simulations to conduct power analyses for linear mixed models, we refer readers to Gelman & Hill (2006) and Galecki & Burzykowski (2013), who provide worked examples of writing simulations for power analysis using the freely available R software. We also refer readers to the freely available MLPowSim software that can be used for similar types of simulation-based power analyses (http://www.bristol.ac.uk/cmm/software/mlpowsim/).

2.11   Chapter Summary

LMMs are flexible tools for the analysis of clustered and repeated-measures/longitudinal data that allow for subject-specific inferences. LMMs extend the capabilities of standard linear models by allowing:

1. Unbalanced and missing data, as long as the missing data are MAR;

2. The fixed effects of time-varying covariates to be estimated in models for repeated-measures or longitudinal data sets;

3. Structured variance-covariance matrices for both the random effects (the D matrix) and the residuals (the R_i matrix).

In building an LMM for a specific data set, we aim to specify a model that is appropriate both for the mean structure and the variance-covariance structure of the observed responses. The variance-covariance structure in an LMM should be specified in light of the observed data and a thorough understanding of the subject matter. From a statistical point of view, we aim to choose a simple (or parsimonious) model with a mean and variance-covariance structure that reflects the basic relationships among observations, and maximizes the likelihood of the observed data. A model with a variance-covariance structure that fits the data well leads to more accurate estimates of the fixed-effect parameters and to appropriate statistical tests of significance.

3

Two-Level Models for Clustered Data: The Rat Pup Example

3.1 Introduction

In this chapter, we illustrate the analysis of a **two-level clustered data set**. Such data sets typically include **randomly sampled clusters** (Level 2) and **units of analysis** (Level 1), which are randomly selected from each cluster. Covariates can measure characteristics of the clusters or of the units of analysis, so they can be either Level 2 or Level 1 variables. The dependent variable, which is measured on each unit of analysis, is always a Level 1 variable.

The models fitted to clustered data sets with two or more levels of data (or to longitudinal data) are often called **multilevel models** (see Subsection 2.2.4). **Two-level models** are the simplest examples of multilevel models and are often used to analyze two-level data sets. In this chapter, we consider **two-level random intercept models** that include only a single random effect associated with the intercept for each cluster. We formally define an example of a two-level random intercept model in Subsection 3.3.2.

Study designs that can result in two-level clustered data sets include **observational studies** on units within clusters, in which characteristics of both the clusters and the units are measured; **cluster-randomized trials**, in which a treatment is randomly assigned to all units within a cluster; and **randomized block design experiments**, in which the blocks represent clusters and treatments are assigned to units within blocks. Examples of two-level data sets and related study designs are presented in Table 3.1.

This is the first chapter in which we illustrate the analysis of a data set using the five software procedures discussed in this book: `proc mixed` in SAS, the `MIXED` command in SPSS, the `lme()` and `lmer()` functions in R, the `mixed` command in Stata, and the HLM2 procedure in HLM. We highlight the SAS software in this chapter. SAS is used for the initial data summary, and for the model diagnostics at the end of the analysis. We also go into the modeling steps in more detail in SAS.

TABLE 3.1: Examples of Two-Level Data in Different Research Settings

Level of Data		Research Setting/Study Design		
		Sociology	Education	Toxicology
		Observational Study	Cluster-Randomized Trial	Cluster-Randomized Trial
Level 2	Cluster (random factor)	City Block	Classroom	Litter
	Covariates	Urban vs. rural indicator, percentage single-family dwellings	Teaching method, teacher years of experience	Treatment, litter size
Level 1	Unit of analysis	Household	Student	Rat Pup
	Dependent variable	Household income	Test score	Birth weight
	Covariates	Number of people in household, own or rent home	Gender, age	Sex

3.2 The Rat Pup Study

3.2.1 Study Description

Jose Pinheiro and Doug Bates, authors of the `lme()` function in R, provide the Rat Pup data in their book *Mixed-Effects Models in S and S-PLUS* (2000). The data come from a study in which 30 female rats were randomly assigned to receive one of three doses (high, low, or control) of an experimental compound. The objective of the study was to compare the birth weights of pups from litters born to female rats that received the high- and low-dose treatments to the birth weights of pups from litters that received the control treatment. Although 10 female rats were initially assigned to receive each treatment dose, three of the female rats in the high-dose group died, so there are no data for their litters. In addition, litter sizes varied widely, ranging from 2 to 18 pups. Because the number of litters per treatment and the number of pups per litter were unequal, the study has an unbalanced design.

The Rat Pup data is an example of a two-level clustered data set obtained from a cluster-randomized trial: each litter (cluster) was randomly assigned to a specific level of treatment, and rat pups (units of analysis) were nested within litters. The birth weights of rat pups within the same litter are likely to be correlated because the pups shared the same maternal environment. In models for the Rat Pup data, we include random litter effects (which imply that observations on the same litter are correlated) and fixed effects associated

TABLE 3.2: Sample of the Rat Pup Data Set

Litter (Level 2)			Rat Pup (Level 1)		
Cluster ID	Covariates		Unit ID	Dependent Variable	Covariate
LITTER	TREATMENT	LITSIZE	PUP_ID	WEIGHT	SEX
1	Control	12	1	6.60	Male
1	Control	12	2	7.40	Male
1	Control	12	3	7.15	Male
1	Control	12	4	7.24	Male
1	Control	12	5	7.10	Male
1	Control	12	6	6.04	Male
1	Control	12	7	6.98	Male
1	Control	12	8	7.05	Male
1	Control	12	9	6.95	Female
1	Control	12	10	6.29	Female
...					
11	Low	16	132	5.65	Male
11	Low	16	133	5.78	Male
...					
21	High	14	258	5.09	Male
21	High	14	259	5.57	Male
21	High	14	260	5.69	Male
21	High	14	261	5.50	Male
...					

Note: "..." indicates that a portion of data is not displayed.

with treatment. Our analysis uses a two-level random intercept model to compare the mean birth weights of rat pups from litters assigned to the three different doses, after taking into account variation both between litters and between pups within the same litter.

A portion of the 322 observations in the Rat Pup data set is shown in Table 3.2, in the "long"[1] format appropriate for a linear mixed model (LMM) analysis using the procedures in SAS, SPSS, R, and Stata. Each data row represents the values for an individual rat pup. The litter ID and litter-level covariates TREATMENT and LITSIZE are included, along with the pup-level variables WEIGHT and SEX. Note that the values of TREATMENT and LITSIZE are the same for all rat pups within a given litter, whereas SEX and WEIGHT vary from pup to pup.

Each variable in this data set is classified as either a Level 2 or Level 1 variable, as follows:

Litter (Level 2) Variables

- **LITTER** = Litter ID number

---

[1]The HLM software requires a different data setup, which will be discussed in Subsection 3.4.5.

- **TREATMENT** = Dose level of the experimental compound assigned to the litter (high, low, control)

- **LITSIZE** = Litter size (i.e., number of pups per litter)

Rat Pup (Level 1) Variables

- **PUP_ID** = Unique identifier for each rat pup

- **WEIGHT** = Birth weight of the rat pup (the dependent variable)

- **SEX** = Sex of the rat pup (male, female)

3.2.2 Data Summary

The data summary for this example was generated using SAS Release 9.3. A link to the syntax and commands that can be used to carry out a similar data summary in the other software packages is included on the web page for the book (see Appendix A).

We first create the `ratpup` data set in SAS by reading in the tab-delimited raw data file, `rat_pup.dat`, assumed to be located in the `C:\temp` directory of a Windows machine. Note that SAS users can optionally import the data directly from a web site (see the second filename statement that has been "commented out"):

```
filename ratpup "C:\temp\rat_pup.dat";
*filename ratpup url "http://www-personal.umich.edu/~bwest/rat_pup.dat";
data ratpup;
    infile ratpup firstobs = 2 dlm = "09"X;
    input pup_id weight sex $ litter litsize treatment $;
    if treatment = "High" then treat = 1;
    if treatment = "Low" then treat = 2;
    if treatment = "Control" then treat = 3;
run;
```

We skip the first row of the raw data file, containing variable names, by using the `firstobs = 2` option in the `infile` statement. The `dlm = "09"X` option tells SAS that tabs, having the hexadecimal code of 09, are the delimiters in the data file.

We create a new numeric variable, TREAT, that represents levels of the original character variable, TREATMENT, recoded into numeric values (High = 1, Low = 2, and Control = 3). This recoding is carried out to facilitate interpretation of the parameter estimates for TREAT in the output from `proc mixed` in later analyses (see Subsection 3.4.1 for an explanation of how this recoding affects the output from `proc mixed`).

Next we create a user-defined format, `trtfmt`, to label the levels of TREAT in the output. Note that assigning a format to a variable can affect the order in which levels of the variable are processed in different SAS procedures; we will provide notes on the ordering of the TREAT variable in each procedure that we use.

```
proc format;
    value trtfmt 1 = "High"
                 2 = "Low"
                 3 = "Control";
run;
```

The following SAS syntax can be used to generate descriptive statistics for the birth weights of rat pups at each level of treatment by sex. The `maxdec = 2` option specifies that values displayed in the output from `proc means` are to have only two digits after the decimal.

Software Note: By default the levels of the `class` variable, TREAT, are sorted by their numeric (unformatted) values in the `proc means` output, rather than by their alphabetic (formatted) values. The values of SEX are ordered alphabetically, because no format is applied.

```
title "Summary statistics for weight by treatment and sex";
proc means data = ratpup maxdec = 2;
   class treat sex;
   var weight;
   format treat trtfmt.;
run;
```

The SAS output displaying descriptive statistics for each level of treatment and sex are shown in the Analysis Variable: weight table below.

```
                    Analysis Variable : weight
      ------------------------------------------------------------
                      N
      Treat    Sex    Obs    N Mean   Std Dev    Minimum    Maximum

      High     Female  32    32  5.85    0.60       4.48       7.68
               Male    33    33  5.92    0.69       5.01       7.70

      Low      Female  65    65  5.84    0.45       4.75       7.73
               Male    61    61  6.03    0.38       5.25       7.13

      Control  Female  54    54  6.12    0.69       3.68       7.57
               Male    77    77  6.47    0.75       4.57       8.33
```

The experimental treatments appear to have a negative effect on mean birth weight: the sample means of birth weight for pups born in litters that received the high- and low-dose treatments are lower than the mean birth weights of pups born in litters that received the control dose. We note this pattern in both female and male rat pups. We also see that the sample mean birth weights of male pups are consistently higher than those of females within all levels of treatment.

We use box plots to compare the distributions of birth weights for each treatment by sex combination graphically. We generate these box plots using the `sgpanel` procedure, creating panels for each treatment and showing box plots for each sex within each treatment:

```
title "Boxplots of rat pup birth weights (Figure 3.1)";
ods listing style = journal2;
proc sgpanel data = ratpup;
  panelby treat / novarname columns=3;
  vbox weight / category = sex;
  format treat trtfmt.;
run;
```

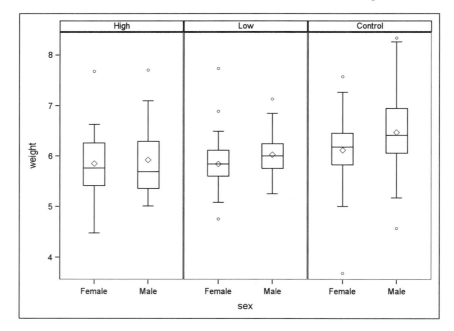

FIGURE 3.1: Box plots of rat pup birth weights for levels of treatment by sex.

The pattern of lower birth weights for the high- and low-dose treatments compared to the control group is apparent in Figure 3.1. Male pups appear to have higher birth weights than females in both the low and control groups, but not in the high group. The distribution of birth weights appears to be roughly symmetric at each level of treatment and sex. The variances of the birth weights are similar for males and females within each treatment but appear to differ across treatments (we will check the assumption of constant variance across the treatment groups as part of the analysis). We also note potential outliers, which are investigated in the model diagnostics (Section 3.10). Because each box plot pools measurements for rat pups from several litters, the possible effects of litter-level covariates, such as litter size, are not shown in this graph.

In Figure 3.2, we use box plots to illustrate the relationship between birth weight and litter size. Each panel shows the distributions of birth weights for all litters ranked by size, within a given level of treatment and sex. We first create a new variable, RANKLIT, to order the litters by size. The smallest litter has a size of 2 pups (RANKLIT = 1), and the largest litter has a size of 18 pups (RANKLIT = 27). After creating RANKLIT, we create box plots as a function of RANKLIT for each combination of TREAT and SEX by using `proc sgpanel`:

```
proc sort data=ratpup;
   by litsize litter;
run;

data ratpup2;
   set ratpup;
   by litsize litter;
   if first.litter then ranklit+1;
   label ranklit = "New Litter ID (Ordered by Size)";
run;
```

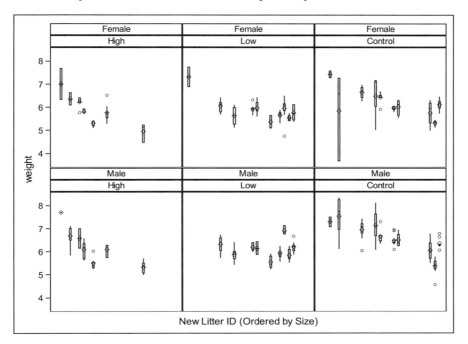

FIGURE 3.2: Litter-specific box plots of rat pup birth weights by treatment level and sex. Box plots are ordered by litter size.

```
/* Box plots for weight by litsize, for each level of treat and sex */

proc sgpanel data=ratpup2;
   panelby sex treat / novarname columns=3 ;
   vbox weight / category = ranklit meanattrs=(size=6) ;
   format treat trtfmt.;
   colaxis display=(novalues noticks);
run;
```

Figure 3.2 shows a strong tendency for birth weights to decrease as a function of litter size in all groups except males from litters in the low-dose treatment. We consider this important relationship in our models for the Rat Pup data.

3.3   Overview of the Rat Pup Data Analysis

For the analysis of the Rat Pup data, we follow the "top-down" modeling strategy outlined in Subsection 2.7.1. In Subsection 3.3.1 we outline the analysis steps, and informally introduce related models and hypotheses to be tested. Subsection 3.3.2 presents a more formal specification of selected models that are fitted to the Rat Pup data, and Subsection 3.3.3 details the hypotheses tested in the analysis. To follow the analysis steps outlined in this section, we refer readers to the schematic diagram presented in Figure 3.3.

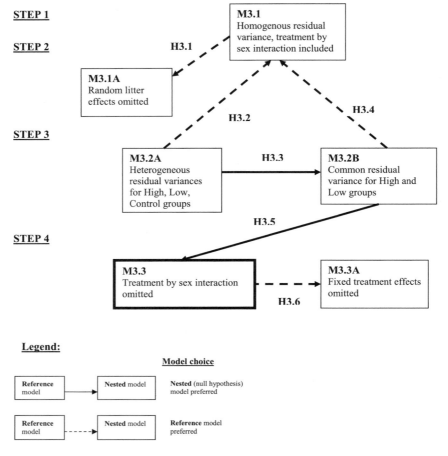

FIGURE 3.3: Model selection and related hypotheses for the analysis of the Rat Pup data.

3.3.1   Analysis Steps

Step 1: Fit a model with a "loaded" mean structure (Model 3.1).

Fit a two-level model with a "loaded" mean structure and random litter-specific intercepts.

Model 3.1 includes the fixed effects of treatment, sex, litter size, and the interaction between treatment and sex. The model also includes a random effect associated with the intercept for each litter and a residual associated with each birth weight observation. The residuals are assumed to be independent and identically distributed, given the random and fixed effects, with constant variance across the levels of treatment and sex.

Step 2: Select a structure for the random effects (Model 3.1 vs. Model 3.1A).

Decide whether to keep the random litter effects in Model 3.1.

In this step we test whether the random litter effects associated with the intercept should be omitted from Model 3.1 (Hypothesis 3.1), by fitting a nested model omitting the random effects (Model 3.1A) and performing a likelihood ratio test. Based on the result of this test, we decide to retain the random litter effects in all subsequent models.

Step 3: Select a covariance structure for the residuals (Models 3.1, 3.2A, or 3.2B).

Decide whether the model should have homogeneous residual variances (Model 3.1), heterogeneous residual variances for each of the treatment groups (Model 3.2A), or grouped heterogeneous residual variances (Model 3.2B).

We observed in Figure 3.2 that the within-litter variance in the control group appears to be larger than the within-litter variance in the high and low treatment groups, so we investigate heterogeneity of residual variance in this step.

In Model 3.1, we assume that the residual variance is homogeneous across all treatment groups. In Model 3.2A, we assume a heterogeneous residual variance structure, i.e., that the residual variance of the birth weight observations differs for each level of treatment (high, low, and control). In Model 3.2B, we specify a common residual variance for the high and low treatment groups, and a different residual variance for the control group.

We test Hypotheses 3.2, 3.3, and 3.4 (specified in Section 3.3) in this step to decide which covariance structure to choose for the residual variance. Based on the results of these tests, we choose Model 3.2B as our preferred model at this stage of the analysis.

Step 4: Reduce the model by removing nonsignificant fixed effects, test the main effects associated with treatment, and assess model diagnostics.

Decide whether to keep the treatment by sex interaction in Model 3.2B (Model 3.2B vs. Model 3.3).

Test the significance of the treatment effects in our final model, Model 3.3 (Model 3.3 vs. Model 3.3A).

Assess the assumptions for Model 3.3.

We first test whether we wish to keep the treatment by sex interaction in Model 3.2B (Hypothesis 3.5). Based on the result of this test, we conclude that the treatment by sex interaction is not significant, and can be removed from the model. Our new model is Model 3.3. The model-building process is complete at this point, and Model 3.3 is our final model.

We now focus on testing the main hypothesis of the study: whether the main effects of treatment are equal to zero (Hypothesis 3.6). We use ML estimation to refit Model 3.3 and to fit a nested model, Model 3.3A, excluding the fixed treatment effects. We then carry out a likelihood ratio test for the fixed effects of treatment. Based on the result of the test, we conclude that the fixed effects of treatment are significant. The estimated fixed-effect parameters indicate that, controlling for sex and litter size, treatment lowers the mean birth weight of rat pups in litters receiving the high and low dose of the drug compared to the birth weight of rat pups in litters receiving the control dose.

Finally, we refit Model 3.3 using REML estimation to reduce the bias of the estimated covariance parameters, and carry out diagnostics for Model 3.3 using SAS (see Section 3.10 for diagnostics).

Software Note: Steps 3 and 4 of the analysis are carried out using `proc mixed` in SAS, the `GENLINMIXED` procedure in SPSS, the `lme()` function in R, and the `mixed` command in Stata only. The other procedures discussed in this book do not allow us to fit models that have heterogeneous residual variance for groups defined by Level 2 (cluster-level) factors.

In Figure 3.3, we summarize the model selection process and hypotheses considered in the analysis of the Rat Pup data. Each box corresponds to a model and contains a brief description of the model. Each arrow corresponds to a hypothesis and connects two models involved in the specification of that hypothesis. The arrow starts at the box representing the reference model for the hypothesis and points to the simpler (nested) model. A dashed arrow indicates that, based on the result of the hypothesis test, we chose the reference model as the preferred one, and a solid arrow indicates that we chose the nested (null) model. The final model is included in a bold box.

3.3.2   Model Specification

In this section we specify selected models considered in the analysis of the Rat Pup data. We summarize the models in Table 3.3.

3.3.2.1   General Model Specification

The general specification of Model 3.1 for an individual birth weight observation (WEIGHT_{ij}) on rat pup i within the j-th litter is shown in (3.1). This specification corresponds closely to the syntax used to fit the model using the procedures in SAS, SPSS, R, and Stata.

$$
\begin{aligned}
\text{WEIGHT}_{ij} = \quad & \beta_0 + \beta_1 \times \text{TREAT1}_j + \beta_2 \times \text{TREAT2}_j + \beta_3 \times \text{SEX1}_{ij} \\
& + \beta_4 \times \text{LITSIZE}_j + \beta_5 \times \text{TREAT1}_j \times \text{SEX1}_{ij} \\
& + \beta_6 \times \text{TREAT2}_j \times \text{SEX1}_{ij} \\
& + u_j + \varepsilon_{ij}
\end{aligned}
\left. \begin{aligned} \\ \\ \\ \end{aligned} \right\} \text{fixed} \\
\left. \right\} \ \text{random} \quad (3.1)
$$

In Model 3.1 WEIGHT_{ij} is the dependent variable, and TREAT1_j and TREAT2_j are Level 2 indicator variables for the high and low levels of treatment, respectively. SEX1_{ij} is a Level 1 indicator variable for female rat pups.

In this model, WEIGHT_{ij} depends on the β parameters (i.e., the fixed effects) in a linear fashion. The fixed intercept parameter, β_0, represents the expected value of WEIGHT_{ij} for the reference levels of treatment and of sex (i.e., males in the control group) when LITSIZE_j is equal to zero. We do not interpret the fixed intercept, because a litter size of zero is outside the range of the data (alternatively, the LITSIZE variable could be centered to make the intercept interpretable; see Subsection 2.9.5).

The parameters β_1 and β_2 represent the fixed effects of the dummy variables (TREAT1_j and TREAT2_j) for the high and low treatment levels vs. the control level, respectively. The β_3 parameter represents the effect of SEX1_{ij} (female vs. male), and β_4 represents the fixed effect of LITSIZE_j. The two parameters, β_5 and β_6, represent the fixed effects associated with the treatment by sex interaction.

The random effect associated with the intercept for litter j is indicated by u_j, which is assumed to have a normal distribution with mean of 0 and constant variance σ^2_{litter}. We write the distribution of these random effects as:

$$
u_j \sim \mathcal{N}(0, \sigma^2_{litter})
$$

where σ^2_{litter} represents the variance of the random litter effects.

In Model 3.1, the distribution of the residual ε_{ij}, associated with the observation on an individual rat pup i within litter j, is assumed to be the same for all levels of treatment:

$$
\varepsilon_{ij} \sim \mathcal{N}(0, \sigma^2)
$$

In Model 3.2A, we allow the residual variances for observations at different levels of treatment to differ:

$$\text{High Treatment:} \quad \varepsilon_{ij} \sim \mathcal{N}(0, \sigma^2_{high})$$
$$\text{Low Treatment:} \quad \varepsilon_{ij} \sim \mathcal{N}(0, \sigma^2_{low})$$
$$\text{Control Treatment:} \quad \varepsilon_{ij} \sim \mathcal{N}(0, \sigma^2_{control})$$

In Model 3.2B, we consider a separate residual variance for the combined high/low treatment group and for the control group:

$$\text{High/Low Treatment:} \quad \varepsilon_{ij} \sim \mathcal{N}(0, \sigma^2_{high/low})$$
$$\text{Control Treatment:} \quad \varepsilon_{ij} \sim \mathcal{N}(0, \sigma^2_{control})$$

In Model 3.3, we include the same residual variance structure as in Model 3.2B, but remove the fixed effects, β_5 and β_6, associated with the treatment by sex interaction from the model. In all models, we assume that the random effects, u_j, associated with the litter-specific intercepts and the residuals, ε_{ij}, are independent.

3.3.2.2 Hierarchical Model Specification

We now present Model 3.1 in the hierarchical form used by the HLM software, with the same notation as in (3.1). The correspondence between this notation and the HLM software notation is defined in Table 3.3.

The hierarchical model has two components, reflecting two sources of variation: namely variation between litters, which we attempt to explain using the **Level 2 model** and variation between pups within a given litter, which we attempt to explain using the **Level 1 model**. We write the Level 1 model as:

Level 1 Model (Rat Pup)

$$\text{WEIGHT}_{ij} = \text{b}_{0j} + \text{b}_{1j} \times \text{SEX1}_{ij} + \varepsilon_{ij} \tag{3.2}$$

where

$$\varepsilon_{ij} \sim \mathcal{N}(0, \sigma^2)$$

The Level 1 model in (3.2) assumes that WEIGHT_{ij}, i.e., the birth weight of rat pup i within litter j, follows a simple ANOVA-type model defined by the litter-specific intercept, b_{0j} and the litter-specific effect of $\text{SEX1}_{ij}, b_{1j}$.

Both b_{0j} and b_{1j} are unobserved quantities that are defined as functions of Level 2 covariates in the Level 2 model:

Level 2 Model (Litter)

$$\begin{aligned} b_{0j} &= \beta_0 + \beta_1 \times \text{TREAT1}_j + \beta_2 \times \text{TREAT1}_j + \beta_4 \times \text{LITSIZE}_j + u_j \\ b_{1j} &= \beta_3 + \beta_5 \times \text{TREAT1}_j + \beta_6 \times \text{TREAT2}_j \end{aligned} \tag{3.3}$$

where

$$u_j \sim \mathcal{N}(0, \sigma^2_{litter})$$

TABLE 3.3: Selected Models Considered in the Analysis of the Rat Pup Data

	Term/Variable	General Notation	HLM Notation	Model 3.1	Model 3.2A[a]	Model 3.2B[a]	Model 3.3[a]
Fixed effects	Intercept	β_0	γ_{00}	✓	✓	✓	✓
	TREAT1 (High vs. control)	β_1	γ_{02}	✓	✓	✓	✓
	TREAT2 (Low vs. control)	β_2	γ_{03}	✓	✓	✓	✓
	SEX1 (Female vs. male)	β_3	γ_{10}	✓	✓	✓	✓
	LITSIZE	β_4	γ_{01}	✓	✓	✓	✓
	TREAT1 × SEX1	β_5	γ_{11}	✓	✓	✓	✓
	TREAT2 × SEX1	β_6	γ_{12}	✓	✓	✓	✓
Random effects	Litter (j) Intercept	u_j	u_j	✓	✓	✓	✓
Residuals	Rat pup (pup i in litter j)	ε_{ij}	r_{ij}	✓	✓	✓	✓
Covariance parameters ($\boldsymbol{\theta}_D$) for $\boldsymbol{D}$ matrix	Litter level Variance of intercepts	σ^2_{litter}	τ	✓	✓	✓	✓

TABLE 3.3: (Continued)

Term/Variable		General Notation	HLM Notation	Model 3.1	3.2A[a]	3.2B[a]	3.3[a]
Covariance parameters ($\boldsymbol{\theta}_R$) for $\boldsymbol{R}_i$ matrix	Rat pup level	Variances of residuals	σ^2_{high} σ^2_{low} $\sigma^2_{control}$	σ^2	σ^2	σ^2_{high} σ^2_{low} $\sigma^2_{control}$	$\sigma^2_{high/low}$ $\sigma^2_{control}$	$\sigma^2_{high/low}$, $\sigma^2_{high/low}$, $\sigma^2_{control}$

[a]Models 3.2A, 3.2B, and 3.3 (with heterogeneous residual variances) can only be fit using selected procedures in SAS (proc mixed), SPSS (GENLINMIXED), R (the lme() function), and Stata (mixed).

The Level 2 model in (3.3) assumes that b_{0j}, the intercept for litter j, depends on the fixed intercept, β_0, and for pups in litters assigned to the high- or low-dose treatments, on the fixed effect associated with their level of treatment vs. control (β_1 or β_2, respectively). The intercept also depends on the fixed effect of litter size, β_4, and a random effect, u_j, associated with litter j.

The effect of SEX1 within each litter, b_{1j}, depends on an overall fixed SEX1 effect, denoted by β_3, and an additional fixed effect of either the high or low treatment vs. control (β_5 or β_6, respectively). Note that the effect of sex varies from litter to litter only through the fixed effect of the treatment assigned to the litter; there is no random effect associated with sex.

By substituting the expressions for b_{0j} and b_{1j} from the Level 2 model into the Level 1 model, we obtain the general LMM specified in (3.1). The fixed treatment effects, β_5 and β_6, for TREAT1$_j$ and TREAT2$_j$ in the Level 2 model for the effect of SEX1 correspond to the interaction effects for treatment by sex (TREAT1$_j$ × SEX1$_{ij}$ and TREAT2$_j$ × SEX1$_{ij}$) in the general model specification.

3.3.3 Hypothesis Tests

Hypothesis tests considered in the analysis of the Rat Pup data are summarized in Table 3.4.

Hypothesis 3.1. The random effects, u_j, associated with the litter-specific intercepts can be omitted from Model 3.1.

We do not directly test the significance of the random litter-specific intercepts, but rather test a hypothesis related to the variance of the random litter effects. We write the null and alternative hypotheses as follows:

$$H_0: \sigma^2_{litter} = 0$$
$$H_A: \sigma^2_{litter} > 0$$

To test Hypothesis 3.1, we use a REML-based likelihood ratio test. The test statistic is calculated by subtracting the –2 REML log-likelihood value for Model 3.1 (the reference model) from the value for Model 3.1A (the nested model, which omits the random litter effects). The asymptotic null distribution of the test statistic is a mixture of χ^2 distributions, with 0 and 1 degrees of freedom, and equal weights of 0.5.

Hypothesis 3.2. The variance of the residuals is the same (homogeneous) for the three treatment groups (high, low, and control).

The null and alternative hypotheses for Hypothesis 3.2 are:

$$H_0: \sigma^2_{high} = \sigma^2_{low} = \sigma^2_{control} = \sigma^2$$
$$H_A: \text{At least one pair of residual variances is not equal to each other}$$

We use a REML-based likelihood ratio test for Hypothesis 3.2. The test statistic is obtained by subtracting the –2 REML log-likelihood value for Model 3.2A (the reference model, with heterogeneous variances) from that for Model 3.1 (the nested model). The asymptotic null distribution of this test statistic is a χ^2 with 2 degrees of freedom, where the 2 degrees of freedom correspond to the 2 additional covariance parameters (i.e., the 2 additional residual variances) in Model 3.2A compared to Model 3.1.

Hypothesis 3.3. The residual variances for the high and low treatment groups are equal.

TABLE 3.4: Summary of Hypotheses Tested in the Rat Pup Analysis

				Hypothesis Test			
Hypothesis Specification				**Models Compared**			
Label	Null (H_0)	Alternative (H_A)	Test	Nested Model (H_0)	Ref. Model (H_A)	Est. Method	Test Stat. Dist. under H_0
3.1	Drop u_j ($\sigma^2_{litter} = 0$)	Retain u_j ($\sigma^2_{litter} > 0$)	LRT	Model 3.1 A	Model 3.1	REML	$0.5\chi^2_0 + 0.5\chi^2_1$
3.2	Homogeneous residual variance ($\sigma^2_{high} = \sigma^2_{low} = \sigma^2_{control} = \sigma^2$)	Residual variances are not all equal	LRT	Model 3.1	Model 3.2A	REML	χ^2_2
3.3	Grouped heterogeneous residual variance ($\sigma^2_{high} = \sigma^2_{low}$)	($\sigma^2_{high} \neq \sigma^2_{low}$)	LRT	Model 3.2B	Model 3.2A	REML	χ^2_1
3.4	Homogeneous residual variance ($\sigma^2_{high/low} = \sigma^2_{control} = \sigma^2$)	Grouped heterogeneous residual variance ($\sigma^2_{high/low} \neq \sigma^2_{control}$)	LRT	Model 3.1	Model 3.2B	REML	χ^2_1
3.5	Drop TREATMENT × SEX effects ($\beta_5 = \beta_6 = 0$)	$\beta_5 \neq 0$, or $\beta_6 \neq 0$	Type III F-test	N/A	Model 3.2B	REML	$F(2,194)^a$
3.6	Drop TREATMENT Effects ($\beta_1 = \beta_2 = 0$)	$\beta_1 \neq 0$, or $\beta_2 \neq 0$	LRT / Type III	Model 3.3A / N/A	Model 3.3	ML / REML	χ^2_2 / F(2,24.3)

[a]Different methods for calculating denominator degrees of freedom are available in the software procedures; we report the Satterthwaite estimate of degrees of freedom calculated by proc mixed in SAS.

The null and alternative hypotheses are as follows:

$$H_0 : \sigma^2_{high} = \sigma^2_{low}$$
$$H_A : \sigma^2_{high} \neq \sigma^2_{low}$$

We test Hypothesis 3.3 using a REML-based likelihood ratio test. The test statistic is calculated by subtracting the –2 REML log-likelihood value for Model 3.2A (the reference model) from the corresponding value for Model 3.2B (the nested model, with pooled residual variance for the high and low treatment groups). The asymptotic null distribution of this test statistic is a χ^2 with 1 degree of freedom, where the single degree of freedom corresponds to the one additional covariance parameter (i.e., the one additional residual variance) in Model 3.2A compared to Model 3.2B.

Hypothesis 3.4. The residual variance for the combined high/low treatment group is equal to the residual variance for the control group.

In this case, the null and alternative hypotheses are:

$$H_0: \sigma^2_{high/low} = \sigma^2_{control} = \sigma^2$$
$$H_A: \sigma^2_{high/low} \neq \sigma^2_{control}$$

We test Hypothesis 3.4 using a REML-based likelihood ratio test. The test statistic is obtained by subtracting the –2 REML log-likelihood value for Model 3.2B (the reference model) from that for Model 3.1 (the nested model). The asymptotic null distribution of this test statistic is a χ^2 with 1 degree of freedom, corresponding to the one additional variance parameter in Model 3.2B compared to Model 3.1.

Hypothesis 3.5. The fixed effects associated with the treatment by sex interaction are equal to zero in Model 3.2B.

The null and alternative hypotheses are:

$$H_0: \beta_5 = \beta_6 = 0$$
$$H_A: \beta_5 \neq 0 \text{ or } \beta_6 \neq 0$$

We test Hypothesis 3.5 using an approximate F-test, based on the results of the REML estimation of Model 3.2B. Because this test is not significant, we remove the treatment by sex interaction term from Model 3.2B and obtain Model 3.3.

Hypothesis 3.6. The fixed effects associated with treatment are equal to zero in Model 3.3.

This hypothesis differs from the previous ones, in that it is not being used to select a model, but is testing the primary research hypothesis. The null and alternative hypotheses are:

$$H_0: \beta_1 = \beta_2 = 0$$
$$H_A: \beta_1 \neq 0 \text{ or } \beta_2 \neq 0$$

We test Hypothesis 3.6 using an ML-based likelihood ratio test. The test statistic is calculated by subtracting the –2 ML log-likelihood value for Model 3.3 (the reference model) from that for Model 3.3A (the nested model excluding the fixed treatment effects). The asymptotic null distribution of this test statistic is a χ^2 with 2 degrees of freedom, corresponding to the two additional fixed-effect parameters in Model 3.3 compared to Model 3.3A.

Alternatively, we can test Hypothesis 3.6 using an approximate F-test for TREAT-MENT, based on the results of the REML estimation of Model 3.3.

For the results of these hypothesis tests see Section 3.5.

3.4 Analysis Steps in the Software Procedures

In this section, we illustrate fitting the LMMs for the Rat Pup example using the software procedures in SAS, SPSS, R, Stata, and HLM. Because we introduce the use of the software procedures in this chapter, we present a more detailed description of the steps and options for fitting each model than we do in Chapters 4 through 8. We compare results for selected models across the software procedures in Section 3.6.

3.4.1 SAS

Step 1: Fit a model with a "loaded" mean structure (Model 3.1).

We assume that the `ratpup` data set has been created in SAS, as illustrated in the data summary (Subsection 3.2.2). The SAS commands used to fit Model 3.1 to the Rat Pup data using `proc mixed` are as follows:

```
ods output fitstatistics = fit1;
title "Model 3.1";
proc mixed data = ratpup order = internal covtest;
class treat sex litter;
model weight = treat sex litsize treat*sex /
    solution ddfm = sat;
random int / subject= litter;
format treat trtfmt.;
run;
```

The `ods` statement is used to create a data set, `fit1`, containing the –2 REML log-likelihood and other fit statistics for Model 3.1. We will use the `fit1` data set later to perform likelihood ratio tests for Hypotheses 3.1, 3.2, and 3.4.

The `proc mixed` statement invokes the analysis, using the default REML estimation method. We use the `covtest` option to obtain the standard errors of the estimated covariance parameters for comparison with the results from the other software procedures. The `covtest` option also causes SAS to display a Wald z-test of whether the variance of the random litter effects equals zero (i.e., Hypothesis 3.1), but we do not recommend using this test (see the discussion of Wald tests for covariance parameters in Subsection 2.6.3.2). The `order = internal` option requests that levels of variables declared in the `class` statement be ordered based on their (unformatted) internal numeric values and not on their formatted values.

The `class` statement includes the two categorical factors, TREAT and SEX, which will be included as fixed predictors in the `model` statement, as well as the classification factor, LITTER, that defines subjects in the `random` statement.

The `model` statement sets up the fixed effects. The dependent variable, WEIGHT, is listed on the left side of the equal sign, and the covariates having fixed effects are included on the right of the equal sign. We include the fixed effects of TREAT, SEX, LITSIZE, and the TREAT × SEX interaction in this model. The `solution` option follows a slash (/), and instructs SAS to display the fixed-effect parameter estimates in the output (they are not displayed by default). The `ddfm =` option specifies the method used to estimate denominator degrees of freedom for F-tests of the fixed effects. In this case, we use `ddfm = sat` for the Satterthwaite approximation (see Subsection 3.11.6 for more details on denominator degrees of freedom options in SAS).

Software Note: By default, SAS generates an indicator variable for each level of a `class` variable included in the `model` statement. This typically results in an overparameterized model, in which there are more columns in the X matrix than there are degrees of freedom for a factor or an interaction term involving a factor. SAS then uses a generalized inverse (denoted by $^-$ in the following formula) to calculate the fixed-effect parameter estimates (see the SAS documentation for `proc mixed` for more information):

$$\widehat{\boldsymbol{\beta}} = (\boldsymbol{X}'\widehat{\boldsymbol{V}}^{-1}\boldsymbol{X})^- \boldsymbol{X}'\widehat{\boldsymbol{V}}^{-1}\mathbf{y}$$

In the output for the fixed-effect parameter estimates produced by requesting the `solution` option as part of the `model` statement, the estimate for the highest level of a class variable is by default set to zero; and the level that is considered to be the highest level for a variable will change depending on whether there is a format associated with the variable or not.

In the analysis of the Rat Pup data, we wish to contrast the effects of the high- and low-dose treatments to the control dose, so we use the `order = internal` option to order levels of TREAT. This results in the parameter for Level 3 of TREAT (i.e., the control dose, which is highest numerically) being set to zero, so the parameter estimates for the other levels of TREAT represent contrasts with TREAT = 3 (control). This corresponds to the specification of Model 3.1 in (3.1). The value of TREAT = 3 is labeled "Control" in the output by the user-defined format.

We refer to TREAT = 3 as the reference category throughout our discussion. In general, we refer to the highest level of a class variable as the "reference" level when we estimate models using `proc mixed` throughout the book. For example, in Model 3.1, the "reference category" for SEX is "Male" (the highest level of SEX alphabetically), which corresponds to our specification in (3.1).

The `random` statement specifies that a random intercept, `int`, is to be associated with each litter, and litters are specified as subjects by using the `subject = litter` option. Alternative syntax for the `random` statement is:

```
random litter ;
```

This syntax results in a model that is equivalent to Model 3.1, but is much less efficient computationally. Because `litter` is specified as a random factor, we get the same block-diagonal structure for the variance-covariance matrix for the random effects, which SAS refers to as the G matrix, as when we used `subject = litter` in the previous syntax (see Subsection 2.2.3 for a discussion of the G matrix). However, all observations are assumed to be from one "subject," and calculations for parameter estimation use much larger matrices and take more time than when `subject = litter` is specified.

The `format` statement attaches the user-defined format, `trtfmt.`, to values of the variable TREAT.

Step 2: Select a structure for the random effects (Model 3.1 vs. Model 3.1A).

To test Hypothesis 3.1, we first fit Model 3.1A without the random effects associated with litter, by using the same syntax as for Model 3.1 but excluding the `random` statement:

```
title "Model 3.1A";
ods output fitstatistics = fitla;
proc mixed data = ratpup order = internal covtest;
```

```
class treat sex litter;
model weight = treat sex treat*sex litsize /
    solution ddfm = sat;
format treat trtfmt.;
run;
```

The SAS code below can be used to calculate the likelihood ratio test statistic for Hypothesis 3.1, compute the corresponding p-value, and display the resulting p-value in the SAS log. To apply this syntax, the user has to manually extract the –2 REML log-likelihood value for the reference model, Model 3.1 (–2 REML log-likelihood = 401.1), and for the nested model, Model 3.1A (–2 REML log-likelihood = 490.5), from the output and include these values in the code. Recall that the asymptotic null distribution of the test statistic for Hypothesis 3.1 is a mixture of χ_0^2 and χ_1^2 distributions, each having equal weight of 0.5. Because the χ_0^2 has all of its mass concentrated at zero, its contribution to the p-value is zero, so it is not included in the following code.

```
title "P-value for Hypothesis 3.1: Simple syntax";
data _null_;
lrtstat = 490.5 - 401.1;
df = 1;
pvalue = 0.5*(1 - probchi(lrtstat,df));
format pvalue 10.8;
put lrtstat = df = pvalue = ;
run;
```

Alternatively, we can use the data sets `fit1` and `fit1a`, containing the fit statistics for Models 3.1 and 3.1A, respectively, to perform this likelihood ratio test. The information contained in these two data sets is displayed below, along with more advanced SAS code to merge the data sets, derive the difference of the –2 REML log-likelihoods, calculate the appropriate degrees of freedom, and compute the p-value for the likelihood ratio test. The results of the likelihood ratio test will be included in the SAS log.

```
title "Fit 1";
proc print data = fit1;
run;


title "Fit 1a";
proc print data = fit1a;
run;
```

```
                            Fit 1
           Obs   Descr                      Value

           1     -2 Res Log Likelihood      401.1
           2     AIC (smaller is better)    405.1
           3     AICC (smaller is better)   405.1
           4     BIC (smaller is better)    407.7

                            Fit 1a
           Obs   Descr                      Value
           1     -2 Res Log Likelihood      490.5
           2     AIC (smaller is better)    492.5
           3     AICC (smaller is better)   492.5
           4     BIC (smaller is better)    496.3
```

```
title "p-value for Hypothesis 3.1: Advanced syntax";
data _null_;
merge fit1(rename = (value = reference)) fit1a(rename = (value = nested));
retain loglik_diff;
if descr = "-2 Res Log Likelihood" then
    loglik_diff = nested - reference;
if descr = "AIC (smaller is better)" then do;
    df = floor((loglik_diff - nested + reference)/2);
    pvalue = 0.5*(1 - probchi(loglik_diff,df));
    put loglik_diff = df = pvalue = ;
    format pvalue 10.8;
end;
run;
```

The `data _null_` statement causes SAS to execute the data step calculations without creating a new data set. The likelihood ratio test statistic for Hypothesis 3.1 is calculated by subtracting the –2 REML log-likelihood value for the reference model (contained in the data set `fit1`) from the corresponding value for the nested model (contained in `fit1a`). To calculate degrees of freedom for this test, we take advantage of the fact that SAS defines the AIC statistic as AIC = –2 REML log-likelihood + 2 × number of covariance parameters.

Software Note: When a covariance parameter is estimated to be on the boundary of a parameter space by `proc mixed` (e.g., when a variance component is estimated to be zero), SAS will not include it when calculating the number of covariance parameters for the AIC statistic. Therefore, the advanced SAS code presented in this section for computing likelihood ratio tests for covariance parameters is only valid if the estimates of the covariance parameters being tested do not lie on the boundaries of their respective parameter spaces (see Subsection 2.5.2).

Results from the likelihood ratio test of Hypothesis 3.1 and other hypotheses are presented in detail in Subsection 3.5.1.

Step 3: Select a covariance structure for the residuals (Models 3.1, 3.2A, or 3.2B).

The following SAS commands can be used to fit Model 3.2A, which allows unequal residual variance for each level of treatment. The only change in these commands from those used for Model 3.1 is the addition of the `repeated` statement:

```
title "Model 3.2A";
proc mixed data = ratpup order = internal covtest;
class treat litter sex;
model weight = treat sex treat*sex litsize /
    solution ddfm = sat;
random int / subject = litter;
repeated / group = treat;
format treat trtfmt.;
run;
```

In Model 3.2A, the option `group = treat` in the `repeated` statement allows a heterogeneous variance structure for the residuals, with each level of treatment having its own residual variance.

> **Software Note:** In general, the `repeated` statement in `proc mixed` specifies the structure of the R_j matrix, which contains the variances and covariances of the residuals for the j-th cluster (e.g., litter). If no `repeated` statement is used, the default covariance structure for the residuals is employed, i.e., $R_j = \sigma^2 I_{n_j}$, where I_{n_j} is an $n_j \times n_j$ identity matrix, with n_j equal to the number of observations in a cluster (e.g., the number of rat pups in litter j). In other words, the default specification is homogeneous residual variance. Because this default R_j matrix is used for Model 3.1, we do not include a `repeated` statement for this model.

To test Hypothesis 3.2, we calculate a likelihood ratio test statistic by subtracting the value of the –2 REML log-likelihood for Model 3.2A (the reference model with heterogeneous variance, –2 REML LL = 359.9) from that for Model 3.1 (the nested model, –2 REML LL = 401.1). The simple SAS syntax used to calculate this likelihood ratio test statistic is similar to that used for Hypothesis 3.1. The p-value is calculated by referring the test statistic to a χ^2 distribution with two degrees of freedom, which correspond to the two additional variance parameters in Model 3.2A compared to Model 3.1. We do not use a mixture of χ_0^2 and χ_1^2 distributions, as in Hypothesis 3.1, because we are not testing a null hypothesis with values of the variances on the boundary of the parameter space:

```
title "p-value for Hypothesis 3.2";
data _null_;
lrtstat = 401.1 - 359.9;
df = 2;
pvalue = 1 - probchi(lrtstat,df);
format pvalue 10.8;
put lrtstat =  df =  pvalue = ;
run;
```

The test result is significant ($p < 0.001$), so we choose Model 3.2A, with heterogeneous residual variance, as our preferred model at this stage of the analysis.

Before fitting Model 3.2B, we create a new variable named TRTGRP that combines the high and low treatment groups, to allow us to test Hypotheses 3.3 and 3.4. We also define a new format, TGRPFMT, for the TRTGRP variable.

```
title "RATPUP3 dataset";
data ratpup3;
set ratpup2;
if treatment in ("High", "Low") then TRTGRP = 1;
if treatment = "Control" then TRTGRP = 2;
run;

proc format;
value tgrpfmt 1 = "High/Low"
              2 = "Control";
run;
```

We now fit Model 3.2B using the new data set, `ratpup3`, and the new group variable in the `repeated` statement (`group = trtgrp`). We also include TRTGRP in the `class` statement so that SAS will properly include it as the grouping variable for the residual variance.

```
title "Model 3.2B";
proc mixed data = ratpup3 order = internal covtest;
class treat litter sex trtgrp;
model weight = treat sex treat*sex litsize / solution ddfm = sat;
random int / subject = litter;
repeated / group = trtgrp;
format treat trtfmt. trtgrp tgrpfmt.;
run;
```

We use a likelihood ratio test for Hypothesis 3.3 to decide if we can use a common residual variance for both the high and low treatment groups (Model 3.2B) rather than different residual variances for each treatment group (Model 3.2A). For this hypothesis Model 3.2A is the reference model and Model 3.2B is the nested model. To calculate the test statistic we subtract the –2 REML log-likelihood for Model 3.2A from that for Model 3.2B (–2 REML LL = 361.1). This test has 1 degree of freedom, corresponding to the one fewer covariance parameter in Model 3.2B compared to Model 3.2A.

```
title "p-value for Hypothesis 3.3";
data _null_;
lrtstat = 361.1 - 359.9;
df = 1;
pvalue = 1 - probchi(lrtstat, df);
format pvalue 10.8;
put lrtstat = df = pvalue = ;
run;
```

The likelihood ratio test statistic for Hypothesis 3.3 is not significant ($p = 0.27$), so we choose the simpler grouped residual variance model, Model 3.2B, as our preferred model at this stage of the analysis.

To test Hypothesis 3.4, and decide whether we wish to have a grouped heterogeneous residual variance structure vs. a homogeneous variance structure, we subtract the –2 REML log-likelihood of Model 3.2B ($= 361.1$) from that of Model 3.1 ($= 401.1$). The test statistic has 1 degree of freedom, corresponding to the 1 additional covariance parameter in Model 3.2B as compared to Model 3.1. The syntax for this comparison is not shown here. Based on the significant result of this likelihood ratio test ($p < 0.001$), we conclude that Model 3.2B (with grouped heterogeneous variances) is our preferred model.

Step 4: Reduce the model by removing nonsignificant fixed effects (Model 3.2B vs. 3.3, and Model 3.3 vs. 3.3A).

We test Hypothesis 3.5 to decide whether we can remove the treatment by sex interaction term, making use of the default Type III F-test for the TREAT × SEX interaction in Model 3.2B. Because the result of this test is not significant ($p = 0.73$), we drop the TREAT × SEX interaction term from Model 3.2B, which gives us Model 3.3.

We now test Hypothesis 3.6 to decide whether the fixed effects associated with treatment are equal to zero, using a likelihood ratio test. This test is not used as a tool for possible model reduction but as a way of assessing the impact of treatment on birth weights. To carry out the test, we first fit the reference model, Model 3.3, using maximum likelihood (ML) estimation:

```
title "Model 3.3 using ML";
proc mixed data = ratpup3 order = internal method = ml;
```

```
class treat litter sex trtgrp;
model weight = treat sex litsize / solution ddfm = sat;
random int / subject = litter;
repeated / group = trtgrp;
format treat trtfmt.;
run;
```

The `method = ml` option in the `proc mixed` statement requests maximum likelihood estimation.

To complete the likelihood ratio test for Hypothesis 3.6, we fit a nested model, Model 3.3A, without the fixed treatment effects, again requesting ML estimation, by making the following modifications to the SAS code for Model 3.3:

```
title "Model 3.3A using ML";
proc mixed data = ratpup3 order = internal method = ml;
...
model weight = sex litsize / solution ddfm = sat;
...
```

The likelihood ratio test statistic used to test Hypothesis 3.6 is calculated by subtracting the –2 ML log-likelihood for Model 3.3 (the reference model) from that for Model 3.3A (the nested model without the fixed effects associated with treatment). The SAS code for this test is not shown here.

Because the result of this test is significant ($p < 0.001$), we conclude that treatment has an effect on rat pup birth weights, after adjusting for the fixed effects of sex and litter size and the random litter effects.

We now refit Model 3.3, our final model, using the default REML estimation method to get unbiased estimates of the covariance parameters. We also add a number of options to the SAS syntax. We request diagnostic plots in the `proc mixed` statement by adding a `plots=` option. We add options to the `model` statement to get output data sets containing the conditional predicted and residual values (`outpred = pdat1`) and we get another data set containing the marginal predicted and residual values (`outpredm = pdat2`). We also request that estimates of the implied marginal variance-covariance and correlation matrices, `v` and `vcorr`, for the third litter be displayed in the output by adding the options `v = 3` and `vcorr = 3` to the `random` statement. Post-hoc tests for all estimated differences among treatment means using the Tukey–Kramer adjustment for multiple comparisons are requested by adding the `lsmeans` statement. Finally, we request that the EBLUPs for the random intercept for each litter be saved in a new file called `eblupsdat`, by using an `ods output` statement. We first sort the `ratpup3` data set by PUP_ID, because the diagnostic plots identify individual points by row numbers in the data set, and the sorting will make the PUP_ID variable equal to the row number in the data set for ease in reading the graphical output.

```
proc sort data = ratpup3;
  by pup_id;
run;

ods graphics on;
ods rtf file = "c:\temp\ratpup_diagnostics.rtf" style = journal;
ods exclude influence;
title "Model 3.3 using REML. Model diagnostics";
proc mixed data = ratpup3 order = internal covtest
```

```
           plots = (residualpanel boxplot influencestatpanel);
    class treat litter sex trtgrp;
    model weight = treat sex litsize / solution ddfm = sat
         residual  outpred = pdat1 outpredm = pdat2
                   influence (iter = 5 effect = litter est) ;
    id pup_id litter  treatment trtgrp ranklit litsize;
    random int / subject=litter solution v = 3 vcorr = 3 ;
    repeated / group=trtgrp ;
    lsmeans treat / adjust = tukey ;
    format treat trtfmt. trtgrp tgrpfmt. ;
    ods output solutionR = eblupsdat ;
run;
ods graphics off;
ods rtf close;
```

Software Note: In earlier versions of SAS, the default output mode was "Listing," which produced text in the Output window. ODS (Output Delivery System) graphics were not automatically produced. However, in Version 9.3, the default output mode has changed to HTML, and ODS graphics are automatically produced. When you use the default HTML output, statistical tables and graphics are included together in the Results Viewer Window. You can change these default behaviors through SAS commands, or by going to **Tools** > **Options** > **Preferences** and clicking on the **Results** tab. By selecting "Create listing," you will add text output to the Output window, as in previous versions of SAS. This will be in addition to the HTML output, which can be turned off by deselecting "Create HTML." To turn off ODS graphics, which can become cumbersome when running large jobs, deselect "Use ODS graphics." Click "OK" to confirm these changes. These same changes can be accomplished by the following commands:

```
  ods listing;
  ods html close;
  ods graphics off;
```

The type of output that is produced can be modified by choosing an output type (e.g., .rtf) and sending the output and SAS ODS graphics to that file. There are a number of possible styles, with Journal being a black-and-white option suitable for a manuscript. If no style is selected, the default will be used, which includes color graphics. The .rtf file can be closed when the desired output has been captured.

```
  ods graphics on;
  ods rtf file="example.rtf" style=journal;
  proc mixed ...;
  run;
  ods rtf close;
```

Any portion of the SAS output can be captured in a SAS data set, or can be excluded from the output by using ODS statements. For example, in the SAS code for Model 3.3, we used `ods exclude influence`. This prevents the influence statistics for each individual observation from being printed in the output, which can get very long if there are a large number of cases, but it still allows the influence diagnostic plots to be displayed. We must also request that the influence statistics be calculated by including an `influence` option as part of the `model` statement.

Output that we wish to save to a SAS data set can be requested by using an `ods output` statement. The statement below requests that the solutions for the random effects (the `SolutionR` table from the output) be output to a new data set called `eblupsdat`.

```
ods output solutionR = eblupsdat;
```

To view the names of SAS output tables so that they can be captured in a data set, use the following code, which will place the names of each portion of the output in the SAS log.

```
ods trace on;
proc mixed ...;
run;
ods trace off;
```

The `proc mixed` statement for Model 3.3 has been modified by the addition of the `plots =` option, with the specific ODS plots that are requested listed within parentheses: (`residualpanel boxplot influencestatpanel`). The `residualpanel` suboption generates diagnostic plots for both conditional and marginal residuals. Although these residual plots were requested, we do not display them here, because these plots are not broken down by TRTGRP. Histograms and normal quantile–quantile (Q–Q) plots of residuals for each treatment group are displayed in Figures 3.4, 3.5, and 3.6. The `model` statement has also been modified by adding the `residual` option, to allow the generation of panels of residual diagnostic plots as part of the ODS graphics output. The `boxplot` suboption requests box plots of the marginal and conditional residuals by the levels of each `class` variable, including class variables specified in the `subject =` and `group =` options, to be created in the ODS graphics output. SAS also generates box plots for levels of the "subject" variable (LITTER), but only if we do not use nesting specification for litter in the `random` statement (i.e., we must use `subject = litter` rather than `subject = litter(treat)`). Box plots showing the studentized residuals for each litter are shown in Figure 3.7. The `influencestatpanel` suboption requests that influence plots for the model fit (REML distance), overall model statistics, covariance parameters, and fixed effects be produced. These plots are illustrated in Figures 3.8 through 3.11. Diagnostic plots generated for Model 3.3 are presented in Section 3.10.

The `influence` option has also been added to the `model` statement for Model 3.3 to obtain influence plots as part of the ODS graphics output (see Subsection 3.10.2). The inclusion of the `iter =` suboption is used to produce iterative updates to the model, by removing the effect of each litter, and then re-estimating all model parameters (including both fixed-effect and random-effect parameters). The option `effect =` specifies an *effect* according to which observations are grouped, i.e., observations sharing the same level of the effect are removed as a group when calculating the influence diagnostics. The effect must contain only class variables, but these variables do not need to be contained in the model. Without the `effect =` suboption, influence statistics would be created for the values for the individual rat pups and not for litters. The influence diagnostics are discussed in Subsection 3.10.2.

We caution readers that running `proc mixed` with these options (e.g., `iter = 5` and `effect = litter`) can cause the procedure to take considerably longer to finish running. In our example, with 27 litters and litter-specific subset deletion, the longer execution time is a result of the fact that `proc mixed` needs to fit 28 models, i.e., the initial model and the model corresponding to each deleted litter, with up to five iterations per model. Clearly, this can become very time-consuming if there are a large number of levels of a variable that is being checked for influence.

The `outpred =` option in the `model` statement causes SAS to output the predicted values and residuals conditional on the random effects in the model. In this example, we output the conditional residuals and predicted values to a data set called `pdat1` by specifying `outpred = pdat1`. The `outpredm =` option causes SAS to output the marginal predicted and residual values to another data set. In this case, we request that these variables be output to the `pdat2` data set by specifying `outpredm = pdat2`. We discuss these residuals in Subsection 3.10.1.

The `id` statement allows us to place variables in the output data sets, `pdat1` and `pdat2`, that identify each observation. We specify that PUP_ID, LITTER, TREATMENT, TRT-GRP, RANKLIT, and LITSIZE be included, so that we can use them later in the residual diagnostics.

The `random` statement has been modified to include the `v =` and `vcorr =` options, so that SAS displays the estimated marginal V_j matrix and the corresponding marginal correlation matrix implied by Model 3.3 for birth weight observations from the third litter ($v = 3$ and $vcorr = 3$). We chose the third litter in this example because it has only four rat pups, to keep the size of the estimated V_j matrix in the output manageable (see Section 3.8 for a discussion of the implied marginal covariance matrix). We also include the `solution` option, to display the EBLUPs for the random litter effects in the output.

The `lsmeans` statement allows us to obtain estimates of the least-squares means of WEIGHT for each level of treatment, based on the fixed-effect parameter estimates for TREAT. The least-squares means are evaluated at the mean of LITSIZE, and assuming that there are equal numbers of rat pups for each level of SEX. We also carry out post-hoc comparisons among all pairs of the least-squares means using the Tukey–Kramer adjustment for multiple comparisons by specifying `adjust = tukey`. Many other adjustments for multiple comparisons can be obtained, such as Dunnett's and Bonferroni. Refer to the SAS documentation for `proc mixed` for more information on post-hoc comparison methods available in the `lsmeans` statement.

Diagnostics for this final model using the REML fit for Model 3.3 are presented in Section 3.10.

3.4.2   SPSS

Most analyses in SPSS can be performed using either the menu system or SPSS syntax. The syntax for LMMs can be obtained by specifying a model using the menu system and then pasting the syntax into the syntax window. We recommend pasting the syntax for any LMM that is fitted using the menu system, and then saving the syntax file for documentation. We present SPSS syntax throughout the example chapters for ease of presentation, although the models were usually set up using the menu system. A link to an example of setting up an LMM using the SPSS menus is included on the web page for this book (see Appendix A).

For the analysis of the Rat Pup data, we first read in the raw data from the tab-delimited file rat_pup.dat (assumed to be located in the `C:\temp` folder) using the following syntax. This SPSS syntax was pasted after reading in the data using the SPSS menu system.

```
* Read in Rat Pup data.
GET DATA
```

```
/TYPE = TXT
/FILE = "C:\temp\rat_pup.dat"
/DELCASE = LINE
/DELIMITERS = "\t"
/ARRANGEMENT = DELIMITED
/FIRSTCASE = 2
/IMPORTCASE = ALL
/VARIABLES =
pup_id F2.1
weight F4.2
sex A6
litter F1.
litsize F2.1
treatment A7
.
CACHE.
EXECUTE.
```

Because the MIXED command in SPSS sets the fixed-effect parameter associated with the highest-valued level of a fixed factor to 0 by default, to prevent overparameterization of models (similar to proc mixed in SAS; see Subsection 3.4.1), the highest-valued levels of fixed factors can be thought of as "reference categories" for the factors. As a result, we recode TREATMENT into a new variable named TREAT, so that the control group (TREAT = 3) will be the reference category.

```
* Recode TREATMENT variable .
RECODE
Treatment
("High"=1) ("Low"=2) ("Control"=3) INTO treat .
EXECUTE .

VARIABLE LABEL treat "Treatment" .
VALUE LABELS treat 1 "High" 2 "Low" 3 "Control" .
```

Step 1: Fit a model with a "loaded" mean structure (Model 3.1).

The following SPSS syntax can be used to fit Model 3.1:

```
* Model 3.1.
MIXED
weight BY treat sex WITH litsize
/CRITERIA = CIN(95) MXITER(100) MXSTEP(5) SCORING(1)
SINGULAR(0.000000000001) HCONVERGE(0, ABSOLUTE)
LCONVERGE(0, ABSOLUTE) PCONVERGE(0.000001, ABSOLUTE)
/FIXED = treat sex litsize treat*sex | SSTYPE(3)
/METHOD = REML
/PRINT = SOLUTION
/RANDOM INTERCEPT | SUBJECT(litter) COVTYPE(VC)
/SAVE = PRED RESID .
```

The first variable listed after invoking the MIXED command is the dependent variable, WEIGHT. The BY keyword indicates that the TREAT and SEX variables are to be considered as categorical factors (they can be either fixed or random). Note that we do not need to include LITTER as a factor, because this variable is identified as a SUBJECT variable later in the code. The WITH keyword identifies continuous covariates, and in this case, we specify LITSIZE as a continuous covariate.

The CRITERIA subcommand specifies default settings for the convergence criteria obtained by specifying the model using the menu system.

In the FIXED subcommand, we include terms that have fixed effects associated with them in the model: TREAT, SEX, LITSIZE and the TREAT × SEX interaction. The SSTYPE(3) option after the vertical bar indicates that the default Type III analysis is to be used when calculating F-statistics. We also use the METHOD = REML subcommand, which requests that the REML estimation method (the default) be used.

The SOLUTION keyword in the PRINT subcommand specifies that the estimates of the fixed-effect parameters, covariance parameters, and their associated standard errors are to be included in the output.

The RANDOM subcommand specifies that there is a random effect in the model associated with the INTERCEPT for each level of the SUBJECT variable (i.e., LITTER). The information about the "subject" variable is specified after the vertical bar (|). Note that because we included LITTER as a "subject" variable, we did not need to list it after the BY keyword (including LITTER after BY does not affect the analysis if LITTER is also indicated as a SUBJECT variable). The COVTYPE(VC) option indicates that the default Variance Components (VC) covariance structure for the random effects (the D matrix) is to be used. We did not need to specify a COVTYPE here because only a single variance associated with the random effects is being estimated.

Conditional predicted values and residuals are saved in the working data set by specifying PRED and RESID in the SAVE subcommand. The keyword PRED saves litter-specific predicted values that incorporate both the estimated fixed effects and the EBLUPs of the random litter effects for each observation. The keyword RESID saves the conditional residuals that represent the difference between the actual value of WEIGHT and the predicted value for each rat pup, based on the estimated fixed effects and the EBLUP of the random effect for each observation. The set of population-averaged predicted values, based only on the estimated fixed-effect parameters, can be obtained by adding the FIXPRED keyword to the SAVE subcommand, as shown later in this chapter (see Section 3.9 for more details):

```
/SAVE = PRED RESID FIXPRED
```

Software Note: There is currently no option to display or save the predicted values of the random litter effects (EBLUPs) in the output in SPSS. However, because all models considered for the Rat Pup data contain a single random intercept for each litter, the EBLUPs can be calculated by simply taking the difference between the "population-averaged" and "litter-specific" predicted values. The values of FIXPRED from the first LMM can be stored in a variable called FIXPRED_1, and the values of PRED from the first model can be stored as PRED_1. We can then compute the difference between these two predicted values and store the result in a new variable that we name EBLUP:

```
COMPUTE eblup = pred_1 - fixpred_1 .
EXECUTE .
```

The values of the EBLUP variable, which are constant for each litter, can then be displayed in the output by using this syntax:

```
SORT CASES BY litter.
SPLIT FILE
```

```
   LAYERED BY litter.
DESCRIPTIVES
VARIABLES = eblup
/STATISTICS=MEAN STDDEV MIN MAX.
SPLIT FILE OFF.
```

Step 2: Select a structure for the random effects (Model 3.1 vs. Model 3.1A).

We now use a likelihood ratio test of Hypothesis 3.1 to decide if the random effects associated with the intercept for each litter can be omitted from Model 3.1. To carry out the likelihood ratio test we first fit a nested model, Model 3.1A, using the same syntax as for Model 3.1 but with the RANDOM subcommand omitted:

```
* Model 3.1A .
MIXED
weight BY treat sex WITH litsize
/CRITERIA = CIN(95) MXITER(100) MXSTEP(5) SCORING(1)
SINGULAR(0.000000000001) HCONVERGE(0, ABSOLUTE) LCONVERGE(0,
ABSOLUTE) PCONVERGE(0.000001, ABSOLUTE)
/FIXED = treat sex litsize treat*sex | SSTYPE(3)
/METHOD = REML
/PRINT = SOLUTION
/SAVE = PRED RESID FIXPRED .
```

The test statistic for Hypothesis 3.1 is calculated by subtracting the –2 REML log-likelihood value associated with the fit of Model 3.1 (the reference model) from that for Model 3.1A (the nested model). These values are displayed in the SPSS output for each model. The null distribution for this test statistic is a mixture of χ_0^2 and χ_1^2 distributions, each with equal weight of 0.5 (see Subsection 3.5.1). Because the result of this test is significant ($p < 0.001$), we choose to retain the random litter effects.

Step 3: Select a covariance structure for the residuals (Models 3.1, 3.2A, or 3.2B).

To fit Model 3.2A, with heterogeneous error variances as a function of the TREAT factor, we need to use the GENLINMIXED command in IBM SPSS Statistics (Version 21). This command can fit linear mixed models in addition to generalized linear mixed models, which are not covered in this book. Models with heterogeneous error variances for different groups of clusters cannot be fitted using the MIXED command.

We use the following syntax to fit Model 3.2A using the GENLINMIXED command:

```
* Model 3.2A.
GENLINMIXED
  /DATA_STRUCTURE SUBJECTS=litter REPEATED_MEASURES=pup_id
    GROUPING=treat COVARIANCE_TYPE=IDENTITY
  /FIELDS TARGET=weight TRIALS=NONE OFFSET=NONE
  /TARGET_OPTIONS DISTRIBUTION=NORMAL LINK=IDENTITY
  /FIXED  EFFECTS=sex litsize treat sex*treat USE_INTERCEPT=TRUE
  /RANDOM USE_INTERCEPT=TRUE SUBJECTS=litter
    COVARIANCE_TYPE=VARIANCE_COMPONENTS
```

```
/BUILD_OPTIONS TARGET_CATEGORY_ORDER=ASCENDING
   INPUTS_CATEGORY_ORDER=ASCENDING MAX_ITERATIONS=100
   CONFIDENCE_LEVEL=95 DF_METHOD=SATTERTHWAITE COVB=MODEL
/EMMEANS_OPTIONS SCALE=ORIGINAL PADJUST=LSD.
```

There are several important notes to consider when fitting this type of model using the GENLINMIXED command:

- First, the LITTER variable (or the random factor identifying clusters of observations, more generally) needs to be declared as a *Nominal* variable in the SPSS Variable View, under the "Measure" column. Continuous predictors with fixed effects included in the model (e.g., LITSIZE) need to be declared as *Scale* variables in Variable View, and categorical fixed factors with fixed effects included in the model should be declared as *Nominal* variables.

- Second, heterogeneous error variance structures for clustered data can only be set up if some type of repeated measures index is defined for each cluster. We arbitrarily defined the variable PUP_ID, which has a unique value for each pup within a litter, as this index (in the REPEATED_MEASURES option of the DATA_STRUCTURE subcommand). This index variable should be a *Scale* variable.

- Third, the grouping factor that defines cases that will have different error variances is defined in the GROUPING option (TREAT).

- Fourth, if we desire a simple error covariance structure for each TREAT group of litters, defined only by a constant error variance, we need to use COVARIANCE_TYPE=IDENTITY. Other error covariance structures are possible, but in models for cross-sectional clustered data that already include random cluster effects, additional covariance among the errors (that is not already accounted for by the random effects) is generally unlikely.

- Fifth, note that the dependent variable is identified as the TARGET variable, the marginal distribution of the dependent variable is identified as NORMAL, and the IDENTITY link is used (in the context of generalized linear mixed models, these options set up a linear mixed model).

- Sixth, the RANDOM subcommand indicates random intercepts for each litter by USE_INTERCEPT=TRUE, with SUBJECTS=litter and COVARIANCE_TYPE=VARIANCE _COMPONENTS.

After submitting this syntax, SPSS will generate output for the model in the output viewer. The output generated by the GENLINMIXED command is fairly unusual relative to the other procedures. By default, most of the output appears in "graphical" format, and users need to double-click on the "Model Viewer" portion of the output to open the full set of output windows. To test Hypothesis 3.2, we first need to find the –2 REML log-likelihood value for this model (Model 3.2A). This value can be found in the very first "Model Summary" window, in the footnote (359.9). We subtract this value from the –2 REML log-likelihood value of 401.1 for Model 3.1 (with constant error variance across the treatment groups), and compute a *p*-value for the resulting chi-square test statistic using the following syntax:

```
COMPUTE hyp32a = SIG.CHISQ(401.1 - 359.9, 2) .
EXECUTE .
```

We use two degrees of freedom given the two additional error variance parameters in Model 3.2A relative to Model 3.1. The resulting *p*-value will be shown in the last column of the SPSS data set, and suggests that we reject the null hypothesis (Model 3.1, with equal error variance across treatment groups) and proceed with Model 3.2A ($p < 0.001$).

Before fitting Model 3.2B, we recode the original TREAT variable into a new variable, TRTGRP, that combines the high and low treatment groups (for testing Hypotheses 3.3 and 3.4):

```
RECODE treat (1 = 1) (2 = 1) (3 = 2) into trtgrp .
EXECUTE .
VALUE LABELS trtgrp 1 "High/Low" 2 "Control".
```

We now fit Model 3.2B using the GENLINMIXED command once again:

```
* Model 3.2B .
GENLINMIXED
  /DATA_STRUCTURE SUBJECTS=litter REPEATED_MEASURES=pup_id
    GROUPING=trtgrp COVARIANCE_TYPE=IDENTITY
  /FIELDS TARGET=weight TRIALS=NONE OFFSET=NONE
  /TARGET_OPTIONS DISTRIBUTION=NORMAL LINK=IDENTITY
  /FIXED  EFFECTS=sex litsize treat sex*treat USE_INTERCEPT=TRUE
  /RANDOM USE_INTERCEPT=TRUE SUBJECTS=litter
    COVARIANCE_TYPE=VARIANCE_COMPONENTS
  /BUILD_OPTIONS TARGET_CATEGORY_ORDER=ASCENDING
    INPUTS_CATEGORY_ORDER=ASCENDING MAX_ITERATIONS=100
    CONFIDENCE_LEVEL=95 DF_METHOD=SATTERTHWAITE COVB=MODEL
  /EMMEANS_OPTIONS SCALE=ORIGINAL PADJUST=LSD.
```

Note that the only difference from the syntax used for Model 3.2A is the use of the newly recoded TRTGRP variable in the GROUPING option, which will define a common error variance for the high and low treatment groups, and a different error variance for the control group. The resulting –2 REML log-likelihood value for this model is 361.1, and we compute a likelihood ratio test p-value to test this model against Model 3.2A:

```
COMPUTE hyp33 = SIG.CHISQ(361.1 - 359.9, 1) .
EXECUTE .
```

The resulting p-value added to the data set ($p = 0.27$) suggests that we choose the simpler model (Model 3.2B) moving forward. To test Hypothesis 3.4, and compare the fit of Model 3.2B with Model 3.1, we perform another likelihood ratio test:

```
COMPUTE hyp34 = SIG.CHISQ(401.1 - 359.9, 1) .
EXECUTE .
```

This test result ($p < 0.001$) indicates that we should reject the null hypothesis of constant error variance across the treatment groups, and proceed with Model 3.2B, allowing for different error variances in the high/low treatment group and the control group.

Software Note: When processing the GENLINMIXED output for Model 3.2B, SPSS users can navigate the separate windows within the "Model Viewer" window. In windows showing the tests for the fixed effects and the estimated fixed effects themselves, we recommend changing the display style to "Table" in the lower left corner of each window. This will show the actual tests and estimates rather than a graphical display.

> In the window showing estimates of the covariance parameters, users can change the "Effect" being shown to examine the estimated variance of the random litter effects or the estimates of the error variance parameters, and when examining the estimates of the error variance parameters, users can toggle the (treatment) group being shown in the lower left corner of the window.

Step 4: Reduce the model by removing nonsignificant fixed effects (Model 3.2B vs. 3.3, and Model 3.3 vs. 3.3A).

We test Hypothesis 3.5 to decide whether we can remove the treatment by sex interaction term, making use of the default Type III F-test for the TREAT × SEX interaction in Model 3.2B. The result of this test can be found in the "Model Viewer" window for Model 3.2B, in the window entitled "Fixed Effects." Because the result of this test is not significant ($p = 0.73$), we drop the TREAT × SEX interaction term from Model 3.2B, which gives us Model 3.3.

We now test Hypothesis 3.6 to decide whether the fixed effects associated with treatment are equal to 0 (in the model omitting the interaction term). While we illustrate the use of likelihood ratio tests based on maximum likelihood estimation to test this hypothesis in the other procedures, we do not have the option of fitting a model using ML estimation when using GENLINMIXED. This is because this command takes the general approach of using penalized quasi-likelihood (PQL) estimation, which can be used for a broad class of generalized linear mixed models. For this reason, we simply refer to the Type III F-test for treatment based on Model 3.3. Here is the syntax to fit Model 3.3:

```
* Model 3.3 .
GENLINMIXED
  /DATA_STRUCTURE SUBJECTS=litter REPEATED_MEASURES=pup_id
    GROUPING=trtgrp COVARIANCE_TYPE=IDENTITY
  /FIELDS TARGET=weight TRIALS=NONE OFFSET=NONE
  /TARGET_OPTIONS DISTRIBUTION=NORMAL LINK=IDENTITY
  /FIXED  EFFECTS=sex litsize treat USE_INTERCEPT=TRUE
  /RANDOM USE_INTERCEPT=TRUE SUBJECTS=litter
    COVARIANCE_TYPE=VARIANCE_COMPONENTS
  /BUILD_OPTIONS TARGET_CATEGORY_ORDER=ASCENDING
    INPUTS_CATEGORY_ORDER=ASCENDING MAX_ITERATIONS=100
    CONFIDENCE_LEVEL=95 DF_METHOD=SATTERTHWAITE COVB=MODEL
  /EMMEANS_OPTIONS SCALE=ORIGINAL PADJUST=LSD.
```

Note that the interaction term has been dropped from the FIXED subcommand. The resulting F-test for TREAT in the "Fixed Effects" window suggests that TREAT is strongly significant. Pairwise comparisons of the marginal means for each treatment group along with fitted values and residuals (for diagnostic purposes) can be generated using the following syntax:

```
* Model 3.3, pairwise comparisons and diagnostics .
GENLINMIXED
  /DATA_STRUCTURE SUBJECTS=litter REPEATED_MEASURES=pup_id
    GROUPING=trtgrp COVARIANCE_TYPE=IDENTITY
  /FIELDS TARGET=weight TRIALS=NONE OFFSET=NONE
  /TARGET_OPTIONS DISTRIBUTION=NORMAL LINK=IDENTITY
  /FIXED  EFFECTS=sex litsize treat USE_INTERCEPT=TRUE
```

```
/RANDOM USE_INTERCEPT=TRUE SUBJECTS=litter
    COVARIANCE_TYPE=VARIANCE_COMPONENTS
/BUILD_OPTIONS TARGET_CATEGORY_ORDER=ASCENDING
    INPUTS_CATEGORY_ORDER=ASCENDING MAX_ITERATIONS=100
    CONFIDENCE_LEVEL=95 DF_METHOD=SATTERTHWAITE COVB=MODEL
/EMMEANS TABLES=treat COMPARE=treat CONTRAST=PAIRWISE
/EMMEANS_OPTIONS SCALE=ORIGINAL PADJUST=SEQBONFERRONI
/SAVE PREDICTED_VALUES(PredictedValue)
    PEARSON_RESIDUALS(PearsonResidual) .
```

Note the two new **EMMEANS** subcommands: the first requests a table showing pairwise comparisons of the means for TREAT, while the second indicates a sequential Bonferroni adjustment for the multiple comparisons. The resulting comparisons can be found in the "Estimated Means" window of the "Model Viewer" output (where we again recommend using a Table style for the display). In addition, the new **SAVE** subcommand generates fitted values and residuals based on this model in the data set, which can be used for diagnostic purposes (see Section 3.10).

3.4.3   R

Before starting the analysis in R, we first import the tab-delimited data file, `rat_pup.dat` (assumed to be located in the `C:\temp` directory), into a data frame object in R named `ratpup`.[2]

```
> ratpup <- read.table("c:\\temp\\rat_pup.dat", h = T)
> attach(ratpup)
```

The `h = T` argument in the `read.table()` function indicates that the first row (the header) in the `rat_pup.dat` file contains variable names. After reading the data, we "attach" the `ratpup` data frame to R's working memory so that the columns (i.e., variables) in the data frame can be easily accessed as separate objects. Note that we show the ">"prompt for each command as it would appear in R, but this prompt is not typed as part of the commands.

To facilitate comparisons with the analyses performed using the other software procedures, we recode the variable SEX into SEX1, which is an indicator variable for females, so that males will be the reference group:

```
> ratpup$sex1[sex == "Female"] <- 1
> ratpup$sex1[sex == "Male"] <- 0
```

We first consider the analysis using the `lme()` function (from the `nlme` package), and then replicate as many steps of the analysis as possible using the newer `lmer()` function (from the `lme4` package).

3.4.3.1   Analysis Using the `lme()` Function

Step 1: Fit a model with a "loaded" mean structure (Model 3.1).

We first load the `nlme` package, so that the `lme()` function will be available for model fitting:

```
> library(nlme)
```

---

[2]The Rat Pup data set is also available as a data frame object in the `nlme` package. After loading the package, the name of the data frame object is `RatPupWeight`.

We next fit the initial LMM, Model 3.1, to the Rat Pup data using the `lme()` function:

```
> # Model 3.1.
> model3.1.fit <- lme(weight ~ treatment + sex1 + litsize +
    treatment:sex1, random = ~ 1 | litter,
    data = ratpup, method = "REML")
```

We explain each part of the syntax used for the `lme()` function below:

- `model3.1.fit` is the name of the object that will contain the results of the fitted model.

- The first argument of the function, `weight ~ treatment + sex1 + litsize + treatment:sex1`, defines a model formula. The response variable, WEIGHT, and the terms that have associated fixed effects in the model (TREATMENT, SEX1, LITSIZE, and the TREATMENT × SEX1 interaction), are listed. The `factor()` function is not necessary for the categorical variable TREATMENT, because the original treatment variable has string values High, Low, and Control, and will therefore be considered as a factor automatically. We also do not need to declare SEX1 as a factor, because it is an indicator variable having only values of 0 and 1.

- The second argument, `random = ~ 1 | litter`, includes a random effect for each level of LITTER in the model. These random effects will be associated with the intercept, as indicated by `~ 1`.

- The third argument of the function, `ratpup`, indicates the name of the data frame object to be used in the analysis.

- The final argument, `method = "REML"`, specifies that the default REML estimation method is to be used.

By default, the `lme()` function treats the lowest level (alphabetically or numerically) of a categorical fixed factor as the reference category. This means that "Control" will be the reference category of TREATMENT because "Control" is the lowest level of treatment alphabetically. The `relevel()` function can also be used to change the reference categories of factors. For example, if one desired "High" to be the reference category of treatment, they could use the following function:

```
> treatment <- relevel(treatment, ref = "High")
```

We obtain estimates from the model fit by using the `summary()` function:

```
> summary(model3.1.fit)
```

Additional results for the fit of Model 3.1 can be obtained by using other functions in conjunction with the `model3.1.fit` object. For example, we can obtain F-tests for the fixed effects in the model by using the `anova()` function:

```
> anova(model3.1.fit)
```

The `anova()` function performs a series of Type I (or sequential) F-tests for the fixed effects in the model, each of which are conditional on the preceding terms in the model specification. For example, the F-test for SEX1 is conditional on the TREATMENT effects, but the F-test for TREATMENT is not conditional on the SEX1 effect.

The `random.effects()` function can be used to display the EBLUPs for the random litter effects:

```
> # Display the random effects (EBLUPs) from the model.
> random.effects(model3.1.fit)
```

Step 2: Select a structure for the random effects (Model 3.1 vs. Model 3.1A).

We now test Hypothesis 3.1 to decide whether the random effects associated with the intercept for each litter can be omitted from Model 3.1, using a likelihood ratio test. We do this indirectly by testing whether the variance of the random litter effects, σ^2_{litter}, is zero vs. the alternative that the variance is greater than zero. We fit Model 3.1A, which is nested within Model 3.1, by excluding the random litter effects.

Because the `lme()` function requires the specification of at least one random effect, we use the `gls()` function, which is also available in the `nlme` package, to fit Model 3.1, excluding the random litter effects. The `gls()` function fits marginal linear models using REML estimation. We fit Model 3.1A using the `gls()` function and then compare the –2 REML log-likelihood values for Models 3.1 and 3.1A using the `anova()` function:

```
> # Model 3.1A.
> model3.la.fit <- gls(weight ~ treatment + sex1 + litsize +
    treatment:sex1, data = ratpup)
> anova(model3.1.fit, model3.la.fit) # Test Hypothesis 3.1.
```

The `anova()` function performs a likelihood ratio test by subtracting the –2 REML log-likelihood value for Model 3.1 (the reference model) from the corresponding value for Model 3.1A (the nested model) and referring the difference to a χ^2 distribution with 1 degree of freedom. The result of this test ($p < 0.001$) suggests that the random litter effects should be retained in this model.

To get the correct p-value for Hypothesis 3.1, however, we need to divide the p-value reported by the `anova()` function by 2; this is because we are testing the null hypothesis that the variance of the random litter effects equals zero, which is on the boundary of the parameter space for a variance. The null distribution of the likelihood ratio test statistic for Hypothesis 3.1 follows a mixture of χ^2_0 and χ^2_1 distributions, with equal weight of 0.5 (see Subsection 3.5.1 for more details). Based on the significant result of this test ($p < 0.0001$), we keep the random litter effects in this model and in all subsequent models.

Step 3: Select a covariance structure for the residuals (Models 3.1, 3.2A, or 3.2B).

We now fit Model 3.2A, with a separate residual variance for each treatment group ($\sigma^2_{high}, \sigma^2_{low},$ and $\sigma^2_{control}$).

```
> # Model 3.2A.
> model3.2a.fit <- lme(weight ~ treatment + sex1 + litsize
+ treatment:sex1, random = ~1 | litter, ratpup, method = "REML",
weights  = varIdent(form = ~1 | treatment))
```

The arguments of the `lme()` function are the same as those used to fit Model 3.1, with the addition of the `weights` argument. The `weights = varIdent(form = ~ 1 | treatment)` argument sets up a heterogeneous residual variance structure, with observations at different levels of TREATMENT having different residual variance parameters. We apply the `summary()` function to review the results of the model fit:

```
> summary(model3.2a.fit)
```

In the **Variance function** portion of the following output, note the convention used by the `lme()` function to display the heterogeneous variance parameters:

```
Random effects:
Formula:  ~1 | litter

        (Intercept) Residual
StdDev: 0.3134714 0.5147866

Variance function:
Structure: Different standard deviations per stratum
Formula:  ~ 1 |  treatment

Parameter estimates:

   Control      Low      High
 1.0000000 0.5650369 0.6393779
```

We first note in the `Random effects` portion of the output that the estimated `Residual` standard deviation is equal to 0.5147866. The `Parameter estimates` specify the values by which the residual standard deviation should be multiplied to obtain the estimated standard deviation of the residuals in each treatment group. This multiplier is 1.0 for the control group (the reference group). The multipliers reported for the low and high treatment groups are very similar (0.565 and 0.639, respectively), suggesting that the residual standard deviation is smaller in these two treatment groups than in the control group. The estimated residual variance for each treatment group can be obtained by squaring their respective standard deviations.

To test Hypothesis 3.2, we subtract the –2 REML log-likelihood for the heterogeneous residual variance model, Model 3.2A, from the corresponding value for the model with homogeneous residual variance for all treatment groups, Model 3.1, by using the `anova()` function.

```
> # Test Hypothesis 3.2.
> anova(model3.1.fit, model3.2a.fit)
```

We do not need to adjust the p-value returned by the `anova()` function for Hypothesis 3.2, because the null hypothesis (stating that the residual variance is identical for each treatment group) does not set a covariance parameter equal to the boundary of its parameter space, as in Hypothesis 3.1. Because the result of this likelihood ratio test is significant ($p < 0.001$), we choose the heterogeneous variances model (Model 3.2A) as our preferred model at this stage of the analysis.

We next test Hypothesis 3.3 to decide if we can pool the residual variances for the high and low treatment groups. To do this, we first create a pooled treatment group variable, TRTGRP:

```
> ratpup$trtgrp[treatment == "Control"] <- 1
> ratpup$trtgrp[treatment == "Low" | treatment == "High"] <- 2
```

We now fit Model 3.2B, using the new TRTGRP variable in the `weights =` argument to specify the grouped heterogeneous residual variance structure:

```
> model3.2b.fit <- lme(weight ~ treatment + sex1 + litsize + treatment:sex1,
    random  = ~ 1 | litter, ratpup, method = "REML",
    weights = varIdent(form = ~1 | trtgrp))
```

We test Hypothesis 3.3 using a likelihood ratio test, by applying the `anova()` function to the objects containing the fits for Model 3.2A and Model 3.2B:

```
> # Test Hypothesis 3.3.
> anova(model3.2a.fit, model3.2b.fit)
```

The null distribution of the test statistic in this case is a χ^2 with one degree of freedom. Because the test is not significant ($p = 0.27$), we select the nested model, Model 3.2B, as our preferred model at this stage of the analysis.

We use a likelihood ratio test for Hypothesis 3.4 to decide whether we wish to retain the grouped heterogeneous error variances in Model 3.2B or choose the homogeneous error variance model, Model 3.1. The `anova()` function is also used for this test:

```
> # Test Hypothesis 3.4.
> anova(model3.1.fit, model3.2b.fit)
```

The result of this likelihood ratio test is significant ($p < 0.001$), so we choose the pooled heterogeneous residual variances model, Model 3.2B, as our preferred model. We can view the parameter estimates from the fit of this model using the `summary()` function:

```
> summary(model3.2b.fit)
```

Step 4: Reduce the model by removing nonsignificant fixed effects (Model 3.2B vs. Model 3.3, and Model 3.3 vs. Model 3.3A).

We test Hypothesis 3.5 to decide whether the fixed effects associated with the treatment by sex interaction are equal to zero in Model 3.2B, using a Type I F-test in R. To obtain the results of this test (along with Type I F-tests for all of the other fixed effects in the model), we apply the `anova()` function to the `model3.2b.fit` object:

```
> # Test Hypothesis 3.5.
> anova(model3.2b.fit)
```

Based on the nonsignificant Type I F-test ($p = 0.73$), we delete the TREATMENT $\times$ SEX1 interaction term from the model and obtain our final model, Model 3.3.

We test Hypothesis 3.6 to decide whether the fixed effects associated with treatment are equal to zero in Model 3.3, using a likelihood ratio test based on maximum likelihood (ML) estimation. We first fit the reference model, Model 3.3, using ML estimation (`method = "ML"`), and then fit a nested model, Model 3.3A, without the TREATMENT term, also using ML estimation.

```
> model3.3.ml.fit <- lme(weight ~ treatment + sex1 + litsize,
    random = ~1 | litter, ratpup, method = "ML",
    weights = varIdent(form = ~1 | trtgrp))
> model3.3a.ml.fit <- lme(weight ~ sex1 + litsize,
    random = ~1 | litter, ratpup, method = "ML",
    weights = varIdent(form = ~1 | trtgrp))
```

We then use the `anova()` function to carry out the likelihood ratio test of Hypothesis 3.6:

```
> # Test Hypothesis 3.6.
> anova(model3.3.ml.fit, model3.3a.ml.fit)
```

The likelihood ratio test result is significant ($p < 0.001$), so we retain the significant fixed treatment effects in the model. We keep the fixed effects associated with SEX1 and LITSIZE without testing them, to adjust for these fixed effects when assessing the treatment effects. See Section 3.5 for a discussion of the results of all hypothesis tests for the Rat Pup data analysis.

We now refit our final model, Model 3.3, using REML estimation to get unbiased estimates of the variance parameters. Note that we now specify TREATMENT as the last term in the fixed-effects portion of the model, so the Type I F-test reported for TREATMENT by the `anova()` function will be comparable to the Type III F-test reported by SAS `proc mixed`.

```
> # Model 3.3: Final Model.
> model3.3.reml.fit <- lme(weight ~ sex1 + litsize + treatment,
    random  = ~1 | litter, ratpup, method = "REML",
    weights = varIdent(form = ~1 | trtgrp))
> summary(model3.3.reml.fit)
> anova(model3.3.reml.fit)
```

3.4.3.2   Analysis Using the `lmer()` Function

Step 1: Fit a model with a "loaded" mean structure (Model 3.1).

We first load the `lme4` package, so that the `lmer()` function will be available for model fitting:

```
> library(lme4)
```

We next fit the initial LMM, Model 3.1, to the Rat Pup data using the `lmer()` function:

```
> # Model 3.1.
> model3.1.fit.lmer <- lmer(weight ~ treatment + sex1 + litsize +
    treatment:sex1 + (1 | litter),
    ratpup, REML = T)
```

We explain each part of the syntax used for the `lmer()` function below:

- `model3.1.fit.lmer` is the name of the object that will contain the results of the fitted model.

- The first portion of the first argument of the function, `weight ~ treatment + sex1 + litsize + treatment:sex1`, partly defines the model formula. The response variable, WEIGHT, and the terms that have associated fixed effects in the model (TREATMENT, SEX1, LITSIZE, and the TREATMENT × SEX1 interaction), are listed. The `factor()` function is not necessary for the categorical variable TREATMENT, because the original treatment variable has string values High, Low, and Control, and will therefore be considered as a factor automatically. We also do not need to declare SEX1 as a factor, because it is an indicator variable having only values of 0 and 1.

- The model formula also includes the term `(1 | litter)`, which includes a random effect associated with the intercept (`1`) for each unique level of LITTER. This is a key difference from the `lme()` function, where there is a separate `random` argument needed to specify the random effects in a given model.

- The third argument of the function, `ratpup`, indicates the name of the data frame object to be used in the analysis.

- The final argument, `REML = T`, specifies that the default REML estimation method is to be used.

Like the `lme()` function, the `lmer()` function treats the lowest level (alphabetically or numerically) of a categorical fixed factor as the reference category by default. This means that "Control" will be the reference category of TREATMENT because "Control" is the lowest level of treatment alphabetically. The `relevel()` function can also be used in conjunction with the `lmer()` function to change the reference categories of factors.

We obtain estimates from the model fit by using the `summary()` function:

```
> summary(model3.1.fit.lmer)
```

In the resulting output, we see that the `lmer()` function only produces *t*-statistics for the fixed effects, with no corresponding *p*-values. This is primarily due to the lack of agreement in the literature over appropriate degrees of freedom for these test statistics. The `anova()` function also does not provide *p*-values for the *F*-statistics when applied to a model fit object generated by using the `lmer()` function. In general, we recommend use of the `lmerTest` package in R for users interested in testing hypotheses about parameters estimated using the `lmer()` function. In this chapter and others, we illustrate likelihood ratio tests using selected functions available in the `lme4` and `lmerTest` packages.

The `ranef()` function can be used to display the EBLUPs for the random litter effects:

```
> # Display the random effects (EBLUPs) from the model.
> ranef(model3.1.fit.lmer)
```

Step 2: Select a structure for the random effects (Model 3.1 vs. Model 3.1A).

For this step, we first load the `lmerTest` package, which enables likelihood ratio tests of hypotheses concerning the variances of random effects in models fitted using the `lmer()` function (including models with only a single random effect, such as Model 3.1). As an alternative to loading the `lme4` package, R users may also simply load the `lmerTest` package first, which includes all related packages required for its use.

```
> library(lmerTest)
```

We once again employ the `lmer()` function and fit Model 3.1 to the Rat Pup data, after loading the `lmerTest` package:

```
> # Model 3.1.
> model3.1.fit.lmer <- lmer(weight ~ treatment + sex1 + litsize +
    treatment:sex1 + (1 | litter),
    ratpup, REML = T)
```

We then apply the `summary()` function to this model fit object:

```
> summary(model3.1.fit.lmer)
```

We note that the `lmer()` function now computes *p*-values for all of the fixed effects included in this model, using a Satterthwaite approximation of the degrees of freedom for this test (similar to the MIXED command in SPSS). For testing Hypothesis 3.1 (i.e., is the variance of the random litter effects greater than zero?), we can use the `rand()` function to perform a likelihood ratio test:

```
> rand(model3.1.fit.lmer)
```

In this case, the `rand()` function fits Model 3.1A (without the random litter effects) and computes the appropriate likelihood ratio test statistic, representing the positive difference in the −2 REML log-likelihood values of the two models (89.4). The corresponding *p*-value based on a mixture of chi-square distributions ($p < 0.001$) suggests a strong rejection of the null hypothesis, and we therefore retain the random litter effects in all subsequent models.

Step 3: Select a covariance structure for the residuals (Models 3.1, 3.2A, or 3.2B).

At the time of this writing, the `lmer()` function does not allow users to fit models with heterogeneous error variance structures. We therefore do not consider Models 3.2A or 3.2B in this analysis. The fixed effects in the model including the random litter effects and assuming constant error variance across the treatment groups (Model 3.1) can be tested using the `lmerTest` package, and the `lme()` function can be used to test for the possibility of heterogeneous error variances, as illustrated in Section 3.4.3.1.

3.4.4   Stata

We begin by importing the tab-delimited version of the Rat Pup data set into Stata, assuming that the `rat_pup.dat` data file is located in the `C:\temp` directory. Note that we present the Stata commands including the prompt (.), which is not entered as part of the commands.

```
. insheet using "C:\temp\rat_pup.dat", tab
```

Alternatively, users of web-aware Stata can import the Rat Pup data set directly from the book's web site:

```
. insheet using http://www-personal.umich.edu/~bwest/rat_pup.dat, tab
```

We now utilize the `mixed` command to fit the models for this example.

Step 1: Fit a model with a "loaded" mean structure (Model 3.1).

Because string variables cannot be used as categorical factor variables in Stata, we first recode the TREATMENT and SEX variables into numeric format:

```
. gen female = (sex == "Female")
. gen treatment2 = 1 if treatment == "Control"
. replace treatment2 = 2 if treatment == "Low"
. replace treatment2 = 3 if treatment == "High"
```

The `mixed` command used to fit Model 3.1 (in Version 13+ of Stata) is then specified as follows:

```
. * Model 3.1 fit .
. mixed weight ib1.treatment2 female litsize ib1.treatment2#c.female
|| litter:, covariance(identity) variance reml
```

The `mixed` command syntax has three parts. The first part specifies the dependent variable and the fixed effects, the second part specifies the random effects, and the third part specifies the covariance structure for the random effects, in addition to miscellaneous options. We note that although we have split the single command onto two lines, readers should attempt to submit the command on a single line in Stata. We discuss these parts of the syntax in detail below.

The first variable listed after the `mixed` command is the continuous dependent variable, WEIGHT. The variables following the dependent variable are the terms that will have associated fixed effects in the model. We include fixed effects associated with TREATMENT (where `ib1.` indicates that the recoded variable TREATMENT2 is a categorical factor, with the category having value 1 (Control) as the reference, or baseline category), the FEMALE indicator, LITSIZE, and the interaction between TREATMENT2 and FEMALE (indicated using #). We note that even though FEMALE is a binary indicator, it needs to be specified as "continuous" in the interaction term using c., because it has not been specified as a categorical factor previously in the variable list using i..

The two vertical bars (||) precede the variable that defines clusters of observations (litter:) in this two-level data set. The absence of additional variables after the colon indicates that there will only be a single random effect associated with the intercept for each level of LITTER in the model.

The covariance option after the comma specifies the covariance structure for the random effects (or the D matrix). Because Model 3.1 includes only a single random effect associated with the intercept (and therefore a single variance parameter associated with the random effects), it has an identity covariance structure. The covariance option is actually not necessary in this simple case.

Finally, the variance option requests that the estimated variances of the random effects and the residuals be displayed in the output, rather than their estimated standard deviations, which is the default. The mixed procedure also uses ML estimation by default, so we also include the reml option to request REML estimation for Model 3.1.

The AIC and BIC information criteria for this model can be obtained by using the following command after the mixed command has finished running:

```
. * Information criteria.
. estat ic
```

By default, the mixed command does not display F-tests for the fixed effects in the model. Instead, omnibus Wald chi-square tests for the fixed effects in the model can be performed using the test command. For example, to test the overall significance of the fixed treatment effects, the following command can be used:

```
. * Test overall significance of the fixed treatment effects.
. test 2.treatment2 3.treatment2
```

The two terms listed after the test command are the dummy variables automatically generated by Stata for the fixed effects of the Low and High levels of treatment (as indicated in the estimates of the fixed-effect parameters). The test command is testing the null hypothesis that the two fixed effects associated with these dummy variables are both equal to zero (i.e., the null hypothesis that the treatment means are all equal for males, given that the interaction between TREATMENT2 and FEMALE has been included in this model). Similar omnibus tests may be obtained for the fixed FEMALE effect, the fixed LITSIZE effect, and the interaction between TREATMENT2 and SEX:

```
. * Omnibus tests for FEMALE, LITSIZE and the
. * TREATMENT2*FEMALE interaction.
. test female
. test litsize
. test 2.treatment2#c.female 3.treatment2#c.female
```

Once a model has been fitted using the mixed command, EBLUPs of the random effects associated with the levels of the random factor (LITTER) can be saved in a new variable (named EBLUPS) using the following command:

```
. predict eblups, reffects
```

The saved EBLUPs can then be used to check for random effects that may be outliers.

Step 2: Select a structure for the random effects (Model 3.1 vs. Model 3.1A).

We perform a likelihood ratio test of Hypothesis 3.1 to decide whether the random effects associated with the intercept for each litter can be omitted from Model 3.1. In the case of two-level models with random intercepts, the `mixed` command performs the appropriate likelihood ratio test automatically. We read the following output after fitting Model 3.1:

```
LR test vs. linear regression: chibar2(01) = 89.41 Prob >= chibar2 = 0.0000
```

Stata reports `chibar2(01)`, indicating that it uses the correct null hypothesis distribution of the test statistic, which in this case is a mixture of χ_0^2 and χ_1^2 distributions, each with equal weight of 0.5 (see Subsection 3.5.1). The likelihood ratio test reported by the `mixed` command is an overall test of the covariance parameters associated with all random effects in the model. In models with a single random effect for each cluster, as in Model 3.1, it is appropriate to use this test to decide if that random effect should be included in the model. The significant result of this test ($p < 0.001$) suggests that the random litter effects should be retained in Model 3.1.

Step 3: Select a covariance structure for the residuals (Models 3.1, 3.2A, or 3.2B).

We now fit Model 3.2A, with a separate residual variance for each treatment group ($\sigma_{high}^2, \sigma_{low}^2$, and $\sigma_{control}^2$).

```
. * Model 3.2A.
. mixed weight ib1.treatment2 female litsize ib1.treatment2#c.female
|| litter:, covariance(identity) variance reml
residuals(independent, by(treatment2))
```

We use the same `mixed` command that was used to fit Model 3.1, with one additional option: `residuals(independent, by(treatment2))`. This option specifies that the residuals in this model are independent of each other (which is reasonable for two-level clustered data if a random effect associated with each cluster has already been included in the model), and that residuals associated with different levels of TREATMENT2 have different variances. The resulting output will provide estimates of the residual variance for each treatment group.

To test Hypothesis 3.2, we subtract the –2 REML log-likelihood for the heterogeneous residual variance model, Model 3.2A, from the corresponding value for the model with homogeneous residual variance for all treatment groups, Model 3.1. We can do this automatically by fitting both models and saving the results from each model in new objects. We first save the results from Model 3.2A in an object named `model32Afit`, refit Model 3.1 and save those results in an object named `model31fit`, and then use the `lrtest` command to perform the likelihood ratio test using the two objects (where Model 3.1 is nested within Model 3.2A, and listed second):

```
. est store model32Afit
. mixed weight ib1.treatment2 female litsize ib1.treatment2#c.female
|| litter:, covariance(identity) variance reml
. est store model31fit
. lrtest model32Afit model31fit
```

We do not need to modify the p-value returned by the `lrtest` command for Hypothesis 3.2, because the null hypothesis (stating that the residual variance is identical for each treatment group) does not set a covariance parameter equal to the boundary of its parameter space, as in Hypothesis 3.1. Because the result of this likelihood ratio test is significant ($p < 0.001$), we choose the heterogeneous variances model (Model 3.2A) as our preferred model at this stage of the analysis.

We next test Hypothesis 3.3 to decide if we can pool the residual variances for the high and low treatment groups. To do this, we first create a pooled treatment group variable, TRTGRP:

```
. gen trtgrp = 1 if treatment2 == 1
. replace trtgrp = 2 if treatment2 == 2 | treatment2 == 3
```

We now fit Model 3.2B, using the new TRTGRP variable to specify the grouped heterogeneous residual variance structure and saving the results in an object named `model32Bfit`:

```
. * Model 3.2B.
. mixed weight ib1.treatment2 female litsize ib1.treatment2#c.female
|| litter:, covariance(identity) variance reml
residuals(independent, by(trtgrp))
. est store model32Bfit
```

We test Hypothesis 3.3 using a likelihood ratio test, by applying the `lrtest` command to the objects containing the fits for Model 3.2A and Model 3.2B:

```
. lrtest model32Afit model32Bfit
```

The null distribution of the test statistic in this case is a χ^2 with one degree of freedom. Because the test is not significant ($p = 0.27$), we select the nested model, Model 3.2B, as our preferred model at this stage of the analysis.

We use a likelihood ratio test for Hypothesis 3.4 to decide whether we wish to retain the grouped heterogeneous error variances in Model 3.2B or choose the homogeneous error variance model, Model 3.1. The `lrtest` command is also used for this test:

```
. * Test Hypothesis 3.4.
. lrtest model32Bfit model31fit
```

The result of this likelihood ratio test is significant ($p < 0.001$), so we choose the pooled heterogeneous residual variances model, Model 3.2B, as our preferred model at this stage.

Step 4: Reduce the model by removing nonsignificant fixed effects (Model 3.2B vs. Model 3.3, and Model 3.3 vs. Model 3.3A).

We test Hypothesis 3.5 to decide whether the fixed effects associated with the treatment by sex interaction are equal to zero in Model 3.2B, using a Type III F-test in Stata. To obtain the results of this test, we execute the `test` command below after fitting Model 3.2B:

```
. * Test Hypothesis 3.5.
. test 2.treatment2#c.female 3.treatment2#c.female
```

Based on the nonsignificant Type III F-test ($p = 0.73$), we delete the TREATMENT2 × FEMALE interaction term from the model and obtain our final model, Model 3.3.

We test Hypothesis 3.6 to decide whether the fixed effects associated with treatment are equal to zero in Model 3.3, using a likelihood ratio test based on maximum likelihood (ML) estimation. We first fit the reference model, Model 3.3, using ML estimation (the default of the `mixed` command, meaning that we drop the `reml` option along with the interaction term), and then fit a nested model, Model 3.3A, without the TREATMENT2 term, also using ML estimation. We then use the `lrtest` command to carry out the likelihood ratio test of Hypothesis 3.6:

```
. * Test Hypothesis 3.6.
. mixed weight ib1.treatment2 female litsize
|| litter:, covariance(identity) variance
residuals(independent, by(trtgrp))
. est store model33fit
. mixed weight female litsize
|| litter:, covariance(identity) variance
residuals(independent, by(trtgrp))
. est store model33Afit
. lrtest model33fit model33Afit
```

The likelihood ratio test result is significant ($p < 0.001$), so we retain the significant fixed treatment effects in the model. We keep the fixed effects associated with FEMALE and LITSIZE without testing them, to adjust for these fixed effects when assessing the treatment effects. See Section 3.5 for a discussion of the results of all hypothesis tests for the Rat Pup data analysis.

We now refit our final model, Model 3.3, using REML estimation to get unbiased estimates of the variance parameters:

```
. * Model 3.3: Final Model.
. mixed weight ib1.treatment2 female litsize
|| litter:, covariance(identity) variance reml
residuals(independent, by(trtgrp))
```

3.4.5   HLM

3.4.5.1   Data Set Preparation

To perform the analysis of the Rat Pup data using the HLM software package, we need to prepare two separate data sets.

1. The **Level 1 (pup-level) data set** contains a single observation (row of data) for each rat pup. This data set includes the Level 2 cluster identifier variable, LITTER, and the variable that identifies the units of analysis, PUP_ID. The response variable, WEIGHT, which is measured for each pup, must also be included, along with any pup-level covariates. In this example, we have only a single pup-level covariate, SEX. In addition, the data set must be sorted by the cluster-level identifier, LITTER.

2. The **Level 2 (litter-level) data set** contains a single observation for each LITTER. The variables in this data set remain constant for all rat pups within a given litter. The Level 2 data set needs to include the cluster identifier, LITTER, and the litter-level covariates, TREATMENT and LITSIZE. This data set must also be sorted by LITTER.

Because the HLM program does not automatically create dummy variables for categorical predictors, we need to create dummy variables to represent the nonreference levels of the categorical predictors prior to importing the data into HLM. We first need to add an indicator variable for SEX to represent female rat pups in the Level 1 data set, and we need to create two dummy variables in the Level 2 data set for TREATMENT, to represent the high and low dose levels. If the input data files were created in SPSS, the SPSS syntax to create these indicator variables in the Level 1 and Level 2 data files would look like this:

Level 1 data

```
COMPUTE sex1 = (sex = "Female") .
EXECUTE .
```

Level 2 data

```
COMPUTE treat1 = (treatment = "High") .
EXECUTE .
COMPUTE treat2 = (treatment = "Low") .
EXECUTE .
```

3.4.5.2 Preparing the Multivariate Data Matrix (MDM) File

We create a new MDM file, using the Level 1 and Level 2 data sets described earlier. In the main HLM menu, click **File**, **Make new MDM file**, and then **Stat package input**. In the window that opens, select **HLM2** to fit a two-level hierarchical linear model, and click **OK**. In the **Make MDM** window that opens, select the **Input File Type** as **SPSS/Windows**.

Now, locate the **Level-1 Specification** area of the MDM window, and **Browse** to the location of the Level 1 SPSS data set. Once the data file has been selected, click on the **Choose Variables** button and select the following variables from the Level 1 file: LITTER (check "ID" for the LITTER variable, because this variable identifies the Level 2 units), WEIGHT (check "in MDM" for this variable, because it is the dependent variable), and the indicator variable for females, SEX1 (check "in MDM").

Next, locate the **Level-2 Specification** area of the MDM window and **Browse** to the location of the Level 2 SPSS data set that has one record per litter. Click on the **Choose Variables** button to include LITTER (check "ID"), TREAT1 and TREAT2 (check "in MDM" for each indicator variable), and finally LITSIZE (check "in MDM").

After making these choices, check the **cross sectional (persons within groups)** option for the MDM file, to indicate that the Level 1 data set contains measures on individual rat pups ("persons" in this context), and that the Level 2 data set contains litter-level information (the litters are the "groups"). Also, select **No** for **Missing Data?** in the Level 1 data set, because we do not have any missing data for any of the litters in this example. Enter a name for the MDM file with an .mdm extension (e.g., ratpup.mdm) in the upper right corner of the MDM window. Finally, save the .mdmt template file under a new name (click **Save mdmt file**), and click the **Make MDM** button.

After HLM has processed the MDM file, click the **Check Stats** button to see descriptive statistics for the variables in the Level 1 and Level 2 data sets (HLM 7+ will show these automatically). This step, which is required prior to fitting a model, allows you to check that the correct number of records has been read into the MDM file and that there are no unusual values for the variables included in the MDM file (e.g., values of 999 that were previously coded as missing data; such values would need to be set to system-missing in SPSS prior to using the data file in HLM). Click **Done** to proceed to the model-building window.

Step 1: Fit a model with a loaded mean structure (Model 3.1).

In the model-building window, select WEIGHT from the list of variables, and click **Outcome variable**. The initial "unconditional" (or "means-only") model for WEIGHT, broken down into Level 1 and Level 2 models, is now displayed in the model-building window. The initial Level 1 model is:

Model 3.1: Level 1 Model (Initial)

$$\text{WEIGHT} = \beta_0 + \boldsymbol{r}$$

To add more informative subscripts to the models (if they are not already shown), click **File** and **Preferences**, and choose **Use level subscripts**. The Level 1 model now includes the subscripts i and j, where i indexes individual rat pups and j indexes litters, as follows:

Model 3.1: Level 1 Model (Initial) With Subscripts

$$\text{WEIGHT}_{ij} = \beta_{0j} + r_{ij}$$

This initial Level 1 model shows that the value of WEIGHT_{ij} for an individual rat pup i, within litter j, depends on the intercept, β_{0j}, for litter j, plus a residual, r_{ij}, associated with the rat pup.

The initial Level 2 model for the litter-specific intercept, β_{0j}, is also displayed in the model-building window.

Model 3.1: Level 2 Model (Initial)

$$\beta_{0j} = \gamma_{00} + u_{0j}$$

This model shows that at Level 2 of the data set, the litter-specific intercept depends on the fixed overall intercept, γ_{00}, plus a random effect, u_{0j}, associated with litter j. In this "unconditional" model, β_{0j} is allowed to vary randomly from litter to litter. After clicking the **Mixed** button for this model (in the lower-right corner of the model-building window), the initial means-only mixed model is displayed.

Model 3.1: Overall Mixed Model (Initial)

$$\text{WEIGHT}_{ij} = \gamma_{00} + u_{0j} + r_{ij}$$

To complete the specification of Model 3.1, we add the pup-level covariate, SEX1. Click the **Level 1** button in the model-building window and then select SEX1. Choose **add variable uncentered**. SEX1 is then added to the Level 1 model along with a litter-specific coefficient, β_{1j}, for the effect of this covariate.

Model 3.1: Level 1 Model (Final)

$$\text{WEIGHT}_{ij} = \beta_{0j} + \beta_{1j}(\text{SEX1}_{ij}) + r_{ij}$$

The Level 2 model now has equations for both the litter-specific intercept, β_{0j}, and for β_{1j}, the litter-specific coefficient associated with SEX1.

Model 3.1: Level 2 Model (Intermediate)

$$\beta_{0j} = \gamma_{00} + u_{0j}$$
$$\beta_{1j} = \gamma_{10}$$

The equation for the litter-specific intercept is unchanged. The value of β_{1j} is defined as a constant (equal to the fixed effect γ_{10}) and does not include any random effects, because we assume that the effect of SEX1 (i.e., the effect of being female) does not vary randomly from litter to litter.

To finish the specification of Model 3.1, we add the uncentered versions of the two litter-level dummy variables for treatment, TREAT1 and TREAT2, to the Level 2 equations for the intercept, β_{0j}, and for the effect of being female, β_{1j}. We add the effect of the uncentered version of the LITSIZE covariate to the Level 2 equation for the intercept only, because we do not wish to allow the effect of being female to vary as a function of litter size. Click the **Level 2** button in the model-building window. Then, click on each Level 2 equation and click on the specific variables (uncentered) to add.

Model 3.1: Level 2 Model (Final)

$$\beta_{0j} = \gamma_{00} + \gamma_{01}(\text{LITSIZE}_j) + \gamma_{02}(\text{TREAT1}_j) + \gamma_{03}(\text{TREAT2}_j) + u_{0j}$$
$$\beta_{1j} = \gamma_{10} + \gamma_{11}(\text{TREAT1}_j) + \gamma_{12}(\text{TREAT2}_j)$$

In this final Level 2 model, the main effects of TREAT1 and TREAT2, i.e., γ_{02} and γ_{03}, enter the model through their effect on the litter-specific intercept, β_{0j}. The interaction between treatment and sex enters the model by allowing the litter-specific effect for SEX1, β_{1j}, to depend on fixed effects associated with TREAT1 and TREAT2 (γ_{11} and γ_{12}, respectively). The fixed effect associated with LITSIZE, γ_{01}, is only included in the equation for the litter-specific intercept and is not allowed to vary by sex (i.e., our model does not include a LITSIZE × SEX1 interaction).

We can view the final LMM by clicking the **Mixed** button in the HLM model-building window:

Model 3.1: Overall Mixed Model (Final)

$$
\begin{aligned}
\text{WEIGHT}_{ij} = \ & \gamma_{00} + \gamma_{01} * \text{LITSIZE}_j + \gamma_{02} * \text{TREAT1}_j \\
& + \gamma_{03} * \text{TREAT2}_j + \gamma_{10} * \text{SEX1}_{ij} \\
& + \gamma_{11} * \text{TREAT1}_j * \text{SEX1}_{ij} + \gamma_{12} * \text{TREAT2}_j * \text{SEX1}_{ij} \\
& + u_{0j} + r_{ij}
\end{aligned}
$$

The final mixed model in HLM corresponds to Model 3.1 as it was specified in (3.1), but with somewhat different notation. Table 3.3 shows the correspondence of this notation with the general LMM notation used in (3.1).

After specifying Model 3.1, click **Basic Settings** to enter a title for this analysis (such as "Rat Pup Data: Model 3.1") and a name for the output (.html) file. Note that the default outcome variable distribution is Normal (Continuous), so we do not need to specify it. The HLM2 procedure automatically creates two residual data files, corresponding to the two levels of the model. The "Level-1 Residual File" contains the conditional residuals, r_{ij}, and the "Level-2 Residual File" contains the EBLUPs of the random litter effects, u_{0j}. To change the names and/or file formats of these residual files, click on either of the two buttons for the files in the **Basic Settings** window. Click **OK** to return to the model-building window.

Click **File ... Save As** to save this model specification to a new .hlm file. Finally, click **Run Analysis** to fit the model. HLM2 by default uses REML estimation for two-level models such as Model 3.1. Click on **File ... View Output** to see the estimates for this model.

Step 2: Select a structure for the random effects (Model 3.1 vs. Model 3.1A).

In this step, we test Hypothesis 3.1 to decide whether the random effects associated with the intercept for each litter can be omitted from Model 3.1. We cannot perform a likelihood ratio test for the variance of the random litter effects in this model because HLM does not allow us to remove the random effects in the Level 2 model (there must be at least one random effect associated with each level of the data set in HLM). Because we cannot use a likelihood ratio test for the variance of the litter-specific intercepts, we instead use the χ^2 tests for the covariance parameters provided by HLM2. These χ^2 statistics are calculated using methodology described in Raudenbush & Bryk (2002) and are displayed near the bottom of the output file.

Step 3: Select a covariance structure for the residuals (Models 3.1, 3.2A, or 3.2B).

Models 3.2A, 3.2B and 3.3, which have heterogeneous residual variance for different levels of treatment, cannot be fitted using HLM2, because this procedure does not allow the Level 1 variance to depend on a factor measured at Level 2 of the data. However, HLM does provide an option labeled **Test homogeneity of Level 1 variance** under the **Hypothesis Testing** settings, which can be used to obtain a test of whether the assumption of homogeneous residual variance is met [refer to Raudenbush & Bryk (2002) for more details].

Step 4: Reduce the model by removing nonsignificant fixed effects (Model 3.2B vs. Model 3.3, and Model 3.3 vs. Model 3.3A).

We can set up general linear hypothesis tests for the fixed effects in a model in HLM by clicking **Other Settings**, and then **Hypothesis Testing** prior to fitting a model. In the hypothesis-testing window, each numbered button corresponds to a test of a null hypothesis that $C\gamma = 0$, where C is a known matrix for a given hypothesis, and γ is a vector of fixed-effect parameters. This specification of the linear hypothesis in HLM corresponds to the linear hypothesis specification, $L\beta = 0$, described in Subsection 2.6.3.1. For each hypothesis, HLM computes a Wald-type test statistic, which has a χ^2 null distribution, with degrees of freedom equal to the rank of C [see Raudenbush & Bryk (2002) for more details].

For example, to test the overall effect of treatment in Model 3.1, which has seven fixed-effect parameters (γ_{00}, associated with the intercept term; γ_{01}, with litter size; γ_{02} and γ_{03}, with the treatment dummy variables TREAT1 and TREAT2; γ_{10}, with sex; and γ_{11} and γ_{12}, with the treatment by sex interaction terms), we would need to set up the following C matrix and γ vector:

$$C = \begin{pmatrix} 0 & 0 & 1 & 0 & 0 & 0 & 0 \\ 0 & 0 & 0 & 1 & 0 & 0 & 0 \end{pmatrix}, \gamma = \begin{pmatrix} \gamma_{00} \\ \gamma_{01} \\ \gamma_{02} \\ \gamma_{03} \\ \gamma_{10} \\ \gamma_{11} \\ \gamma_{12} \end{pmatrix}$$

This specification of C and γ corresponds to the null hypothesis H_0: $\gamma_{02} = 0$ and $\gamma_{03} = 0$. Each row in the C matrix corresponds to a column in the HLM Hypothesis Testing window.

To set up this hypothesis test, click on the first numbered button under **Multivariate Hypothesis Tests** in the **Hypothesis Testing** window. In the first column of zeroes, corresponding to the first row of the C matrix, enter a 1 for the fixed effect γ_{02}. In the second column of zeroes, enter a 1 for the fixed effect γ_{03}. To complete the specification of the hypothesis, click on the third column, which will be left as all zeroes, and click **OK**. Additional hypothesis tests can be obtained for the fixed effects associated with other terms in Model 3.1 (including the interaction terms) by entering additional C matrices under different numbered buttons in the **Hypothesis Testing** window. After setting up all hypothesis tests of interest, click **OK** to return to the main model-building window.

3.5 Results of Hypothesis Tests

The hypothesis test results reported in Table 3.5 were derived from output produced by SAS `proc mixed`. See Table 3.4 and Subsection 3.3.3 for more information about the specification of each hypothesis.

3.5.1 Likelihood Ratio Tests for Random Effects

Hypothesis 3.1. The random effects, u_{0j}, associated with the litter-specific intercepts can be omitted from Model 3.1.

To test Hypothesis 3.1, we perform a likelihood ratio test. The test statistic is calculated by subtracting the –2 REML log-likelihood value of the reference model, Model 3.1, from the corresponding value for a nested model omitting the random effects, Model 3.1A. Because a variance cannot be less than zero, the null hypothesis value of $\sigma^2_{litter} = 0$ is at the boundary of the parameter space, and the asymptotic null distribution of the likelihood ratio test statistic is a mixture of χ^2_0 and χ^2_1 distributions, each with equal weight of 0.5 (Verbeke & Molenberghs, 2000). We illustrate calculation of the p-value for the likelihood ratio test statistic:

$$p\text{-value} = 0.5 \times P(\chi^2_0 > 89.4) + 0.5 \times P(\chi^2_1 > 89.4) < 0.001$$

The resulting test statistic is significant ($p < 0.001$), so we retain the random effects associated with the litter-specific intercepts in Model 3.1 and in all subsequent models. As noted in Subsection 3.4.1, the χ^2_0 distribution has all of its mass concentrated at zero, so its contribution to the p-value is zero and the first term can be omitted from the p-value calculation.

3.5.2 Likelihood Ratio Tests for Residual Variance

Hypothesis 3.2. The variance of the residuals is the same (homogeneous) for the three treatment groups (high, low, and control).

We use a REML-based likelihood ratio test for Hypothesis 3.2. The test statistic is calculated by subtracting the value of the –2 REML log-likelihood for Model 3.2A (the reference model) from that for Model 3.1 (the nested model). Under the null hypothesis, the variance parameters are not on the boundary of their parameter space (i.e., the null hypothesis does not specify that they are equal to zero). The test statistic has a χ^2 distribution with 2 degrees of freedom because Model 3.2A has 2 more covariance parameters (i.e., the 2 additional residual variances) than Model 3.1. The test result is significant ($p < 0.001$). We therefore reject the null hypothesis and decide that the model with heterogeneous residual variances, Model 3.2A, is our preferred model at this stage of the analysis.

Hypothesis 3.3. The residual variances for the high and low treatment groups are equal.

To test Hypothesis 3.3, we again carry out a REML-based likelihood ratio test. The test statistic is calculated by subtracting the value of the –2 REML log-likelihood for Model 3.2A (the reference model) from that for Model 3.2B (the nested model).

Under the null hypothesis, the test statistic has a χ^2 distribution with 1 degree of freedom. The nested model, Model 3.2B, has one fewer covariance parameter (i.e., one less residual variance) than the reference model, Model 3.2A, and the null hypothesis value of

TABLE 3.5: Summary of Hypothesis Test Results for the Rat Pup Analysis

Hypo-thesis Label	Test	Estima-tion Method	Models Compared (Nested vs. Reference)	Test Statistic Values (Calculation)	p-value
3.1	LRT	REML	**3.1A vs. 3.1**	$\chi^2(0:1) = 89.4$ $(490.5 - 401.1)$	$< .001$
3.2	LRT	REML	**3.1 vs. 3.2A**	$\chi^2(2) = 41.2$ $(401.1 - 359.9)$	$< .001$
3.3	LRT	REML	**3.2B vs. 3.2A**	$\chi^2(1) = 1.2$ $(361.1 - 359.9)$	0.27
3.4	LRT	REML	**3.1 vs. 3.2B**	$\chi^2(1) = 40.0$ $(401.1 - 361.1)$	$< .001$
3.5	Type III F-test	REML	**3.2B**[a]	$F(2, 194) = 0.3$	0.73
3.6	LRT	ML	**3.3A vs. 3.3**	$\chi^2(2) = 18.6$ $(356.4 - 337.8)$	$< .001$
	Type III F-test	REML	**3.3**[b]	$F(2, 24.3) = 11.4$	$< .001$

Note: See Table 3.4 for null and alternative hypotheses, and distributions of test statistics under H_0.

[a] We use an F-test for the fixed effects associated with TREATMENT × SEX based on the fit of Model 3.2B only.

[b] We use an F-test for the fixed effects associated with TREATMENT based on the fit of Model 3.3 only.

the parameter does not lie on the boundary of the parameter space. The test result is not significant ($p = 0.27$). We therefore do not reject the null hypothesis, and decide that Model 3.2B, with pooled residual variance for the high and low treatment groups, is our preferred model.

Hypothesis 3.4. The residual variance for the combined high/low treatment group is equal to the residual variance for the control group.

To test Hypothesis 3.4, we carry out an additional REML-based likelihood ratio test. The test statistic is calculated by subtracting the value of the –2 REML log-likelihood for Model 3.2B (the reference model) from that of Model 3.1 (the nested model).

Under the null hypothesis, the test statistic has a χ^2 distribution with 1 degree of freedom: the reference model has 2 residual variances, and the nested model has 1. The test result is significant ($p < 0.001$). We therefore reject the null hypothesis and choose Model 3.2B as our preferred model.

3.5.3   *F*-tests and Likelihood Ratio Tests for Fixed Effects

Hypothesis 3.5. The fixed effects, β_5 and β_6, associated with the treatment by sex interaction are equal to zero in Model 3.2B.

We test Hypothesis 3.5 using a Type III F-test for the treatment by sex interaction in Model 3.2B. The results of the test are not significant ($p = 0.73$). Therefore, we drop the fixed effects associated with the treatment by sex interaction and select Model 3.3 as our final model.

Hypothesis 3.6. The fixed effects associated with treatment, β_1 and β_2, are equal to zero in Model 3.3.

Hypothesis 3.6 is not part of the model selection process but tests the primary hypothesis of the study. We would not remove the effect of treatment from the model even if it proved to be nonsignificant because it is the main focus of the study.

The Type III F-test, reported by SAS for treatment in Model 3.3, is significant ($p < 0.001$), and we conclude that the mean birth weights differ by treatment group, after controlling for litter size, sex, and the random effects associated with litter.

We also carry out an ML-based likelihood ratio test for Hypothesis 3.6. To do this, we refit Model 3.3 using ML estimation. We then fit a nested model without the fixed effects of treatment (Model 3.3A), again using ML estimation. The test statistic is calculated by subtracting the –2 log-likelihood value for Model 3.3 from the corresponding value for Model 3.3A. The result of this test is also significant ($p < 0.001$).

3.6 Comparing Results across the Software Procedures

3.6.1 Comparing Model 3.1 Results

Table 3.6 shows selected results generated using each of the six software procedures to fit Model 3.1 to the Rat Pup data. This model is "loaded" with fixed effects, has random effects associated with the intercept for each litter, and has a homogeneous residual variance structure.

All five procedures agree in terms of the estimated fixed-effect parameters and their estimated standard errors for Model 3.1. They also agree on the estimated variance components (i.e., the estimates of σ^2_{litter} and σ^2) and their respective standard errors, when they are reported.

Portions of the model fit criteria differ across the software procedures. Reported values of the –2 REML log-likelihood are the same for the procedures in SAS, SPSS, R, and Stata. However, the reported value in HLM (indicated as the **deviance** statistic in the HLM output) is lower than the values reported by the other four procedures. According to correspondence with HLM technical support staff, the difference arising in this illustration is likely due to differences in default convergence criteria between HLM and the other procedures, or in implementation of the iterative REML procedure within HLM. This minor difference is not critical in this case.

We also note that the values of the AIC and BIC statistics vary because of the different computing formulas being used (the HLM2 procedure does not compute these information criteria). SAS and SPSS compute the AIC as –2 REML log-likelihood + 2 × (# covariance parameters in the model). Stata and R compute the AIC as –2 REML log-likelihood + 2 × (# fixed effects + # covariance parameters in the model). Although the AIC and BIC statistics are not always comparable across procedures, they can be used to compare the fits of models within any given procedure. For details on how the computation of the BIC criteria varies from the presentation in Subsection 2.6.4 across the different software procedures, refer to the documentation for each procedure.

TABLE 3.6: Comparison of Results for Model 3.1

Estimation Method	SAS: proc mixed REML	SPSS: MIXED REML	R: lme() function REML	R: lmer() function REML	Stata: mixed REML	HLM2 REML
Fixed-Effect Parameter	*Estimate (SE)*	*Estimate (SE)*	*Estimate (SE)*	*Estimate (SE)*	*Estimate (SE)*	*Estimate (SE)*[a]
β_0 (Intercept)	8.32(0.27)	8.32(0.27)	8.32(0.27)	8.32(0.27)	8.32(0.27)	8.32(0.27)
β_1 (High vs. Control)	−0.91(0.19)	−0.91(0.19)	−0.91(0.19)	−0.91(0.19)	−0.91(0.19)	−0.91(0.19)
β_2 (Low vs. Control)	−0.47(0.16)	−0.47(0.16)	−0.47(0.16)	−0.47(0.16)	−0.47(0.16)	−0.47(0.16)
β_3 (Female vs. male)	−0.41(0.07)	−0.41(0.07)	−0.41(0.07)	−0.41(0.07)	−0.41(0.07)	−0.41(0.07)
β_4 (Litter size)	−0.13(0.02)	−0.13(0.02)	−0.13(0.02)	−0.13(0.02)	−0.13(0.02)	−0.13(0.02)
β_5 (High × Female)	0.11(0.13)	0.11(0.13)	0.11(0.13)	0.11(0.13)	0.11(0.13)	0.11(0.13)
β_6 (Low × Female)	0.08(0.11)	0.08(0.11)	0.08(0.11)	0.08(0.11)	0.08(0.11)	0.08(0.11)
Covariance Parameter	*Estimate (SE)*	*Estimate (SE)*	*Estimate (n.c.)*	*Estimate (SE)*	*Estimate (SE)*	*Estimate (n.c.)*
σ^2_{litter}	0.10(0.03)	0.10(0.03)	0.10[b]	0.10	0.10(0.03)	0.10
σ^2 (Residual variance)	0.16(0.01)	0.16(0.01)	0.16	0.16	0.16(0.01)	0.16
Model Information Criteria						
−2 REML log-likelihood	401.1	401.1	401.1	401.1	401.1	399.3
AIC	405.1[c]	405.1[c]	419.1[d]	419.1[d]	419.1[d]	n.c.
BIC	407.7	412.6	452.9	453.1	453.1	n.c.
Tests for Fixed Effects	**Type III F-Tests**	**Type III F-Tests**	**Type I F-Tests**	**Type I F-Tests**	**Wald χ²-Tests**	**Wald χ²-Tests**
Intercept	$t(32.9) = 30.5$ $p < .01$	$F(1, 34.0) = 1076.2$ $p < .01$	$F(1, 292) = 9093.8$ $p < .01$	N/A[e]	$Z = 30.5$ $p < .01$	$\chi^2(1) = 927.3$ $p < .01$
TREATMENT	$F(2, 24.3) = 11.5$ $p < .01$	$F(2, 24.3) = 11.5$ $p < .01$	$F(2, 23.0) = 5.08$ $p = 0.01$	$F(2, xx) = 5.08^e$	$\chi^2(2) = 23.7$ $p < .01$	$\chi^2(2) = 23.7$ $p < .01$

TABLE 3.6: (Continued)

Estimation Method	SAS: proc mixed REML	SPSS: MIXED REML	R: lme() function REML	R: lmer() function REML	Stata: mixed REML	HLM2 REML
SEX	$F(1,303) = 47.0$ $p < .01$	$F(1,302.9) = 47.0$ $p < .01$	$F(1,292) = 52.6$ $p < .01$	$F(1,\mathrm{xx}) = 52.6^{e}$	$\chi^2(1) = 31.7$ $p < .01$	$\chi^2(l) = 31.7$ $p < .01$
LITSIZE	$F(1,31.8) = 46.9$ $p < .01$	$F(1,31.8) = 46.9$ $p < .01$	$F(1,23.0) = 47.4$ $p < .01$	$F(1,\mathrm{xx}) = 47.4^{e}$	$\chi^2(l) = 46.9$ $p < .01$	$\chi^2(1) = 46.9$ $p < .01$
TREATMENT × SEX	$F(2,302.0) = 0.5$ $p = .63$	$F(2,302.3) = 0.5$ $p = .63$	$F(2,292.0) = 0.5$ $p = .63$	$F(2,\mathrm{xx}) = 0.5^{e}$	$\chi^2(2) = 0.9$ $p = .63$	$\chi^2(2) = 0.9$ $p > .50$
EBLUPs	Output (w/ sig. tests)	Computed (Subsection 3.3.2)	Can be saved	Can be saved	Can be saved	Saved by default

Note: (n.c.) = not computed

Note: 322 Rat Pups at Level 1; 27 Litters at Level 2

[a]HLM2 also reports "robust" standard errors for the estimated fixed effects in the output by default. We report the model-based standard errors here.

[b]Users of the lme() function in R can use the function intervals(model3.1.ratpup) to obtain approximate 95% confidence intervals for covariance parameters. The estimated standard deviations reported by the summary() function have been squared to obtain variances.

[c]SAS and SPSS compute the AIC as −2 REML log-likelihood + 2 × (# covariance parameters in the model).

[d]Stata and R compute the AIC as −2 REML log-likelihood + 2 × (# fixed effects + # covariance parameters in the model).

[e]Likelihood ratio tests are recommended for making inferences about fixed effects when using the lmer() function in R, given that denominator degrees of freedom for the test statistics are not computed by default. Satterthwaite approximations of the denominator degrees of freedom are computed when using the lmerTest package.

TABLE 3.7: Comparison of Results for Model 3.2B

	SAS: proc mixed	SPSS: GENLINMIXED	R: lme() function	Stata: mixed
Estimation Method	REML	REML	REML	REML
Fixed-Effect Parameter	*Estimate (SE)*	*Estimate (SE)*	*Estimate (SE)*	*Estimate (SE)*
β_0(Intercept)	8.35(0.28)	8.35(0.28)	8.35(0.28)	8.35(0.28)
β_1(High vs. control)	−0.90(0.19)	−0.90(0.19)	−0.90(0.19)	−0.90(0.19)
β_2(Low vs. control)	−0.47(0.16)	−0.47(0.16)	−0.47(0.16)	−0.47(0.16)
β_3(Female vs. male)	−0.41(0.09)	−0.41(0.09)	−0.41(0.09)	−0.41(0.09)
β_4(Litter size)	−0.13(0.02)	−0.13(0.02)	−0.13(0.02)	−0.13(0.02)
β_5(High × female)	0.09(0.12)	0.09(0.12)	0.09(0.12)	0.09(0.12)
β_6(Low × female)	0.08(0.11)	0.08(0.11)	0.08(0.11)	0.08(0.11)
Covariance Parameter	*Estimate (SE)*	*Estimate (SE)*	*Estimate (n.c.)*	*Estimate (SE)*
σ^2_{litter}	0.10(0.03)	0.10(0.03)	0.10	0.10(0.03)
$\sigma^2_{high/low}$	0.09(0.01)	0.09(0.01)	0.09	0.09(0.01)
$\sigma^2_{control}$	0.27(0.03)	0.27(0.03)	0.27	0.27(0.03)
Tests for Fixed Effects	**Type III F-Tests**	**Type III F-Tests**	**Type I F-Tests**	**Wald χ^2-Tests**
Intercept	$t(34) = 30.29$, $p < .001$	$t(34) = 30.29$, $p < .001$	$F(1, 292) = 9027.94$, $p < .001$	$t(34) = 30.30$, $p < .001$
TREATMENT	$F(2, 24.4) = 11.18$, $p < .001$	$F(2, 24.4) = 11.18$, $p < .001$	$F(2, 23.0) = 4.24$, $p = .027$	$\chi^2(2) = 22.9$, $p < .01$

TABLE 3.7: (Continued)

Estimation Method	SAS: proc mixed REML	SPSS: GENLINMIXED REML	R: lme() function REML	Stata: mixed REML
SEX	$F(1, 29.6) = 59.17$, $p < .001$	$F(1, 296) = 59.17$, $p < .001$	$F(1, 292) = 61.57$, $p < .001$	$\chi^2(1) = 19.3$, $p < .01$
LITSIZE	$F(1, 31.2) = 49.33$, $p < .001$	$F(1, 31.0) = 49.33$, $p < .001$	$F(1, 23.0) = 49.58$, $p < .001$	$\chi^2(1) = 49.33$, $p < .01$
TREATMENT $\times$ SEX	$F(2, 194) = 0.32$, $p = .73$	$F(2, 194) = 0.32$, $p = .73$	$F(2, 292) = 0.32$, $p = .73$	$\chi^2(2) = 0.63$, $p = .73$

Model Information Criteria

-2 REML log-likelihood	361.1	361.1	361.1	361.1
AIC	367.1	367.2	381.1	381.1
BIC	371.0	378.3	418.6	418.8

Note: (n.c.) = not computed
Note: 322 Rat Pups at Level 1; 27 Litters at Level 2

The significance tests for the fixed intercept are also different across the software procedures. SPSS and R report F-tests and t-tests for the fixed intercept, whereas SAS only reports a t-test by default (an F-test can be obtained for the intercept in SAS by specifying the `intercept` option in the `model` statement of `proc mixed`), Stata reports a z-test, and the HLM2 procedure can optionally report a Wald chi-square test in addition to the default t-test.

The procedures that report F-tests for fixed effects (other than the intercept) differ in the F-statistics that they report. Type III F-tests are reported in SAS and SPSS, and Type I (sequential) F-tests are reported in R. There are also differences in the denominator degrees of freedom used for the F-tests (see Subsection 3.11.6 for a discussion of different denominator degrees of freedom options in SAS). Using the `test` command for fixed effects in Stata or requesting general linear hypothesis tests for fixed effects in HLM (as illustrated in Subsection 3.4.5) results in Type III Wald chi-square tests being reported for the fixed effects. The p-values for all tests of the fixed effects are similar for this model across the software packages, despite differences in the tests being used.

3.6.2 Comparing Model 3.2B Results

Table 3.7 shows selected results for Model 3.2B, which has the same fixed effects as Model 3.1, random effects associated with the intercept for each litter, and heterogeneous residual variances for the combined high/low treatment group and the control group.

We report results for Model 3.2B generated by `proc mixed` in SAS, `GENLINMIXED` in SPSS, the `lme()` function in R, and the `mixed` command in Stata, because these are the only procedures that currently accommodate models with heterogeneous residual (i.e., Level 1) variances in different groups defined by categorical Level 2 variables.

The estimated variance of the random litter effects and the estimated residual variances for the pooled high/low treatment group and the control treatment group are the same across these procedures. However, these parameters are displayed differently in R; R displays multipliers of a single parameter estimate (see Subsection 3.4.3 for an example of the R output for covariance parameters, and Table 3.7 for an illustration of how to calculate the covariance parameters based on the R output in Subsection 3.4.3).

In terms of tests for fixed effects, only the F-statistics reported for the treatment by sex interaction are similar in R. This is because the other procedures are using Type III tests by default (considering all other terms in the model), and R only produces Type I F-tests. The Type I and Type III F-test results correspond only for the last term entered into the model formula. Type I F-tests can also be obtained in `proc mixed` by using the `htype = 1` option in the `model` statement. Stata uses Wald chi-square tests when the `test` command is used to perform these tests.

The values of the –2 REML log-likelihoods for the fitted models are the same across these procedures. Again, we note that the information criteria (AIC and BIC) differ because of different calculation formulas being used by the two procedures.

3.6.3 Comparing Model 3.3 Results

Table 3.8 shows selected results from the fit of Model 3.3 (our final model) using the procedures in SAS, SPSS, R, and Stata. This model has fixed effects associated with sex, treatment and litter size, and heterogeneous residual variances for the high/low vs. control treatment groups.

These four procedures agree on the reported values of the estimated fixed-effect parameters and their standard errors, as well as the estimated covariance parameters and their standard errors. We note the same differences between the procedures in Table 3.8 that were discussed for Table 3.7, for the model fit criteria (AIC and BIC) and the F-tests for the fixed effects.

3.7 Interpreting Parameter Estimates in the Final Model

The results discussed in this section were obtained from the fit of Model 3.3 using SAS `proc mixed`. Similar results can be obtained when using the `GENLINMIXED` procedure in SPSS, the `lme()` function in R, or the `mixed` command in Stata.

3.7.1 Fixed-Effect Parameter Estimates

The SAS output below displays the fixed-effect parameter estimates and their corresponding t-tests.

```
                        Solution for Fixed Effects
    ------------------------------------------------------------------
                                      Standard
    Effect      Sex       Treat    Estimate   Error      DF    t Value  Pr<|t|

    Intercept                        8.3276   0.27410   33.3    30.39   <.0001
    treat                 High     - 0.8623   0.18290   25.7   - 4.71   <.0001
    treat                 Low      - 0.4337   0.15230   24.3   - 2.85    .0088
    treat                 Control   0.0000
    sex         Female            - 0.3434   0.04204   256.0   - 8.17   <.0001
    sex         Male               0.0000
    litsize                       - 0.1307   0.01855   31.1   - 7.04   <.0001
```

The main effects of treatment (high vs. control and low vs. control) are both negative and significant ($p < 0.01$). We estimate that the mean birth weights of rat pups in the high and low treatment groups are 0.86 g and 0.43 g lower, respectively, than the mean birth weight of rat pups in the control group, adjusting for sex and litter size. Note that the degrees of freedom (DF) reported for the t-statistics are computed using the Satterthwaite method (requested by using the `ddfm = sat` option in `proc mixed`).

The main effect of sex on birth weight is also negative and significant ($p < 0.001$). We estimate that the mean birth weight of female rat pups is 0.34 g less than that of male rat pups, after adjusting for treatment level and litter size.

The main effect of litter size is also negative and significant ($p < 0.001$). We estimate that the mean birth weight of a rat pup in a litter with one additional pup is decreased by 0.13 g, adjusting for treatment and sex. The relationship between litter size and birth weight confirms our impression based on Figure 3.2, and corresponds with our intuition that larger litters would on average tend to have smaller pups.

Post-hoc comparisons of the least-squares means, using the Tukey–Kramer adjustment method to compute p-values for all pairwise comparisons, are shown in the SAS output below. These comparisons are obtained by including the `lsmeans` statement in `proc mixed`:

```
lsmeans treat / adjust = tukey;
```

```
                    Differences of Least-Squares Means
    -----------------------------------------------------------------------------
    Effect  treat  _treat   Estimate  Std Error  DF   t Value  Pr t  Adjustment    Adj P

    treat   High   Low     - 0.4286   0.1755   23.4    2.44    .0225  Tukey--Kramer  .0558
    treat   High   Control - 0.8623   0.1829   25.7    4.71   <.0001  Tukey--Kramer  .0002
    treat   Low    Control - 0.4337   0.1523   24.3    2.85    .0088  Tukey--Kramer  .0231
```

TABLE 3.8: Comparison of Results for Model 3.3

Estimation Method	SAS: proc mixed REML	SPSS: GENLINMIXED REML	R: lme() function REML	Stata: mixed REML
	Estimate (SE)	Estimate (SE)	Estimate (SE)	Estimate (SE)
Fixed-Effect Parameter				
β_0 (Intercept)	8.33(0.27)	8.33(0.27)	8.33(0.27)	8.33(0.27)
β_1 (High vs. Control)	−0.86(0.18)	−0.86(0.18)	−0.86(0.18)	−0.86(0.18)
β_2 (Low vs. Control)	−0.43(0.15)	−0.43(0.15)	−0.43(0.15)	−0.43(0.15)
β_3 (Female vs. Male)	−0.34(0.04)	−0.34(0.04)	−0.34(0.04)	−0.34(0.04)
β_4 (Litter size)	−0.13(0.02)	−0.13(0.02)	−0.13(0.02)	−0.13(0.02)
Covariance Parameter	Estimate (SE)	Estimate (SE)	Estimate (n.c.)	Estimate (SE)
σ^2_{litter}	0.10(0.03)	0.10(0.03)	0.10	0.10(0.03)
$\sigma^2_{high/low}$	0.09(0.01)	0.09(0.01)	0.09	0.09(0.01)
$\sigma^2_{control}$	0.26(0.03)	0.26(0.03)	0.26	0.26(0.03)
Model Information Criteria				
−2 REML log-likelihood	356.3	356.3	356.3	356.3
AIC	362.3	362.3	372.3	372.3
BIC	366.2	373.6	402.3	402.5
Tests for Fixed Effects	Type III F-Tests	Type III F-Tests	Type I F-Tests	Wald χ^2-Tests
Intercept	$t(33.3) = 30.4$, $p < .01$	$t(33.0) = 30.4$, $p < .01$	$F(1, 294) = 9029.6$, $p < .01$	$Z = 30.39$, $p < .01$

TABLE 3.8: (Continued)

Estimation Method	SAS: proc mixed REML	SPSS: GENLINMIXED REML	R: lme() function REML	Stata: mixed REML
SEX	$F(1, 256) = 66.7,$ $p < .01$	$F(1, 256) = 66.7,$ $p < .01$	$F(1, 294) = 63.6,$ $p < .01$	$\chi^2(1) = 66.7,$ $p < .01$
LITSIZE	$F(1, 31.1) = 49.6,$ $p < .01$	$F(1, 31.1) = 49.6,$ $p < .01$	$F(1, 23) = 33.7,$ $p < .01$	$\chi^2(1) = 49.6,$ $p < .01$
TREATMENT	$F(2, 24.3) = 11.4,$ $p < .01$	$F(2, 24) = 11.4,$ $p < .01$	$F(2, 23) = 11.4,$ $p < .01$	$\chi^2(2) = 22.8,$ $p < .01$

Note: (n.c.) = not computed
Note: 322 Rat Pups at Level 1; 27 Litters at Level 2

Both the high and low treatment means of birth weight are significantly different from the Control mean at $\alpha = 0.05$ ($p = .0002$ and $p = .0231$, respectively), and the high and low treatment means are not significantly different ($p = .0558$).

3.7.2   Covariance Parameter Estimates

The SAS output below displays the covariance parameter estimates reported by `proc mixed` for the REML-based fit of Model 3.3.

```
                    Covariance Parameter Estimates
-----------------------------------------------------------------------------
Cov Parm    Subject      Group        Estimate   Std Error   Z Value   Pr Z
Intercept   litter                    0.09900    0.03288     3.01      .0013
Residual                 TRTGRP High/low  0.09178  0.009855   9.31      <.0001
Residual                 TRTGRP control   0.2646   0.03395    7.79      <.0001
```

The estimated variance of the random effects associated with the intercept for litters is 0.099. The estimated residual variances for the high/low and control treatment groups are 0.092 and 0.265, respectively. That is, the estimated residual variance for the combined high/low treatment group is only about one-third as large as that of the control group. Based on this result, it appears that treatment not only lowers the mean birth weights of rat pups, but reduces their within-litter variance as well. The difference in variability based on treatment groups is apparent in the box plots in Figure 3.2.

We do not consider the Wald z-tests for the covariance parameters included in this output, but rather recommend the use of likelihood ratio tests for testing the covariance parameters, as discussed in Subsections 3.5.1 and 3.5.2.

3.8   Estimating the Intraclass Correlation Coefficients (ICCs)

In the context of a two-level model specification with random intercepts, the **intraclass correlation coefficient** (ICC) is a measure describing the similarity (or homogeneity) of the responses within a cluster (e.g., litter) and can be defined as a function of the variance components in the model. For the models investigated in this chapter, the litter-level ICC is defined as the proportion of the total random variation in the responses (the denominator in (3.4)) that is due to the variance of the random litter effects (the numerator in (3.4)):

$$\text{ICC}_{litter} = \frac{\sigma_{litter}^2}{\sigma_{litter}^2 + \sigma^2} \tag{3.4}$$

ICC_{litter} will be high if the total random variation in the denominator is dominated by the variance of the random litter effects. Because variance components are by definition greater than or equal to zero, the ICCs derived using (3.4) are also greater than or equal to zero.

Although the classical definition of an ICC arises from a **variance components** model, in which there are only random intercepts and no fixed effects, we illustrate the calculation of **adjusted** ICCs in this chapter (Raudenbush & Bryk, 2002). The definition of an adjusted ICC is based on the variance components used in the model containing both random intercepts and fixed effects.

We obtain an estimate of the ICC by substituting the estimated variance components from a fitted model into (3.4). The variance component estimates are clearly labeled in the output provided by each software procedure and can be used for this calculation.

Another way to think about the intraclass correlation is in the context of a **marginal model** implied by a mixed model with random intercepts, which has a marginal variance-covariance matrix with a compound symmetry structure (see Subsection 2.3.2). In the marginal model, the ICC can be defined as a correlation coefficient between responses, common for any pair of individuals within the same cluster. For the Rat Pup example, the marginal ICC is defined as the correlation of observed birth weight responses on any two rat pups i and i' within the same litter j, i.e., $corr(\text{WEIGHT}_{ij}, \text{WEIGHT}_{i'j})$.

Unlike the ICC calculated in (3.4), an ICC calculated using the marginal model approach could be either positive or negative. A negative marginal correlation between the weights of pairs of rat pups within the same litter would result in a negative intraclass correlation. This could occur, for example, in the context of competition for maternal resources in the womb, where rat pups that grow to be large might do so at the expense of other pups in the same litter.

In practice, there are arguments for and against using each form of the ICC. The ICC in (3.4) is easy to understand conceptually, but is restricted to be positive by the definition of variance components. The ICC estimated from the marginal model allows for negative correlations, but loses the nice interpretation of the variance components made possible with random intercepts. In general, models with explicitly specified random effects and the corresponding ICC definition in (3.4) are preferred when the pairwise correlations within clusters are positive. However, we advise fitting a marginal model without random effects and with a compound symmetry variance-covariance structure for the residuals (see Subsection 2.3.1) to check for a negative ICC. If the ICC is in fact negative, a marginal model must be used. We illustrate fitting marginal models in Chapters 5 and 6, and discuss marginal models in more detail in Section 2.3.

In this section, we calculate ICCs using the estimated variance components from Model 3.3 and also show how to obtain the intraclass correlations from the variance-covariance matrix of the marginal model implied by Model 3.3. Recall that Model 3.3 has a heterogeneous residual variance structure, with one residual variance for the control group and another for the pooled high/low treatment group. As a result, the observations on rat pups in the control group also have a different estimated intraclass correlation than observations on rat pups in the high/low treatment group. Equation (3.5) shows the estimated ICC for the control group, and (3.6) shows the estimated ICC for the pooled high/low treatment group. Because there is more within-litter variation in the control group, as noted in Figure 3.2, the estimated ICC is lower in the control group than in the pooled high/low group.

$$\text{Control: } \widehat{\text{ICC}}_{litter} = \frac{\widehat{\sigma}^2_{litter}}{\widehat{\sigma}^2_{litter} + \widehat{\sigma}^2_{control}} = \frac{0.0990}{0.0990 + 0.2646} = 0.2722 \qquad (3.5)$$

$$\text{High/Low: } \widehat{\text{ICC}}_{litter} = \frac{\widehat{\sigma}^2_{litter}}{\widehat{\sigma}^2_{litter} + \widehat{\sigma}^2_{high/low}} = \frac{0.0990}{0.0990 + 0.0918} = 0.5189 \qquad (3.6)$$

We can also obtain estimates of the ICC for a single litter from the marginal variance-covariance matrix ($\boldsymbol{V}_j$) implied by Model 3.3. When we specify v = 3 in the random statement in proc mixed in SAS, the estimated marginal variance-covariance matrix for the observations on the four rat pups in litter 3 ($\boldsymbol{V}_3$) is displayed in the following output. Note that this litter was assigned to the control treatment.

```
  Estimated V Correlation Matrix for litter 3

  Row    Col1     Col2     Col3     Col4
  ---------------------------------------------
   1    0.3636   0.0990   0.0990   0.0990
   2    0.0990   0.3636   0.0990   0.0990
   3    0.0990   0.0990   0.3636   0.0990
   4    0.0990   0.0990   0.0990   0.3636
```

The marginal variance-covariance matrices for observations on rat pups within every litter assigned to the control dose would have the same structure, with constant variance on the diagonal and constant covariance in the off-diagonal cells. The size of the matrix for each litter would depend on the number of pups in that litter. Observations on rat pups from different litters would have zero marginal covariance because they are assumed to be independent.

The estimated **marginal correlations** of observations collected on rat pups in litter 3 can be derived using the SAS output obtained from the vcorr = 3 option in the **random** statement. The estimated marginal correlation matrix for rat pups within litter 3, which received the control treatment, is as follows:

```
  Estimated V Correlation Matrix for litter 3

  Row    Col1     Col2     Col3     Col4
  ---------------------------------------------
   1    1.0000   0.2722   0.2722   0.2722
   2    0.2722   1.0000   0.2722   0.2722
   3    0.2722   0.2722   1.0000   0.2722
   4    0.2722   0.2722   0.2722   1.0000
```

We see the same estimates of the marginal within-litter correlation for control litters, 0.2722, that was found when we calculated the ICC using the estimated variance components.

Similar estimates of the marginal variance-covariance matrices implied by Model 3.3 can be obtained in R using the getVarCov() function:

```
> getVarCov(model3.3.reml.fit, individual="3", type="marginal")
```

Estimates for litters in different treatment groups can be obtained by changing the value of the "individual" argument in the preceding function in R, or by specifying different litter numbers in the v = and vcorr = options in SAS **proc mixed**. The option of displaying these marginal variance-covariance matrices is currently not available in the other three software procedures.

3.9 Calculating Predicted Values

3.9.1 Litter-Specific (Conditional) Predicted Values

Using the estimates obtained from the fit of Model 3.3, a formula for the predicted birth weight of an individual observation on rat pup i from litter j is written as:

$$\text{WEI}\widehat{\text{G}}\text{HT}_{ij} = \quad 8.33 - 0.36 \times \text{TREAT1}_j - 0.43 \times \text{TREAT2}_j$$
$$- 0.34 \times \text{SEX}_{ij} - 0.13 \times \text{LITSIZE}_j + \hat{u}_j \quad (3.7)$$

Recall that TREAT1_j and TREAT2_j in (3.7) are dummy variables, indicating whether or not an observation is for a rat pup in a litter that received the high or low dose treatment, SEX1_{ij} is a dummy variable indicating whether or not a particular rat pup is a female, and LITSIZE_j is a continuous variable representing the size of litter j. The predicted value of u_j is the realization (or EBLUP) of the random litter effect for the j-th litter. This equation can also be used to write three separate equations for predicting birth weight in the three treatment groups:

High: $\quad \text{WEI}\widehat{\text{G}}\text{HT}_{ij} = 7.47 - 0.34 \times \text{SEX1}_{ij} - 0.13 \times \text{LITSIZE}_j + \hat{u}_j$

Low: $\quad \text{WEI}\widehat{\text{G}}\text{HT}_{ij} = 7.90 - 0.34 \times \text{SEX1}_{ij} - 0.13 \times \text{LITSIZE}_j + \hat{u}_j$

Control: $\quad \text{WEI}\widehat{\text{G}}\text{HT}_{ij} = 8.33 - 0.34 \times \text{SEX1}_{ij} - 0.13 \times \text{LITSIZE}_j + \hat{u}_j$

For example, after fitting Model 3.3 using `proc mixed`, we see the EBLUPs of the random litter effects displayed in the SAS output. The EBLUP of the random litter effect for litter 7 (assigned to the control dose) is 0.3756, suggesting that pups in this litter tend to have higher birth weights than expected. The formula for the predicted birth weight values in this litter is:

$$\text{WEI}\widehat{\text{G}}\text{HT}_{ij} = 8.33 - 0.34 \times \text{SEX1}_{ij} - 0.13 \times \text{LITSIZE}_j + 0.3756$$
$$= 8.71 - 0.34 \times \text{SEX1}_{ij} - 0.13 \times 18$$
$$= 6.37 - 0.34 \times \text{SEX1}_{ij}$$

We present this calculation for illustrative purposes only. The conditional predicted values for each rat pup can be placed in an output data set by using the `outpred =` option in the `model` statement in `proc mixed`, as illustrated in the final syntax for Model 3.3 in Subsection 3.4.1.

3.9.2 Population-Averaged (Unconditional) Predicted Values

Population-averaged (unconditional) predicted values for birth weight in the three treatment groups, based only on the estimated fixed effects in Model 3.3, can be calculated using the following formulas:

High: $\quad \text{WEI}\widehat{\text{G}}\text{HT}_{ij} = 7.47 - 0.34 \times \text{SEX1}_{ij} - 0.13 \times \text{LITSIZE}_j$

Low: $\quad \text{WEI}\widehat{\text{G}}\text{HT}_{ij} = 7.90 - 0.34 \times \text{SEX1}_{ij} - 0.13 \times \text{LITSIZE}_j$

Control: $\quad \text{WEI}\widehat{\text{G}}\text{HT}_{ij} = 8.33 - 0.34 \times \text{SEX1}_{ij} - 0.13 \times \text{LITSIZE}_j$

Note that these formulas, which can be used to determine the expected values of birth weight for rat pups of a given sex in a litter of a given size, simply omit the random litter effects. These population-averaged predicted values represent expected values of birth weight based on the marginal distribution implied by Model 3.3 and are therefore sometimes referred to as **marginal predicted values**.

The marginal predicted values for each rat pup can be placed in an output data set by using the `outpredm =` option in the `model` statement in `proc mixed`, as illustrated in the final syntax for Model 3.3 in Subsection 3.4.1.

3.10   Diagnostics for the Final Model

The model diagnostics considered in this section are for the final model, Model 3.3, and are based on REML estimation of this model using SAS `proc mixed`. The syntax for this model is shown in Subsection 3.4.1.

3.10.1   Residual Diagnostics

We first examine conditional raw residuals in Subsection 3.10.1.1, and then discuss studentized conditional raw residuals in Subsection 3.10.1.2. The conditional raw residuals are not as well suited to detecting outliers as are the **studentized conditional residuals** (Schabenberger, 2004).

3.10.1.1   Conditional Residuals

In this section, we examine residual plots for the **conditional raw residuals**, which are the realizations of the residuals, ε_{ij}, based on the REML fit of Model 3.3. They are the differences between the observed values and the litter-specific predicted values for birth weight, based on the estimated fixed effects and the EBLUPs of the random effects. Similar plots for the conditional residuals are readily obtained for this model using the other software procedures. Instructions on how to obtain these diagnostic plots in the other software procedures are included on the book's web page (see Appendix A).

The conditional raw residuals for Model 3.3 were saved in a new SAS data set (`pdat1`) by using the `outpred =` option in the `model` statement of `proc mixed` in Subsection 3.4.1. To check the assumption of normality for the conditional residuals, a histogram and a normal quantile–quantile plot (Q–Q plot) are obtained separately for the high/low treatment group and the control treatment group. To do this, we first use `proc sgpanel` with the `histogram` keyword, and then use `proc univariate` with the `qqplot` keyword:

```
/* Figure 3.4: Histograms of residuals by treatment group */
title;
proc sgpanel data=pdat1;
  panelby trtgrp  / novarname rows=2;
  histogram resid;
  format trtgrp tgrpfmt.;
run;

/* Figure 3.5: Normal Q-Q plots of residuals by treatment group */
proc univariate data=pdat1;
  class trtgrp ;
  var resid;
  qqplot / normal(mu=est sigma=est);
  format trtgrp tgrpfmt.;
run;
```

We look at separate graphs for the high/low and control treatment groups. It would be inappropriate to combine these graphs for the conditional raw residuals, because of the differences in the residual variances for levels of TRTGRP in Model 3.3.

The plots in Figures 3.4 and 3.5 show that there are some outliers at the low end of the distribution of the conditional residuals and potentially negatively skewed distributions for both the high/low treatment and control groups. However, the skewness is not severe.

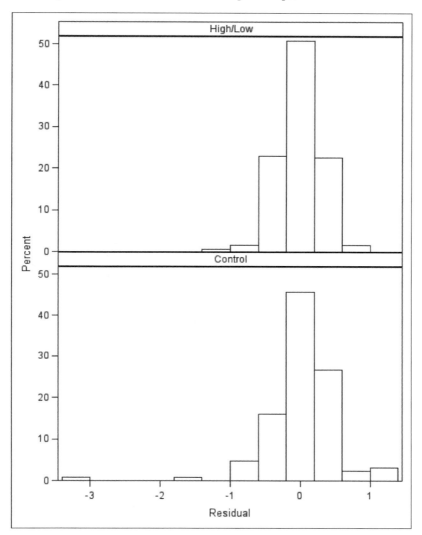

FIGURE 3.4: Histograms for conditional raw residuals in the pooled high/low and control treatment groups, based on the fit of Model 3.3.

The Shapiro–Wilk test of normality for the conditional residuals reported by `proc univariate` (with the `normal` option) reveals that the assumption of normality for the conditional residuals is violated in both the pooled high/low treatment group (Shapiro–Wilk $W = 0.982$, $p = 0.015$)[3] and the control group ($W = 0.868$, $p < 0.001$). These results suggest that the model still requires some additional refinement (e.g., transformation of the response variable, or further investigation of outliers).

A plot of the conditional raw residuals vs. the predicted values by levels of TRTGRP can be generated using the following syntax:

---

[3]SAS `proc univariate` also reports three additional tests for normality (including Kolmogorov–Smirnov), the results of which indicate that we would not reject a null hypothesis of normality for the conditional residuals in the high/low group (KS $p > .15$).

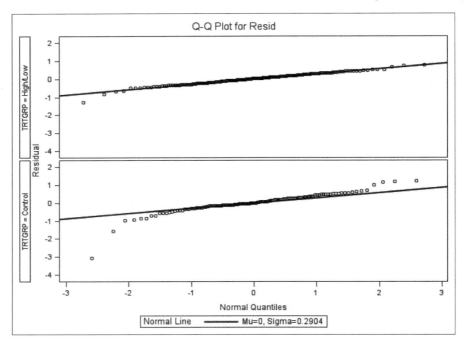

FIGURE 3.5: Normal Q–Q plots for the conditional raw residuals in the pooled high/low and control treatment groups, based on the fit of Model 3.3.

```
/* Figure 3.6: Plot of conditional raw residuals vs. predicted values */
proc sgpanel data = pdat1;
  panelby trtgrp / novarname rows=2;
  scatter y = resid x = pred ;
  format trtgrp tgrpfmt.;
run;
```

We do not see strong evidence of nonconstant variance for the pooled high/low group in Figure 3.6. However, in the control group, there is evidence of an outlier, which after some additional investigation was identified as being for PUP_ID = 66. An analysis without the values for this rat pup resulted in similar estimates for each of the fixed effects included in Model 3.3, suggesting that this observation, though poorly fitted, did not have a great deal of influence on these estimates.

3.10.1.2   Conditional Studentized Residuals

The **conditional studentized residual** for an observation is the difference between the observed value and the predicted value, based on both the fixed and random effects in the model, divided by its estimated standard error. The standard error of a residual depends on both the residual variance for the treatment group (high/low or control), and the leverage of the observation. The conditional studentized residuals are more appropriate to examine model assumptions and to detect outliers and potentially influential points than are the raw residuals because the raw residuals may not come from populations with equal variance, as was the case in this analysis. Furthermore, if an observation has a large conditional residual, we cannot know if it is large because it comes from a population with a larger variance, or if it is just an unusual observation (Schabenberger, 2004).

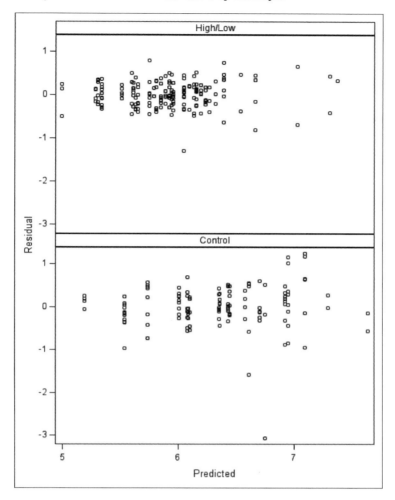

FIGURE 3.6: Scatter plots of conditional raw residuals vs. predicted values in the pooled high/low and control treatment groups, based on the fit of Model 3.3.

We first examine the distribution of the conditional studentized residuals by obtaining box plots of the distribution of these residuals for each litter. These box plots are produced as a result of the `boxplot` option being used in the `plots =` specification of `proc mixed`, or they can be obtained by using the following syntax:

```
proc sgplot data = pdat1 noautolegend;
  vbox studentresid / category=litter;
run;
```

In Figure 3.7, we notice that the rat pup in row 66 of the data set (corresponding to PUP_ID = 66, from litter 6, because we sorted the `ratpup3` data set by PUP_ID before fitting Model 3.3) is an outlier. These box plots also show that the distribution of the conditional studentized residuals is approximately homogeneous across litters, unlike the raw birth weight observations displayed in Figure 3.2. Note that there is no need to plot studentized residuals separately for different treatments as there was for the raw conditional residuals. This plot gives us some confidence in the appropriateness of Model 3.3.

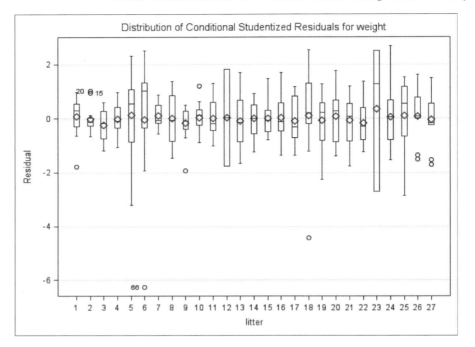

FIGURE 3.7: Box plot of conditional studentized residuals by new litter ID, based on the fit of Model 3.3.

3.10.2   Influence Diagnostics

In this section, we examine influence diagnostics generated by `proc mixed` for the rat pups and their litters.

3.10.2.1   Overall Influence Diagnostics

The graphs displayed here were obtained by using the `influence (iter = 5 effect = litter est)` option in the `model` statement. The `iter =` suboption controls the maximum number of additional iterations that will be performed by `proc mixed` to update the parameter estimates when a given litter is deleted. The `est` suboption causes SAS to display a graph of the effects of each litter on the estimates of the fixed effects (shown in Figure 3.11).

```
model weight = treat sex litsize / solution ddfm=sat
       influence(iter=5 effect=litter est) ;
```

Figure 3.8 shows the effect of deleting one litter at a time on the restricted (REML) likelihood distance, a measure of the overall fit of the model. Based on this graph, litter 6 is shown to have a very large influence on the REML likelihood. In the remaining part of the model diagnostics, we will investigate which aspect of the model fit is influenced by individual litters (and litter 6 in particular).

Figure 3.9 shows the effect of deleting each litter on selected model summary statistics (Cook's D and Covratio). The upper left panel, **Cook's D Fixed Effects**, shows the overall effect of the removal of each litter on the vector of fixed-effect parameter estimates. Litter 6 does not have a high value of Cook's D statistic, indicating that it does not have a large effect on the fixed-effect parameter estimates, even though it has a large influence on the

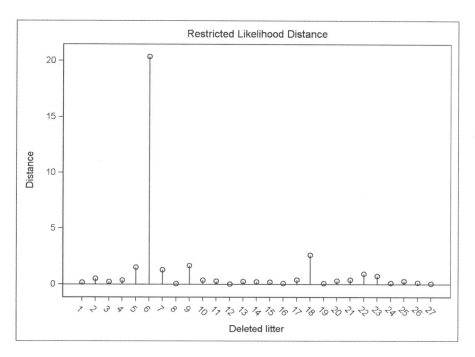

FIGURE 3.8: Effect of deleting each litter on the REML likelihood distance for Model 3.3.

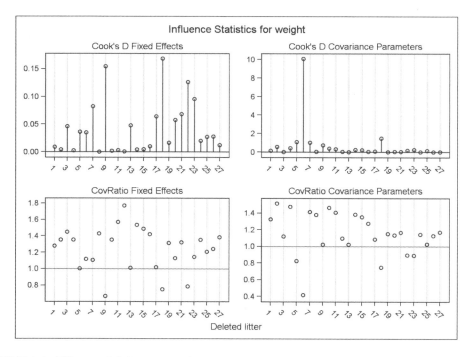

FIGURE 3.9: Effects of deleting one litter at a time on summary measures of influence for Model 3.3.

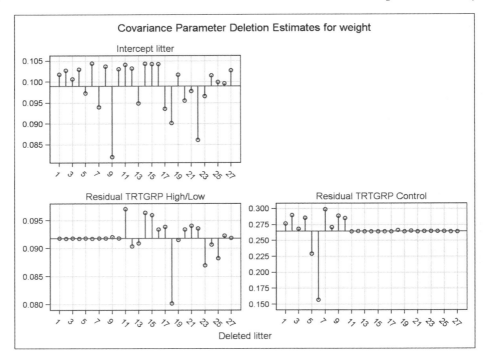

FIGURE 3.10: Effects of deleting one litter at a time on measures of influence for the covariance parameters in Model 3.3.

REML likelihood. However, deletion of litter 6 has a very large influence on the estimated covariance parameters, as shown in the **Cook's D Covariance Parameters** panel. The **Covratio Fixed Effects** panel illustrates the change in the precision of the fixed-effect parameter estimates brought about by deleting the j-th litter. A reference line is drawn at 1.0 for this graph. A litter with a value of covratio = 1.0 indicates that deletion of that litter has "no effect" on the precision of the fixed-effect parameter estimates, and a value of covratio substantially below 1.0 indicates that deleting that litter will improve the precision of the fixed-effect parameter estimates. A litter that greatly affects the covratio for the fixed effects could have a large influence on statistical tests (such as t-tests and F-tests). There do not appear to be any litters that have a large influence on the precision of the estimated fixed effects (i.e., much lower covratio than the others). However, the very small value of covratio = 0.4 for litter 6 in the panel of **Covratio for Covariance Parameters** indicates that the precision of the estimated covariance parameters can be substantially improved by removing litter 6.

3.10.2.2   Influence on Covariance Parameters

The ODS graphics output from SAS also produces influence diagnostics for the covariance parameter estimates. These can be useful in identifying litters that may have a large effect on the estimated between-litter variance (σ^2_{litter}), or on the residual variance estimates for the two treatment groups (high/low and control). We have included the panel of plots displaying the influence of each litter on the covariance parameters in Figure 3.10.

Each panel of Figure 3.10 depicts the influence of individual litters on one of the covariance parameters in the model. There is a horizontal line showing the estimated value of the covariance parameter when all litters are used in the model. For example, in the first panel

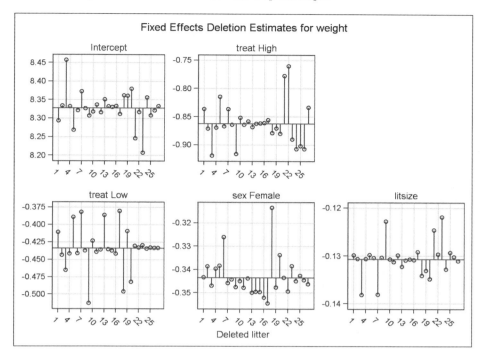

FIGURE 3.11: Effect of deleting each litter on measures of influence for the fixed effects in Model 3.3.

of the graph (labeled **Intercept litter**), the between-litter variance is displayed, with a horizontal line at 0.099, the estimated between-litter variance from Model 3.3. The change in the between-litter variance that would result from deletion of each litter is depicted. Litter 6 has little influence on the between-litter variance estimate. The two lower panels depict the residual variance estimates for the high/low treatment group (labeled **Residual TRTGRP HIGH/LOW**), and for the control group (labeled **Residual TRTGRP Control**). This graph shows that deletion of litter 6 would decrease the residual variance estimated for the control group from 0.2646 (as in Model 3.3) to about 0.155. We refitted Model 3.3 excluding litter 6, and found that this was in fact the case, with the residual variance for the control group decreasing from 0.2646 to 0.1569.

3.10.2.3 Influence on Fixed Effects

Figure 3.11 shows the effect of the deletion of each litter on the fixed-effect parameter estimates, with one panel for each fixed effect. A horizontal line is drawn to represent the value of the parameter estimate when all cases are used in the analysis, and the effect of deleting each litter is depicted. For example, there is very little effect of litter 6 on the estimated intercept in the model (because litter 6 is a control litter, its effect would be included in the intercept).

We refitted Model 3.3 excluding litter 6, and found that the estimated fixed effects were very similar (litter 6 received the control treatment). The residual variance for the control group is much smaller without the presence of this litter, as discussed in Subsection 3.10.2.2. The likelihood ratio test for Hypothesis 3.4 was still significant after deleting litter 6 ($p < 0.001$), which suggested retaining the heterogeneous residual variance model (Model 3.2B) rather than assuming homogeneous residual variance (Model 3.1).

3.11 Software Notes and Recommendations

3.11.1 Data Structure

SAS `proc mixed`, the `MIXED` and `GENLINMIXED` commands in SPSS, the `lme()` and `lmer()` functions in R, and the `mixed` command in Stata all require a single data set in the "long" format, containing both Level 1 and Level 2 variables. See Table 3.2 for an illustration of the Rat Pup data set in the "long" format.

The HLM2 procedure requires the construction of two data sets for a two-level analysis: a Level 1 data set containing information on the dependent variable, and covariates measured at Level 1 of the data (e.g., rat pups); and a Level 2 data set containing information on the clusters (e.g., litters). See Subsection 3.4.5 for more details on setting up the two data sets for HLM.

3.11.2 Syntax vs. Menus

SAS `proc mixed` and the `lme()` and `lmer()` functions in R require users to create syntax to specify an LMM. The `MIXED` and `GENLINMIXED` procedures in SPSS allows users to specify a model using the menu system and then paste the syntax into a syntax window, where it can be saved or run later. Stata allows users to specify a model either through the menu system or by typing syntax for the `mixed` command. The HLM program works mainly by menus (although batch processing of syntax for fitting HLM models is possible; see the HLM documentation for more details).

3.11.3 Heterogeneous Residual Variances for Level 2 Groups

In the Rat Pup analysis we considered models (e.g., Model 3.2A, Model 3.2B, and Model 3.3) in which rat pups from different treatment groups were allowed to have different residual variances. In other words, the Level 1 (residual) variance was allowed to vary as a function of a Level 2 variable (TREATMENT, which is a characteristic of the litter). The option of specifying a heterogeneous structure for the residual variance matrix, R_j, based on groups defined by Level 2 variables, is currently only available in SAS `proc mixed`, the `GENLINMIXED` command in SPSS, the `lme()` function in R, and the `mixed` command in Stata. The HLM2 procedure currently allows the residual variance at Level 1 of a two-level data set to be a function of Level 1 variables, but not of Level 2 variables.

The ability to fit models with heterogeneous residual variance structures for different groups of observations can be very helpful in certain settings and was found to improve the fits of the models in the Rat Pup example. In some analyses, fitting a heterogeneous residual variance structure could make a critical difference in the results obtained and in the conclusions for the fixed effects in a model.

3.11.4 Display of the Marginal Covariance and Correlation Matrices

SAS `proc mixed` and the `lme()` function in R allow users to request that the marginal covariance matrix implied by a fitted model be displayed. The option of displaying the implied marginal covariance matrix is currently not available in the other three software procedures. We have found that it is helpful to investigate the implied marginal variance-covariance and correlation matrices for a linear mixed model to see whether the estimated variance-covariance structure is appropriate for the data set being analyzed.

3.11.5   Differences in Model Fit Criteria

The –2 REML log-likelihood values obtained by fitting the two-level random intercept models discussed in this chapter are nearly identical across the software procedures (See Tables 3.6 to 3.8). However, there are some minor differences in this statistic, as calculated by the HLM2 procedure, which are perhaps due to computational differences. The AIC statistic differs across the software procedures that calculate it, because of differences in the formulas used (see Subsection 3.6.1 for a discussion of the different calculation formulas for AIC).

3.11.6   Differences in Tests for Fixed Effects

SAS `proc mixed`, the `anova()` function in R, and the `MIXED` and `GENLINMIXED` commands in SPSS all compute F-tests for the overall effects of fixed factors and covariates in the model, in addition to t-tests for individual fixed-effect parameters. In contrast, the `mixed` command in Stata calculates z-tests for individual fixed-effect parameters, or alternatively, Wald chi-square tests for fixed effects via the use of the `test` command. The HLM2 program calculates t-tests for fixed-effect parameters, or alternatively, Wald chi-square tests for sets of fixed effects.

By default, the procedures in SAS and SPSS both calculate Type III F-tests, in which the significance of each term is tested conditional on the fixed effects of all other terms in the model. When the `anova()` function in R is applied to a model fit object, it returns Type I (sequential) F-tests, which are conditional on just the fixed effects listed in the model prior to the effects being tested. The Type III and Type I F-tests are comparable only for the term entered last in the model (except for certain models for balanced data). When making comparisons between the F-test results for SAS and R, we used the F-tests only for the last term entered in the model to maintain comparability. Note that SAS `proc mixed` can calculate Type I F-tests by specifying the `htype = 1` option in the model statement.

The programs that calculate F-tests and t-tests use differing methods to calculate denominator degrees of freedom (or degrees of freedom for t-tests) for these approximate tests. SAS allows the most flexibility in specifying methods for calculating denominator degrees of freedom, as discussed below.

The `MIXED` command in SPSS and the `lmerTest` package in R both use the Satterthwaite approximation when calculating denominator degrees of freedom for these tests and do not provide other options.

A note on denominator degrees of freedom (df) methods for F-tests in SAS

The containment method is the default denominator degrees of freedom method for `proc mixed` when a `random` statement is used, and no denominator degrees of freedom method is specified. The containment method can be explicitly requested by using the `ddfm = contain` option. The containment method attempts to choose the correct degrees of freedom for fixed effects in the model, based on the syntax used to define the random effects. For example, the df for the variable TREAT would be the df for the random effect that contains the word "treat" in it. The syntax

```
random int / subject = litter(treat);
```

would cause SAS to use the degrees of freedom for `litter(treat)` as the denominator degrees of freedom for the F-test of the fixed effects of TREAT. If no random effect syntactically includes the word "treat," the residual degrees of freedom would be used.

The Satterthwaite approximation (`ddfm = sat`) is intended to produce a more accurate F-test approximation, and hence a more accurate p-value for the F-test. The

SAS documentation warns that the small-sample properties of the Satterthwaite approximation have not been thoroughly investigated for all types of models implemented in `proc mixed`.

The Kenward–Roger method (`ddfm = kr`) is an attempt to make a further adjustment to the F-statistics calculated by SAS, to take into account the fact that the REML estimates of the covariance parameters are in fact estimates and not known quantities. This method inflates the marginal variance-covariance matrix and then uses the Satterthwaite method on the resulting matrix.

The between-within method (`ddfm = bw`) is the default for repeated-measures designs, and may also be specified for analyses that include a random statement. This method divides the residual degrees of freedom into a between-subjects portion and a within-subjects portion. If levels of a covariate change within a subject (e.g., a time-dependent covariate), then the degrees of freedom for effects associated with that covariate are the within-subjects df. If levels of a covariate are constant within subjects (e.g., a time-independent covariate), then the degrees of freedom are assigned to the between-subjects df for F-tests.

The residual method (`ddfm = resid`) assigns $n - \text{rank}(\boldsymbol{X})$ as the denominator degrees of freedom for all fixed effects in the model. The rank of $\boldsymbol{X}$ is the number of linearly independent columns in the $\boldsymbol{X}$ matrix for a given model. This is the same as the degrees of freedom used in ordinary least squares (OLS) regression (i.e., $n-p$, where n is the total number of observations in the data set and p is the number of fixed-effect parameters being estimated).

In R, when applying the `summary()` or `anova()` functions to the object containing the results of an `lme()` fit, denominator degrees of freedom for tests associated with specific fixed effects are calculated as follows:

Denominator degrees of freedom (df) methods for F-tests in R

> **Level 1:** df = # of Level 1 observations – # of Level 2 clusters – # of Level 1 fixed effects
>
> **Level 2:** df = # of Level 2 clusters – # of Level 2 fixed effects

---

For example, in the Type I F-test for TREATMENT from R reported in Table 3.8, the denominator degrees of freedom is 23, which is calculated as 27 (the number of litters) minus 4 (the number of fixed-effect parameters associated with the Level 2 variables). In the Type I F-test for SEX, the denominator degrees of freedom is 294, which is calculated as 322 (the number of pups in the data set) minus 27 (the number of litters), minus 1 (the number of fixed-effect parameters associated with the Level 1 variables).

In general, the differences in the denominator degrees of freedom calculations are not critical for the t-tests and F-tests when working with data sets and relatively large (> 100) samples at each level (e.g., rat pups at Level 1 and litters at Level 2) of the data set. However, more attention should be paid when working with small samples, and these cases are best handled using the different options provided by `proc mixed` in SAS. We recommend using the Kenward–Roger method in these situations (Verbeke & Molenberghs, 2000).

We also remind readers that denominator degrees of freedom are not computed by default when using the `lme4` package version of the `lmer()` function in R, and p-values for computed test statistics are not displayed as a result. Users of the `lmer()` function can consider loading the `lmerTest` package to compute Satterthwaite approximations of these degrees of freedom (and p-values for the corresponding test statistics).

3.11.7 Post-Hoc Comparisons of Least Squares (LS) Means (Estimated Marginal Means)

A very useful feature of SAS `proc mixed` and the `MIXED` and `GENLINMIXED` commands in SPSS is the ability to obtain post-hoc pairwise comparisons of the least-squares means (called *estimated marginal means* in SPSS) for different levels of a fixed factor, such as TREATMENT. This can be accomplished by the use of the `lsmeans` statement in SAS and the `EMMEANS` subcommand in SPSS. Different adjustments are available for multiple comparisons in these two software procedures. SAS has the wider array of choices (e.g., Tukey–Kramer, Bonferroni, Dunnett, Sidak, and several others). SPSS currently offers only the Bonferroni and Sidak methods. Post-hoc comparisons can also be computed in R with some additional programming (Faraway, 2005), and users of Stata can apply the `margins` command after fitting a model using `mixed` to explore alternative differences in marginal means (see `help margins` in Stata).

3.11.8 Calculation of Studentized Residuals and Influence Statistics

Whereas each software procedure can calculate both conditional and marginal raw residuals, SAS `proc mixed` is currently the only program that automatically provides studentized residuals, which are preferred for model diagnostics (detection of outliers and identification of influential observations).

SAS has also developed powerful and flexible influence diagnostics that can be used to explore the potential influence of individual observations or clusters of observations on both fixed-effect parameter estimates and covariance parameter estimates, and we have illustrated the use of these diagnostic tools in this chapter.

3.11.9 Calculation of EBLUPs

The procedures in SAS, R, Stata, and HLM all allow users to generate and save the EBLUPs for any random effects (and not just random intercepts, as in this example) included in an LMM. In general, the current version of SPSS does not allow users to extract the values of the EBLUPs. However, for models with a single random intercept for each cluster, the EBLUPs can be easily calculated in SPSS, as illustrated for Model 3.1 in Subsection 3.4.2.

3.11.10 Tests for Covariance Parameters

By default, the procedures in SAS and SPSS do not report Wald tests for covariance parameters, but they can be requested as a means of obtaining the standard errors of the estimated covariance parameters. We do not recommend the use of Wald tests for testing hypotheses about the covariance parameters.

REML-based likelihood ratio tests for covariance parameters can be carried out (with a varying amount of effort by the user) in `proc mixed` in SAS, the `MIXED` and `GENLINMIXED` commands in SPSS, and by using the `anova()` function after fitting an LMM with the either the `lme()` or the `lmer()` function in R. The syntax to perform these likelihood ratio tests is most easily implemented using R, although the user may need to adjust the p-value obtained, to take into account mixtures of χ^2 distributions when appropriate.

The default likelihood ratio test reported by the `mixed` command in Stata for the covariance parameters is an overall test of the covariance parameters associated with all random effects in the model. In models with a single random effect for each cluster, as in Model 3.1, it is appropriate to use this test to decide if that effect should be included in the model. However, the likelihood ratio test reported by Stata should be considered conservative in models including multiple random effects.

The HLM procedures utilize alternative chi-square tests for covariance parameters associated with random effects, the definition of which depend on the number of levels of data (e.g., whether one is studying a two-level or three-level data set). For details on these tests in HLM, refer to Raudenbush & Bryk (2002).

3.11.11   Reference Categories for Fixed Factors

By default, `proc mixed` in SAS and the `MIXED` and `GENLINMIXED` commands in SPSS both prevent overparameterization of LMMs with categorical fixed factors by setting the fixed effects associated with the highest-valued levels of the factors to zero. This means that when using these procedures, the highest-valued level of a fixed factor can be thought of as the "reference" category for the factor. The `lme()` and `lmer()` functions in R and the `mixed` command in Stata both use the lowest-valued levels of factors as the reference categories, by default. These differences will result in different parameterizations of the model being fitted in the different software procedures unless recoding is carried out prior to fitting the model. However, the choice of parameterization will not affect the overall model fit criteria, predicted values, hypothesis tests, or residuals.

Unlike the other four procedures, the HLM procedures in general (e.g., HLM2 for two-level models, HLM3 for three-level models, etc.) do not automatically generate dummy variables for levels of categorical fixed factors to be included in a mixed model. Indicator variables for the nonreference levels of fixed factors need to be generated first in another software package before importing the data sets into HLM. This leaves the choice of the reference category to the user.

4

Three-Level Models for Clustered Data: The Classroom Example

4.1 Introduction

In this chapter, we illustrate models for clustered study designs having three levels of data. In three-level clustered data sets, the units of analysis (Level 1) are nested within randomly sampled clusters (Level 2), which are in turn nested within other larger randomly sampled clusters (Level 3). Such study designs allow us to investigate whether covariates measured at each level of the data hierarchy have an impact on the dependent variable, which is always measured on the units of analysis at Level 1.

Designs that lead to three-level clustered data sets can arise in many fields of research, as illustrated in Table 4.1. For example, in the field of education research, as in the Classroom data set analyzed in this chapter, students' math achievement scores are studied by first randomly selecting a sample of schools, then sampling classrooms within each school, and finally sampling students from each selected classroom (Figure 4.1). In medical research, a study to evaluate treatments for blood pressure may be carried out in multiple clinics, with several doctors from each clinic selected to participate and multiple patients treated by each doctor participating in the study. In a laboratory research setting, a study of birth weights in rat pups similar to the Rat Pup study that we analyzed in Chapter 3 might have replicate experimental runs, with several litters of rats involved in each run, and several rat pups in each litter.

In Table 4.1, we provide possible examples of three-level clustered data sets in different research settings. In these studies, the dependent variable is measured on one occasion for each Level 1 unit of analysis, and covariates may be measured at each level of the data hierarchy. For example, in a study of student achievement, the dependent variable, math achievement score, is measured once for each student. In addition, student characteristics such as age, classroom characteristics such as class size, and school characteristics such as neighborhood poverty level are all measured. In the multilevel models that we fit to three-level data sets, we relate the effects of covariates measured at each level of the data to the dependent variable. Thus, in an analysis of the study of student achievement, we might use student characteristics to help explain between-student variability in math achievement scores, classroom characteristics to explain between-classroom variability in the classroom-specific average math scores, and school-level covariates to help explain between-school variation in school-specific mean math achievement scores.

Figure 4.1 illustrates the hierarchical structure of the data for an educational study similar to the Classroom study analyzed in this chapter. This figure depicts the data structure for a single (hypothetical) school with two randomly selected classrooms.

Models applicable for data from such sampling designs are known as **three-level hierarchical linear models (HLMs)** and are extensions of the two-level models introduced in

TABLE 4.1: Examples of Three-Level Data in Different Research Settings

Level of Data		Research Setting		
		Education	Medicine	Biology
Level 3	Cluster of clusters (random factor)	School	Clinic	Replicate
	Covariates	School size, poverty level of neighborhood surrounding the school	Number of doctors in the clinic, clinic type (public or private)	Experimental run, instrument calibration, ambient temperature, run order
Level 2	Cluster of units (random factor)	Classroom	Doctor	Litter
	Covariates	Class size, years of experience of teacher	Specialty, years of experience	Litter size, weight of mother rat
Level 1	Unit of analysis	Student	Patient	Rat Pup
	Dependent variable	Test scores	Blood pressure	Birth weight
	Covariates	Sex, age	Age, illness severity	Sex

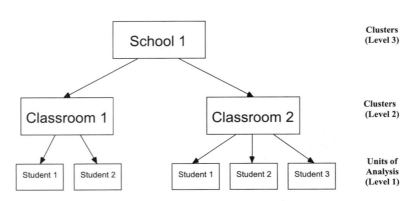

FIGURE 4.1: Nesting structure of a clustered three-level data set in an educational setting.

Chapter 3. Two-level, three-level, and higher-level models are generally referred to as **multilevel models**. Although multilevel models in general may have random effects associated with both the intercept and with other covariates, in this chapter we restrict our discussion to **random intercept models**, in which random effects at each level of clustering are associated with the intercept. The random effects associated with the clusters at Level 2 of a three-level data set are often referred to as **nested random effects**, because the Level 2 clusters are nested within the clusters at Level 3. In the absence of fixed effects associated with covariates, a three-level HLM is also known as a **variance components model**. The HLM software (Raudenbush et al., 2005; Raudenbush & Bryk, 2002), which we highlight in this chapter, was developed specifically for analyses involving these and related types of models.

4.2 The Classroom Study

4.2.1 Study Description

The Study of Instructional Improvement,[1] or SII (Anderson et al., 2009), was carried out by researchers at the University of Michigan to study the math achievement scores of first- and third-grade students in randomly selected classrooms from a national U.S. sample of elementary schools. In this example, we analyze data for 1,190 first-grade students sampled from 312 classrooms in 107 schools. The dependent variable, MATHGAIN, measures change in student math achievement scores from the spring of kindergarten to the spring of first grade.

The SII study design resulted in a three-level data set, in which students (Level 1) are nested within classrooms (Level 2), and classrooms are nested within schools (Level 3).

We examine the contribution of selected student-level, classroom-level and school-level covariates to the variation in students' math achievement gain. Although one of the original study objectives was to compare math achievement gain scores in schools participating in comprehensive school reforms (CSRs) to the gain scores from a set of matched comparison schools not participating in the CSRs, this comparison is not considered here.

A sample of the Classroom data is shown in Table 4.2. The first two rows in the table are included to distinguish between the different types of variables; in the actual electronic data set, the variable names are defined in the first row. Each row of data in Table 4.2 contains the school ID, classroom ID, student ID, the value of the dependent variable, and values of three selected covariates, one at each level of the data.

The layout of the data in the table reflects the hierarchical nature of the data set. For instance, the value of MATHPREP, a classroom-level covariate, is the same for all students in the same classroom, and the value of HOUSEPOV, a school-level covariate, is the same for all students within a given school. Values of student-level variables (e.g., the dependent variable, MATHGAIN, and the covariate, SEX) vary from student to student (row to row) in the data.

---

[1]Work on this study was supported by grants from the U.S. Department of Education to the Consortium for Policy Research in Education (CPRE) at the University of Pennsylvania (Grant # OERI-R308A60003), the National Science Foundation's Interagency Educational Research Initiative to the University of Michigan (Grant #s REC-9979863 & REC-0129421), the William and Flora Hewlett Foundation, and the Atlantic Philanthropies. Opinions expressed in this book are those of the authors and do not reflect the views of the U.S. Department of Education, the National Science Foundation, the William and Flora Hewlett Foundation, or the Atlantic Philanthropies.

TABLE 4.2: Sample of the Classroom Data Set

School (Level 3)		Classroom (Level 2)		Student (Level 1)		
Cluster ID	Covariate	Cluster ID	Covariate	Unit ID	Dependent Variable	Covariate
SCHOOLID	HOUSEPOV	CLASSID	MATHPREP	CHILDID	MATHGAIN	SEX
1	0.0817	160	2.00	1	32	1
1	0.0817	160	2.00	2	109	0
1	0.0817	160	2.00	3	56	1
1	0.0817	217	3.25	4	83	0
1	0.0817	217	3.25	5	53	0
1	0.0817	217	3.25	6	65	1
1	0.0817	217	3.25	7	51	0
1	0.0817	217	3.25	8	66	0
1	0.0817	217	3.25	9	88	1
1	0.0817	217	3.25	10	7	0
1	0.0817	217	3.25	11	60	0
2	0.0823	197	2.50	12	2	1
2	0.0823	197	2.50	13	101	0
2	0.0823	211	2.33	14	30	0
2	0.0823	211	2.33	15	65	0
...						

Note: "..." indicates portion of the data not displayed.

The following variables are considered in the analysis of the Classroom data:

School (Level 3) Variables

- **SCHOOLID** = School ID number[a]

- **HOUSEPOV** = Percentage of households in the neighborhood of the school below the poverty level

Classroom (Level 2) Variables

- **CLASSID** = Classroom ID number[a]

- **YEARSTEA**[b] = First-grade teacher's years of teaching experience

- **MATHPREP** = First-grade teacher's mathematics preparation: number of mathematics content and methods courses

- **MATHKNOW**[b] = First-grade teacher's mathematics content knowledge; based on a scale based composed of 30 items (higher values indicate higher content knowledge)

Student (Level 1) Variables

- **CHILDID** = Student ID number[a]

- **MATHGAIN** = Student's gain in math achievement score from the spring of kindergarten to the spring of first grade (the dependent variable)

- **MATHKIND**[b] = Student's math score in the spring of their kindergarten year

- **SEX** = Indicator variable (0 = boy, 1 = girl)

- **MINORITY**[b] = Indicator variable (0 = non-minority student, 1 = minority student)

- **SES**[b] = Student socioeconomic status

[a]The original ID numbers in the study were randomly reassigned for the data set used in this example.
[b]Not shown in Table 4.2.

4.2.2   Data Summary

The descriptive statistics and plots for the Classroom data presented in this section are obtained using the HLM software (Version 7). Syntax to carry out these descriptive analyses using the other software packages is included on the web page for the book (see Appendix A).

4.2.2.1   Data Set Preparation

To perform the analyses for the Classroom data set using the HLM software, we need to prepare three separate data sets:

1. The *Level 1 (student-level) Data Set* has one record per student, and contains variables measured at the student level, including the response variable, MATHGAIN, the student-level covariates, and the identifying variables: SCHOOLID, CLASSID, and CHILDID. The Level 1 data set is sorted in ascending order by SCHOOLID, CLASSID, and CHILDID.

2. The *Level 2 (classroom-level) Data Set* has one record per classroom, and contains classroom-level covariates, such as YEARSTEA and MATHKNOW, that are measured for the teacher. Each record contains the identifying variables SCHOOLID and CLASSID, and the records are sorted in ascending order by these variables.

3. The *Level 3 (school-level) Data Set* has one record per school, and contains school-level covariates, such as HOUSEPOV, and the identifying variable, SCHOOLID. The data set is sorted in ascending order by SCHOOLID.

The Level 1, Level 2, and Level 3 data sets can easily be derived from a single data set having the "long" structure illustrated in Table 4.2. All three data sets should be stored in a format readable by HLM, such as ASCII (i.e., raw data in text files), or a data set specific to a statistical software package. For ease of presentation, we assume that the three data sets for this analysis have been set up in SPSS format (Version 21+).

4.2.2.2   Preparing the Multivariate Data Matrix (MDM) File

We prepare two MDM files for this initial data summary. One includes only variables that do not have any missing values (i.e., it excludes MATHPREP and MATHKNOW), and consequently, all students ($n = 1190$) are included. The second MDM file contains all variables, including MATHPREP and MATHKNOW, and thus includes complete cases ($n = 1081$) only.

We create the first **Multivariate Data Matrix (MDM)** file using the Level 1, Level 2, and Level 3 data sets defined earlier. After starting HLM, locate the main menu, and click on **File**, **Make new MDM file**, and then **Stat package input**. In the dialog box that opens, select **HLM3** to fit a three-level hierarchical linear model with nested random effects, and click **OK**. In the next window, choose the **Input File Type** as SPSS/Windows.

In the **Level-1 Specification** area of this window, we select the Level 1 data set. **Browse** to the location of the student-level file. Click **Choose Variables**, and select the following variables: CLASSID (check "L2id," because this variable identifies clusters of students at Level 2 of the data), SCHOOLID (check "L3id," because this variable identifies clusters of classrooms at Level 3 of the data), MATHGAIN (check "in MDM," because this is the dependent variable), and the student-level covariates MATHKIND, SEX, MINORITY, and SES (check "in MDM" for all of these variables). Click **OK** when finished selecting these variables.

In the **Missing Data?** box, select "Yes," because some students may have missing values on some of the variables. Then choose **Delete missing level-1 data when: running analyses**, so that observations with missing data will be deleted from individual analyses and not from the MDM file.

In the **Level-2 Specification** area of this window, select the Level 2 (classroom-level) data set defined above. **Browse** to the location of the classroom-level data set. In the **Choose Variables** dialog box, choose the CLASSID variable (check "L2id") and the SCHOOLID variable (check "L3id"). In addition, select the classroom-level covariate that has complete data for all classrooms,[2] which is YEARSTEA (check "in MDM"). Click **OK** when finished selecting these variables.

In the **Level-3 Specification** area of this window, select the Level 3 (school-level) data set defined earlier. **Browse** to the location of the school-level data set, and choose the SCHOOLID variable (click on "L3id") and the HOUSEPOV variable, which has complete data for all schools (check "in MDM"). Click **OK** when finished.

Once all three data sets have been identified and the variables have been selected, go to the **MDM template file** portion of the window, where you will see a white box for an **MDM File Name** (with an .mdm suffix). Enter a name for the MDM file (such as classroom.mdm), including the .mdm suffix, and then click **Save mdmt file** to save an MDM Template file that can be used later when creating the second MDM file. Finally, click on **Make MDM** to create the MDM file using the three input files.

After HLM finishes processing the MDM file, click **Check Stats** to view descriptive statistics for the selected variables at each level of the data, as shown in the following HLM output:

---

[2] We do not select the classroom-level variables MATHPREP and MATHKNOW when setting up the first MDM file, because information on these two variables is missing for some classrooms, which would result in students from these classrooms being omitted from the initial data summary.

```
                      LEVEL-1 DESCRIPTIVE STATISTICS

   VARIABLE NAME      N       MEAN       SD      MINIMUM    MAXIMUM
     SEX            1190      0.51      0.50       0.00       1.00
     MINORITY       1190      0.68      0.47       0.00       1.00
     MATHKIND       1190    466.66     41.46     290.00     629.00
     MATHGAIN       1190     57.57     34.61    -110.00     253.00
     SES            1190     -0.01      0.74      -1.61       3.21


                      LEVEL-2 DESCRIPTIVE STATISTICS


   VARIABLE NAME      N       MEAN       SD      MINIMUM    MAXIMUM
   YEARSTEA           312     12.28      9.65      0.00      40.00


                      LEVEL-3 DESCRIPTIVE STATISTICS


   VARIABLE NAME      N       MEAN       SD      MINIMUM    MAXIMUM
   HOUSEPOV           107      0.19      0.14      0.01       0.56
```

Because none of the selected variables in the MDM file have any missing data, this is a descriptive summary for all 1190 students, within the 312 classrooms and 107 schools. We note that 51% of the 1190 students are female and that 68% of them are of minority status. The average number of years teaching for the 312 teachers is 12.28, and the mean proportion of households in poverty in the neighborhoods of the 107 schools is 0.19 (19%).

We now construct the second MDM file. After closing the window containing the descriptive statistics, click **Choose Variables** in the **Level-2 Specification** area. At this point, we add the MATHKNOW and MATHPREP variables to the MDM file by checking "in MDM" for each of these variables; click **OK** after selecting them. Then, save the .mdm file and the .mdmt template file under different names and click **Make MDM** to generate a new MDM file containing these additional Level 2 variables. After clicking **Check Stats**, the following output is displayed:

```
                      LEVEL-1 DESCRIPTIVE STATISTICS

   VARIABLE NAME      N       MEAN       SD      MINIMUM    MAXIMUM
       SEX          1081      0.50      0.50       0.00       1.00
     MINORITY       1081      0.67      0.47       0.00       1.00
     MATHKIND       1081    467.15     42.00     290.00     629.00
     MATHGAIN       1081     57.84     34.70     -84.00     253.00
       SES          1081     -0.01      0.75      -1.61       3.21


                      LEVEL-2 DESCRIPTIVE STATISTICS

   VARIABLE NAME      N       MEAN       SD      MINIMUM    MAXIMUM
     YEARSTEA        285     12.28      9.80       0.00      40.00
     MATHKNOW        285     -0.08      1.00      -2.50       2.61
     MATHPREP        285      2.58      0.96       1.00       6.00


                      LEVEL-3 DESCRIPTIVE STATISTICS


   VARIABLE NAME      N       MEAN       SD      MINIMUM    MAXIMUM
   HOUSEPOV           105      0.20      0.14      0.01       0.56
```

Note that 27 classrooms have been dropped from the Level 2 file because of missing data on the additional Level 2 variables, MATHKNOW and MATHPREP. Consequently, there were two schools (SCHOOLID = 48 and 58) omitted from the **Check Stats** summary, because none of the classrooms in these two schools had information on MATHKNOW and MATHPREP. This resulted in only 1081 students being analyzed in the data summary based on the second MDM file, which is 109 fewer than were included in the first summary. We close the window containing the descriptive statistics and click **Done** to proceed to the HLM model-building window.

In the model-building window we continue the data summary by examining box plots of the observed MATHGAIN responses for students in the selected classrooms from the first eight schools in the first (complete) MDM file. Select **File**, **Graph Data**, and **box-whisker plots**. To generate these box plots, select MATHGAIN as the **Y-axis** variable and select **Group at level-3** to generate a separate panel of box plots for each school. For **Number of groups**, we select **First ten groups** (corresponding to the first 10 schools). Finally, click **OK**. HLM produces box plots for the first 10 schools, and we display the plots for the first eight schools in Figure 4.2.

In Figure 4.2, we note evidence of both between-classroom and between-school variability in the MATHGAIN responses. We also see differences in the within-classroom variability, which may be explained by student-level covariates. By clicking **Graph Settings** in the HLM graph window, one can select additional Level 2 (e.g., YEARSTEA) or Level 3 (e.g., HOUSEPOV) variables as **Z-focus** variables that color-code box plots, based on values of the classroom- and school-level covariates. Readers should refer to the HLM manual for more information on additional graphing features.

4.3   Overview of the Classroom Data Analysis

For the analysis of the Classroom data, we follow the "step-up" modeling strategy outlined in Subsection 2.7.2. This approach differs from the strategy used in Chapter 3, in that it starts with a simple model, containing only a single fixed effect (the overall intercept), random effects associated with the intercept for classrooms and schools, and residuals, and then builds the model by adding fixed effects of covariates measured at the various levels.

In Subsection 4.3.1 we outline the analysis steps, and informally introduce related models and hypotheses to be tested. In Subsection 4.3.2 we present the specification of selected models that will be fitted to the Classroom data, and in Subsection 4.3.3 we detail the hypotheses tested in the analysis. To follow the analysis steps outlined in this section, we refer readers to the schematic diagram presented in Figure 4.3.

4.3.1   Analysis Steps

Step 1: Fit the initial "unconditional" (variance components) model (Model 4.1).

Fit a three-level model with a fixed intercept, and random effects associated with the intercept for classrooms (Level 2) and schools (Level 3), and decide whether to keep the random intercepts for classrooms (Model 4.1 vs. Model 4.1A).

Because Model 4.1 does not include fixed effects associated with any covariates, it is referred to as a "means-only" or "unconditional" model in the HLM literature and is known as a variance components model in the classical ANOVA context. We include the term

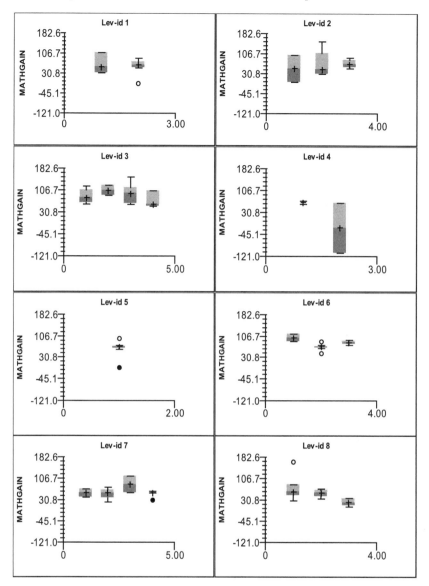

FIGURE 4.2: Box plots of the MATHGAIN responses for students in the selected classrooms in the first eight schools for the Classroom data set.

"unconditional" in quotes because it indicates a model that is not conditioned on any fixed effects other than the intercept, although it is still conditional on the random effects. The model includes a fixed overall intercept, random effects associated with the intercept for classrooms within schools, and random effects associated with the intercept for schools.

After fitting Model 4.1, we obtain estimates of the initial variance components, i.e., the variances of the random effects at the school level and the classroom level, and the residual variance at the student level. We use the variance component estimates from the Model 4.1 fit to estimate intraclass correlation coefficients (ICCs) of MATHGAIN responses at the school level and at the classroom level (see Section 4.8).

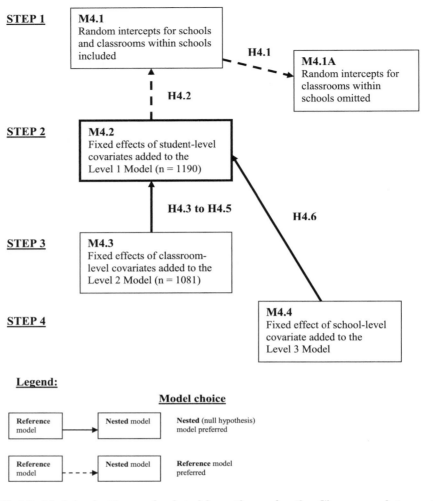

FIGURE 4.3: Model selection and related hypotheses for the Classroom data analysis.

We also test Hypothesis 4.1 to decide whether the random effects associated with the intercepts for classrooms nested within schools can be omitted from Model 4.1. We fit a model without the random classroom effects (Model 4.1A) and perform a likelihood ratio test. We decide to retain the random intercepts associated with classrooms nested within schools, and we also retain the random intercepts associated with schools in all subsequent models, to preserve the hierarchical structure of the data. Model 4.1 is preferred at this stage of the analysis.

Step 2: Build the Level 1 Model by adding Level 1 Covariates (Model 4.1 vs. Model 4.2).

Add fixed effects associated with covariates measured on the students to the Level 1 Model to obtain Model 4.2, and evaluate the reduction in the residual variance.

In this step, we add fixed effects associated with the four student-level covariates (MATHKIND, SEX, MINORITY, and SES) to Model 4.1 and obtain Model 4.2. We informally assess the related reduction in the between-student variance (i.e., the residual variance).

We also test Hypothesis 4.2, using a likelihood ratio test, to decide whether we should add the fixed effects associated with all of the student-level covariates to Model 4.1. We decide to add these fixed effects and choose Model 4.2 as our preferred model at this stage of the analysis.

Step 3: Build the Level 2 Model by adding Level 2 covariates (Model 4.3).

Add fixed effects associated with the covariates measured on the Level 2 clusters (classrooms) to the Level 2 model to create Model 4.3, and decide whether to retain the effects of the Level 2 covariates in the model.

We add fixed effects associated with the three classroom-level covariates (YEARSTEA, MATHPREP, and MATHKNOW) to Model 4.2 to obtain Model 4.3. At this point, we would like to assess whether the Level 2 component of variance (i.e., the variance of the nested random classroom effects) is reduced when we include the effects of these Level 2 covariates in the model. However, Models 4.2 and 4.3 are fitted using different sets of observations, owing to the missing values on the MATHPREP and MATHKNOW covariates, and a simple comparison of the classroom-level variance components obtained from these two models is therefore not appropriate.

We also test Hypotheses 4.3, 4.4, and 4.5, to decide whether we should keep the fixed effects associated with YEARSTEA, MATHPREP, and MATHKNOW in Model 4.3, using individual t-tests for each hypothesis. Based on the results of the t-tests, we decide that none of the fixed effects associated with these Level 2 covariates should be retained and choose Model 4.2 as our preferred model at this stage. We do not use a likelihood ratio test for the fixed effects of all Level 2 covariates at once, as we did for all Level 1 covariates in Step 2, because different sets of observations were used to fit Models 4.2 and 4.3. A likelihood ratio test is only possible if both models were fitted using the same set of observations. In Subsection 4.11.4, we illustrate syntax that could be used to construct a "complete case" data set in each software package.

Step 4: Build the Level 3 Model by adding the Level 3 covariate (Model 4.4).

Add a fixed effect associated with the covariate measured on the Level 3 clusters (schools) to the Level 3 model to create Model 4.4, and evaluate the reduction in the variance component associated with the Level 3 clusters.

In this last step, we add a fixed effect associated with the only school-level covariate, HOUSEPOV, to Model 4.2 and obtain Model 4.4. We assess whether the variance component at the school level (i.e., the variance of the random school effects) is reduced when we include this fixed effect at Level 3 of the model. Because the same set of observations was used to fit Models 4.2 and 4.4, we informally assess the relative reduction in the between-school variance component in this step.

We also test Hypothesis 4.6 to decide whether we should add the fixed effect associated with the school-level covariate to Model 4.2. Based on the result of a t-test for the fixed effect of HOUSEPOV in Model 4.4, we decide not to add this fixed effect, and choose Model 4.2 as our final model. We consider diagnostics for Model 4.2 in Section 4.9, using residual files generated by the HLM software.

Figure 4.3 provides a schematic guide to the model selection process and hypotheses considered in the analysis of the Classroom data. See Subsection 3.3.1 for a detailed interpretation of the elements of this figure. Table 4.3 provides a summary of the various models considered in the Classroom data analyses.

4.3.2   Model Specification

4.3.2.1   General Model Specification

We specify Model 4.3 in (4.1), because it is the model with the most fixed effects that we consider in the analysis of the Classroom data. Models 4.1, 4.1A, and 4.2 are simplifications of this more general model. Selected models are summarized in Table 4.3.

The general specification for Model 4.3 is:

$$
\begin{aligned}
\mathrm{MATHGAIN}_{ijk} = \quad & \left. \begin{aligned} & \beta_0 + \beta_1 \times \mathrm{MATHKIND}_{ijk} + \beta_2 \times \mathrm{SEX}_{ijk} \\ & + \beta_3 \times \mathrm{MINORITY}_{ijk} + \beta_4 \times \mathrm{SES}_{ijk} + \beta_5 \times \mathrm{YEARSTEA}_{jk} \\ & + \beta_6 \times \mathrm{MATHPREP}_{jk} + \beta_7 \times \mathrm{MATHKNOW}_{jk} \end{aligned} \right\} \text{fixed} \\
& \qquad\qquad\qquad\quad \left. + u_k + u_{j|k} + \varepsilon_{ijk} \quad\right\} \text{random}
\end{aligned}
$$
$$(4.1)$$

In this specification, $\mathrm{MATHGAIN}_{ijk}$ represents the value of the dependent variable for student i in classroom j nested within school k; β_0 through β_7 represent the fixed intercept and the fixed effects of the covariates (e.g., MATHKIND,…, MATHKNOW); u_k is the random effect associated with the intercept for school k; $u_{j|k}$ is the random effect associated with the intercept for classroom j within school k; and ε_{ijk} represents the residual. To obtain Model 4.4, we add the fixed effect (β_8) of the school-level covariate HOUSEPOV and omit the fixed effects of the classroom-level covariates (β_5 through β_7).

The distribution of the random effects associated with the schools in Model 4.3 is written as

$$u_k \sim \mathcal{N}(0, \sigma^2_{int:school})$$

where $\sigma^2_{int:school}$ represents the variance of the school-specific random intercepts.

The distribution of the random effects associated with classrooms nested within a given school is

$$u_{j|k} \sim \mathcal{N}(0, \sigma^2_{int:classroom})$$

where $\sigma^2_{int:classroom}$ represents the variance of the random classroom-specific intercepts at any given school. This between-classroom variance is assumed to be constant for all schools.

The distribution of the residuals associated with the student-level observations is

$$\varepsilon_{ijk} \sim \mathcal{N}(0, \sigma^2)$$

where σ^2 represents the residual variance.

We assume that the random effects, u_k, associated with schools, the random effects, $u_{j|k}$, associated with classrooms nested within schools, and the residuals, ε_{ijk}, are all mutually independent.

The general specification of Model 4.3 corresponds closely to the syntax that is used to fit the model in SAS, SPSS, Stata, and R. In Subsection 4.3.2.2 we provide the hierarchical specification that more closely corresponds to the HLM setup of Model 4.3.

4.3.2.2   Hierarchical Model Specification

We now present a hierarchical specification of Model 4.3. The following model is in the form used in the HLM software, but employs the notation used in the general specification of Model 4.3 in (4.1), rather than the HLM notation. The correspondence between the notation used in this section and that used in the HLM software is shown in Table 4.3.

TABLE 4.3: Summary of Selected Models Considered for the Classroom Data

		Term/Variable	General Notation	HLM Notation	Model 4.1	Model 4.2	Model 4.3	Model 4.4
Fixed effects		**Intercept**	β_0	γ_{000}	√	√	√	√
		MATHKIND	β_1	γ_{300}		√	√	√
		SEX	β_2	γ_{100}		√	√	√
		MINORITY	β_3	γ_{200}		√	√	√
		SES	β_4	γ_{400}		√	√	√
		YEARSTEA	β_5	γ_{010}			√	
		MATHPREP	β_6	γ_{030}			√	
		MATHKNOW	β_7	γ_{020}			√	
		HOUSEPOV	β_8	γ_{001}				√
Random effects	**Classroom (j)**	**Intercept**	$u_{j\|k}$	r_{jk}	√	√	√	√
	School (k)	**Intercept**	u_k	u_{00k}	√	√	√	√
Residuals	**Student (i)**		ε_{ijk}	e_{ijk}	√	√	√	√
Covariance parameters ($\boldsymbol{\theta}_D$) for $\boldsymbol{D}$ matrix	**Classroom level**	**Variance of intercepts**	$\sigma^2_{int:class}$	τ_π	√	√	√	√
	School level	**Variance of intercepts**	$\sigma^2_{int:school}$	τ_β	√	√	√	√
Covariance parameters (θ_R) for R_i matrix	**Student level**	**Residual variance**	σ^2	σ^2	√	√	√	√

The hierarchical model has three components, reflecting contributions from the three levels of data shown in Table 4.2. First, we write the **Level 1** component as

Level 1 Model (Student)

$$\text{MATHGAIN}_{ijk} = b_{0j|k} + \beta_1 \times \text{MATHKIND}_{ijk} + \beta_2 \times \text{SEX}_{ijk}$$
$$+ \beta_3 \times \text{MINORITY}_{ijk} + \beta_4 \times \text{SES}_{ijk} + \varepsilon_{ijk} \qquad (4.2)$$

where

$$\varepsilon_{ijk} \sim \mathcal{N}(0, \sigma^2)$$

The **Level 1** model (4.2) shows that at the student level of the data, we have a set of simple classroom-specific linear regressions of MATHGAIN on the student-level covariates. The unobserved classroom-specific intercepts, $b_{0j|k}$, are related to several other fixed and random effects at the classroom level, and are defined in the following **Level 2** model.

The **Level 2** model for the classroom-specific intercepts can be written as

Level 2 Model (Classroom)

$$b_{0j|k} = b_{0k} + \beta_5 \times \text{YEARSTEA}_{jk} + \beta_6 \times \text{MATHPREP}_{jk}$$
$$+ \beta_7 \times \text{MATHKNOW}_{jk} + u_{j|k} \qquad (4.3)$$

where

$$u_{j|k} \sim \mathcal{N}(0, \sigma^2_{int:classroom})$$

The **Level 2** model (4.3) assumes that the intercept, $b_{0j|k}$, for classroom j nested within school k, depends on the unobserved intercept specific to the k-th school, b_{0k}, the classroom-specific covariates associated with the teacher for that classroom (YEARSTEA, MATH-PREP, and MATHKNOW), and a random effect, $u_{j|k}$, associated with classroom j within school k.

The **Level 3** model for the school-specific intercepts in Model 4.3 is:

Level 3 Model (School)

$$b_{0k} = \beta_0 + u_k \qquad (4.4)$$

where

$$u_k \sim \mathcal{N}(0, \sigma^2_{int:school})$$

The **Level 3** model in (4.4) shows that the school-specific intercept in Model 4.3 depends on the overall fixed intercept, β_0, and the random effect, u_k, associated with the intercept for school k.

By substituting the expression for b_{0k} from the Level 3 model into the Level 2 model, and then substituting the resulting expression for $b_{0j|k}$ from the Level 2 model into the Level 1 model, we recover Model 4.3 as it was specified in (4.1).

We specify Model 4.4 by omitting the fixed effects of the classroom-level covariates from Model 4.3, and adding the fixed effect of the school-level covariate, HOUSEPOV, to the **Level 3** model.

4.3.3   Hypothesis Tests

Hypothesis tests considered in the analysis of the Classroom data are summarized in Table 4.4.

TABLE 4.4: Summary of Hypotheses Tested in the Classroom Analysis

| | **Hypothesis Specification** | | | **Hypothesis Test** | | | |
| | | | | **Models Compared** | | | |
Label	**Null (H_0)**	**Alternative (H_A)**	**Test**	**Nested Model (H_0)**	**Ref. Model (H_A)**	**Est. Method**	**Test Stat. Dist. under H_0**
4.1	Drop $u_{j|k}$ ($\sigma^2_{int:classroom} = 0$)	Retain $u_{j|k}$ ($\sigma^2_{int:classroom} > 0$)	LRT	Model 4.1A	Model 4.1	REML	$0.5\chi^2_0 + 0.5\chi^2_1$
4.2	Fixed effects of student-level covariates are all zero ($\beta_1 = \beta_2 = \beta_3 = \beta_4 = 0$)	At least one fixed effect at the student level is different from zero	LRT	Model 4.1	Model 4.2	ML	χ^2_4
4.3	Fixed effect of YEARSTEA is zero ($\beta_5 = 0$)	$\beta_5 \neq 0$	t-test	N/A	Model 4.3	REML/ML	t^a_{177}
4.4	Fixed effect of MATHPREP is zero ($\beta_6 = 0$)	$\beta_6 \neq 0$	t-test	N/A	Model 4.3	REML/ML	t^a_{177}
4.5	Fixed effect of MATHKNOW is zero ($\beta_7 = 0$)	$\beta_7 \neq 0$	t-test	N/A	Model 4.3	REML/ML	t^a_{177}
4.6	Fixed effect of HOUSEPOV is zero ($\beta_8 = 0$)	$\beta_8 \neq 0$	t-test	N/A	Model 4.4	REML/ML	t^a_{105}

[a]Degrees of freedom for the t-statistics are those reported by the HLM3 procedure.

Hypothesis 4.1. The random effects associated with the intercepts for classrooms nested within schools can be omitted from Model 4.1.

We do not directly test the significance of the random classroom-specific intercepts, but rather test null and alternative hypotheses about the *variance* of the classroom-specific intercepts. The null and alternative hypotheses are:

$$H_0 : \sigma^2_{int:classroom} = 0$$
$$H_A : \sigma^2_{int:classroom} > 0$$

We use a REML-based likelihood ratio test for Hypothesis 4.1 in SAS, SPSS, R, and Stata. The test statistic is calculated by subtracting the –2 REML log-likelihood for Model 4.1 (the reference model, including the nested random classroom effects) from the corresponding value for Model 4.1A (the nested model). To obtain a p-value for this test statistic, we refer it to a mixture of χ^2 distributions, with 0 and 1 degrees of freedom and equal weight 0.5.

The HLM3 procedure does not use REML estimation and is not able to fit a model without any random effects at a given level of the model, such as Model 4.1A. Therefore, an LRT cannot be performed, so we consider an alternative chi-square test for Hypothesis 4.1 provided by HLM in Subsection 4.7.2.

We decide that the random effects associated with the intercepts for classrooms nested within schools should be retained in Model 4.1. We do not explicitly test the variance of the random school-specific intercepts, but we retain them in Model 4.1 and all subsequent models to reflect the hierarchical structure of the data.

Hypothesis 4.2. The fixed effects associated with the four student-level covariates should be added to Model 4.1.

The null and alternative hypotheses for the fixed effects associated with the student-level covariates, MATHKIND, SEX, MINORITY, and SES, are:

$$H_0 : \beta_1 = \beta_2 = \beta_3 = \beta_4 = 0$$
$$H_A : \text{At least one fixed effect is not equal to zero}$$

We test Hypothesis 4.2 using a likelihood ratio test, based on maximum likelihood (ML) estimation. The test statistic is calculated by subtracting the –2 ML log-likelihood for Model 4.2 (the reference model with the fixed effects of all student-level covariates included) from that for Model 4.1 (the nested model). Under the null hypothesis, the distribution of this test statistic is asymptotically a χ^2 with 4 degrees of freedom.

We decide that the fixed effects associated with the Level 1 covariates should be added and select Model 4.2 as our preferred model at this stage of the analysis.

Hypotheses 4.3, 4.4, and 4.5. The fixed effects associated with the classroom-level covariates should be retained in Model 4.3.

The null and alternative hypotheses for the fixed effects associated with the classroom-level covariates YEARSTEA, MATHPREP, and MATHKNOW are written as follows:

- **Hypothesis 4.3** for the fixed effect associated with YEARSTEA:

$$H_0 : \beta_5 = 0$$
$$H_A : \beta_5 \neq 0$$

- **Hypothesis 4.4** for the fixed effect associated with MATHPREP:

$$H_0 : \beta_6 = 0$$
$$H_A : \beta_6 \neq 0$$

- **Hypothesis 4.5** for the fixed effect associated with MATHKNOW:

$$H_0 : \beta_7 = 0$$
$$H_A : \beta_7 \neq 0$$

We test each of these hypotheses using *t*-tests based on the fit of Model 4.3. We decide that none of the fixed effects associated with the classroom-level covariates should be retained, and keep Model 4.2 as our preferred model at this stage of the analysis.

Because there are classrooms with missing values for the classroom-level covariate MATHKNOW, there are different sets of observations used to fit Models 4.2 and 4.3. In this case, we cannot use a likelihood ratio test to decide if we should keep the fixed effects of all covariates at Level 2 of the model, as we did for the fixed effects of the covariates at Level 1 of the model when we tested Hypothesis 4.2.

To perform a likelihood ratio test for the effects of all of the classroom-level covariates, we would need to fit both Models 4.2 and 4.3 using the same cases (i.e., those cases with no missing data on all covariates would need to be considered for both models). See Subsection 4.11.4 for a discussion of how to set up the necessary data set, with complete data for all covariates, in each of the software procedures.

Hypothesis 4.6. The fixed effect associated with the school-level covariate HOUSEPOV should be added to Model 4.2.

The null and alternative hypotheses for the fixed effect associated with HOUSEPOV are specified as follows:

$$H_0 : \beta_8 = 0$$
$$H_A : \beta_8 \neq 0$$

We test Hypothesis 4.6 using a *t*-test for the significance of the fixed effect of HOUSE-POV in Model 4.4. We decide that the fixed effect of this school-level covariate should not be added to the model, and choose Model 4.2 as our final model. For the results of these hypothesis tests, see Section 4.5.

4.4 Analysis Steps in the Software Procedures

We compare results for selected models across the software procedures in Section 4.6.

4.4.1 SAS

We begin by reading the comma-delimited data file, classroom.csv (assumed to have the "long" data structure displayed in Table 4.2, and located in the `C:\temp` directory) into a temporary SAS data set named `classroom`:

```
proc import out = WORK.classroom
   datafile = "C:\temp\classroom.csv";
   dbms = csv replace;
   getnames = YES;
   datarow = 2;
   guessingrows = 20;
run;
```

Step 1: Fit the initial "unconditional" (variance components) model (Model 4.1), and decide whether to omit the random classroom effects (Model 4.1 vs. Model 4.1A).

Prior to fitting the models in SAS, we sort the data set by SCHOOLID and CLASSID. Although not necessary for fitting the models, this sorting makes it easier to interpret elements in the marginal variance-covariance and correlation matrices that we display later.

```
proc sort data = classroom;
   by schoolid classid;
run;
```

The SAS code for fitting Model 4.1 using REML estimation is shown below. Because the only fixed effect in Model 4.1 is the intercept, we use the `model` statement with no covariates on the right-hand side of the equal sign; the intercept is included by default. The two `random` statements specify that a random intercept should be included for each school and for each classroom nested within a school.

```
title "Model 4.1";
proc mixed data = classroom noclprint covtest;
   class classid schoolid;
   model mathgain = / solution;
   random intercept / subject = schoolid v vcorr;
   random intercept / subject = classid(schoolid);
run;
```

We specify two options in the `proc mixed` statement. The `noclprint` option suppresses listing of levels of the `class` variables in the output to save space; we recommend using this option only after an initial examination of the levels of the `class` variables to be sure the data set is being read correctly. The `covtest` option is specified to request that the standard errors of the estimated variance components be included in the output. The Wald tests of covariance parameters that are reported when the `covtest` option is specified should not be used to test hypotheses about the variance components.

The `solution` option in the `model` statement causes SAS to print estimates of the fixed-effect parameters included in the model, which for Model 4.1 is simply the overall intercept.

We use the `v` option to request that a single block (corresponding to the first school) of the estimated V matrix for the marginal model implied by Model 4.1 be displayed in the output (see Subsection 2.2.3 for more details on the V and Z matrices in SAS). In addition, we specify the `vcorr` option to request that the corresponding marginal correlation matrix be displayed. The `v` and `vcorr` options can be added to one or both of the random statements, with the same results. We display the SAS output from these commands in Section 4.8.

> **Software Note:** When the `proc mixed` syntax above is used, the overall subject is considered by SAS to be SCHOOLID. Because CLASSID is specified in the `class` statement, SAS sets up the same number of columns in the Z matrix for each school. The SAS output below, obtained by fitting Model 4.1 to the Classroom data, indicates

that there are 107 subjects included in this model (corresponding to the 107 schools). SAS sets up 10 columns in the Z matrix for each school, which corresponds to the maximum number of classrooms in any given school (9) plus a single column for the school. The maximum number of students in any school is 31.

```
                  Dimensions

    Covariance Parameters        3
    Columns in X                 1
    Columns in Z Per Subject     10
    Subjects                     107
    Max Obs Per Subject          31
```

This syntax for Model 4.1 allows SAS to fit the model efficiently, by taking blocks of the marginal V matrix into account. However, it also means that if EBLUPs are requested by specifying the `solution` option in the corresponding `random` statements, as illustrated below, SAS will display output for 10 classrooms within each school, even for schools that have fewer than 10 classrooms. This results in extra unwanted output being generated. We present alternative syntax to avoid this potential problem below.

Alternative syntax for Model 4.1 may be specified by omitting the `subject =` option in the `random` statement for SCHOOLID. By including the `solution` option in one or both `random` statements, we obtain the predicted values of the EBLUPs for each school, followed by the EBLUPs for the classrooms, with no extra output being produced.

```
title "Model 4.1: Alternative Syntax";
proc mixed data = classroom noclprint covtest;
   class classid schoolid;
   model mathgain = / solution;
   random schoolid / solution;
   random int / subject = classid(schoolid);
run;
```

The resulting "Dimensions" output for this model specification is as follows:

```
                  Dimensions

    Covariance Parameters        3
    Columns in x                 1
    Columns in Z Per Subject     419
    Subjects                     1
    Max obs Per Subject          1190
```

SAS now considers there to be only one subject in this model, even though we specified a `subject =` option in the second `random` statement. In this case, the Z matrix has 419 columns, which correspond to the total number of schools (107) plus classrooms (312).

A portion of the resulting output for the EBLUPs is as follows:

```
                        Solution for Random Effects
-------------------------------------------------------------------------
                                         Std Err
    Effect    Classid  Schoolid  Estimate  Pred      DF    t Value   pr > |t|
-------------------------------------------------------------------------
    schoolid             1        0.94110  7.1657    878    0.13     0.8955
    schoolid             2        2.54710  7.0625    878    0.36     0.7184
    schoolid             3       13.67300  6.6390    878    2.06     0.0397

    ...
    Intercept   160      1        1.63800  8.9188    878    0.18     0.8543
    Intercept   217      1       -0.43260  8.1149    878   -0.05     0.9575
    Intercept   197      2       -1.69390  9.1905    878   -0.18     0.8538
    Intercept   211      2        2.51290  8.6881    878    0.29     0.7725
    Intercept   307      2        2.44330  8.6881    878    0.28     0.7786
    Intercept    11      3        3.86990  8.6607    878    0.45     0.6551
    Intercept   137      3        5.56300  9.1818    878    0.61     0.5448
    Intercept   145      3        8.10160  8.4622    878    0.96     0.3386
    Intercept   228      3       -0.02247  8.8971    878   -0.00     0.9980
```

Note that the standard error, along with a t-test, is displayed for each EBLUP. The test statistic is calculated by dividing the predicted EBLUP by its estimated standard error, with the degrees of freedom being the same as indicated by the `ddfm =` option. If no `ddfm =` option is specified, SAS will use the default "containment" method (see Subsection 3.11.6 for a discussion of different methods of calculating denominator degrees of freedom in SAS). Although it may not be of particular interest to test whether the predicted random effect for a given school or classroom is equal to zero, large values of the t-statistic may indicate an outlying value for a given school or classroom.

To suppress the display of the EBLUPs in the output, but obtain them in a SAS data set, we use the following ODS statements prior to invoking `proc mixed`:

```
ods exclude solutionr;
ods output solutionr=eblupdat;
```

The first `ods` statement prevents the EBLUPs from being displayed in the output. The second `ods` statement requests that the EBLUPs be placed in a new SAS data set named `eblupdat`. The distributions of the EBLUPs contained in the `eblupdat` data set can be investigated graphically to check for possible outliers (see Subsection 4.10.1 for diagnostic plots of the EBLUPs associated with Model 4.2). Note that the `eblupdat` data set is only created if the `solution` option is added to either of the `random` statements.

We recommend that readers be cautious about including options, such as `v` and `vcorr`, for displaying the marginal variance-covariance and correlation matrices when using random statements with no `subject =` option, as shown in the first `random` statement in the alternative syntax for Model 4.1. In this case, SAS will display matrices that are of the same dimension as the total number of observations in the entire data set. In the case of the Classroom data, this would result in a 1190×1190 matrix being displayed for both the marginal variance-covariance and correlation matrices.

We now fit Model 4.1A using REML estimation (the default), so that we can perform a likelihood ratio test of Hypothesis 4.1 to decide if we need the nested random classroom effects in Model 4.1. We omit the random classroom-specific intercepts in Model 4.1A by adding an asterisk at the beginning of the second `random` statement, to comment it out:

```
title "Model 4.1A";
proc mixed data = classroom noclprint covtest;
   class classid schoolid;
   model mathgain = / solution;
   random intercept / subject = schoolid;
   *random intercept / subject = classid(schoolid);
run;
```

Results of the REML-based likelihood ratio test for Hypothesis 4.1 are discussed in detail in Subsection 4.5.1.

We decide to retain the nested random classroom effects in Model 4.1 based on the significant ($p < 0.001$) result of this test. We also keep the random school-specific intercepts in the model to reflect the hierarchical structure of the data.

Step 2: Build the Level 1 Model by adding Level 1 Covariates (Model 4.1 vs. Model 4.2).

To fit Model 4.2, again using REML estimation, we modify the `model` statement to add the fixed effects of the student-level covariates MATHKIND, SEX, MINORITY, and SES:

```
title "Model 4.2";
proc mixed data = classroom noclprint covtest;
   class classid schoolid;
   model mathgain = mathkind sex minority ses / solution;
   random intercept / subject = schoolid;
   random intercept / subject = classid(schoolid);
run;
```

Because the covariates SEX and MINORITY are indicator variables, having values of 0 and 1, they do not need to be identified as categorical variables in the `class` statement.

We formally test Hypothesis 4.2 (whether any of the fixed effects associated with the student-level covariates are different from zero) by performing an ML-based likelihood ratio test, subtracting the –2 ML log-likelihood of Model 4.2 (the reference model) from that of Model 4.1 (the nested model, excluding the four fixed effects being tested), and referring the difference to a χ^2 distribution with 4 degrees of freedom. Note that the `method = ML` option is used to request maximum likelihood estimation of both models (see Subsection 2.6.2.1 for a discussion of likelihood ratio tests for fixed effects):

```
title "Model 4.1: ML Estimation";
proc mixed data = classroom noclprint covtest method = ML;
   class classid schoolid;
   model mathgain = / solution;
   random intercept / subject = schoolid;
   random intercept / subject = classid(schoolid);
run;
```

```
title "Model 4.2: ML Estimation";
proc mixed data = classroom noclprint covtest method = ML;
   class classid schoolid;
   model mathgain = mathkind sex minority ses / solution;
   random intercept / subject = schoolid;
   random intercept / subject = classid(schoolid);
run;
```

We reject the null hypothesis that all fixed effects associated with the student-level covariates are equal to zero, based on the result of this test ($p < 0.001$), and choose Model 4.2 as our preferred model at this stage of the analysis.

The significant result for the test of Hypothesis 4.2 suggests that the fixed effects of the student-level covariates explain at least some of the variation at the student level of the data. Informal examination of the estimated residual variance in Model 4.2 suggests that the Level 1 (student-level) residual variance is substantially reduced compared to that for Model 4.1 (see Section 4.6).

Step 3: Build the Level 2 Model by adding Level 2 Covariates (Model 4.3).

To fit Model 4.3, we add the fixed effects of the classroom-level covariates YEARSTEA, MATHPREP, and MATHKNOW to the `model` statement that we specified for Model 4.2:

```
title "Model 4.3";
proc mixed data = classroom noclprint covtest;
   class classid schoolid;
   model mathgain = mathkind sex minority ses yearstea mathprep
      mathknow / solution;
   random intercept / subject = schoolid;
   random intercept / subject = classid(schoolid);
run;
```

We consider t-tests of Hypotheses 4.3, 4.4, and 4.5, for the individual fixed effects of the classroom-level covariates added to the model in this step. The results of these t-tests indicate that none of the fixed effects associated with the classroom-level covariates are significant, so we keep Model 4.2 as the preferred model at this stage of the analysis. Note that results of the t-tests are based on an analysis with 109 observations omitted.

Step 4: Build the Level 3 Model by adding the Level 3 Covariate (Model 4.4).

To fit Model 4.4, we add the fixed effect of the school-level covariate, HOUSEPOV, to the `model` statement for Model 4.2:

```
title "Model 4.4";
proc mixed data = classroom noclprint covtest;
   class classid schoolid;
   model mathgain = mathkind sex minority ses housepov / solution;
   random intercept / subject = schoolid;
   random intercept / subject = classid(schoolid);
run;
```

To test Hypothesis 4.6, we carry out a t-test for the fixed effect of HOUSEPOV, based on the REML fit of Model 4.4. The result of the t-test indicates that the fixed effect for this school-level covariate is not significant ($p = 0.25$). We also note that the estimated variance of the random school effects in Model 4.4 has not been reduced compared to that of Model 4.2. We therefore do not retain the fixed effect of the school-level covariate and choose Model 4.2 as our final model.

We now refit our final model, using REML estimation:

```
title "Model 4.2 (Final)";
proc mixed data = classroom noclprint covtest;
   class classid schoolid;
```

```
      model mathgain = mathkind sex minority ses / solution outpred = pdatl;
      random intercept / subject = schoolid solution;
      random intercept / subject = classid(schoolid);
   run;
```

The data set `pdat1` (requested with the `outpred =` option in the `model` statement) contains the conditional predicted values for each student (based on the estimated fixed effects in the model, and the EBLUPs for the random school and classroom effects) and the conditional residuals for each observation. This data set can be used to visually assess model diagnostics (see Subsection 4.10.2).

4.4.2 SPSS

We first import the comma-delimited data file named classroom.csv, which has the "long" format displayed in Table 4.2, using the following SPSS syntax:

```
GET DATA /TYPE = TXT
  /FILE = "C:\temp\classroom.csv"
  /DELCASE = LINE
  /DELIMITERS = ","
  /ARRANGEMENT = DELIMITED
  /FIRSTCASE = 2
  /IMPORTCASE = ALL
  /VARIABLES =
    sex F1.0
    minority F1.0
    mathkind F3.2
    mathgain F4.2
    ses F5.2
    yearstea F5.2
    mathknow F5.2
    housepov F5.2
    mathprep F4.2
    classid F3.2
    schoolid F1.0
    childid F2.1
    .
  CACHE.
  EXECUTE.
```

Now that we have a data set in SPSS in the "long" form appropriate for fitting linear mixed models, we proceed with the analysis.

Step 1: Fit the initial "unconditional" (variance components) model (Model 4.1), and decide whether to omit the random classroom effects (Model 4.1 vs. Model 4.1A).

There are two ways to set up a linear mixed model with nested random effects using SPSS syntax. We begin by illustrating how to set up Model 4.1 without specifying any "subjects" in the `RANDOM` subcommands:

```
* Model 4.1 (more efficient syntax).
MIXED
mathgain BY classid schoolid
  /CRITERIA = CIN(95) MXITER(100) MXSTEP(5) SCORING(1)
    SINGULAR(0.000000000001) HCONVERGE(0, ABSOLUTE) LCONVERGE (0, ABSOLUTE)
    PCONVERGE(0.000001, ABSOLUTE)
  /FIXED = | SSTYPE(3)
  /METHOD = REML
  /PRINT = SOLUTION
  /RANDOM classid(schoolid) | COVTYPE(VC)
  /RANDOM schoolid | COVTYPE(VC) .
```

The first variable listed after invoking the MIXED command is the dependent variable, MATHGAIN. The two random factors (CLASSID and SCHOOLID) have been declared as categorical factors by specifying them after the BY keyword. The CRITERIA subcommand specifies the estimation criteria to be used when fitting the model, which are the defaults.

The FIXED subcommand has no variables on the right-hand side of the equal sign before the vertical bar (|), which indicates that an intercept-only model is requested (i.e., the only fixed effect in the model is the overall intercept). The SSTYPE(3) option after the vertical bar specifies that SPSS should use the default "Type 3" analysis, in which the tests for the fixed effects are adjusted for all other fixed effects in the model. This option is not critical for Model 4.1, because the only fixed effect in this model is the intercept.

We use the default REML estimation method by specifying REML in the METHOD subcommand.

The PRINT subcommand specifies that the printed output should include the estimated fixed effects (SOLUTION).

Note that there are two RANDOM subcommands: the first identifies CLASSID nested within SCHOOLID as a random factor, and the second specifies SCHOOLID as a random factor. No "subject" variables are specified in either of these RANDOM subcommands. The COVTYPE(VC) option specified after the vertical bar indicates that a variance components (VC) covariance structure for the random effects is desired.

In the following syntax, we show an alternative specification of Model 4.1 that is less efficient computationally for larger data sets. We specify INTERCEPT before the vertical bar and a SUBJECT variable after the vertical bar for each RANDOM subcommand. This syntax means that for each level of the SUBJECT variables, we add a random effect to the model associated with the INTERCEPT:

```
* Model 4.1 (less efficient syntax).
MIXED
mathgain
  /CRITERIA = CIN(95) MXITER(100) MXSTEP(5) SCORING(1)
    SINGULAR(0.000000000001) HCONVERGE(0, ABSOLUTE) LCONVERGE (0, ABSOLUTE)
    PCONVERGE(0.000001, ABSOLUTE)
  /FIXED = | SSTYPE(3)
  /METHOD = REML
  /PRINT = SOLUTION
  /RANDOM INTERCEPT | SUBJECT(classid*schoolid) COVTYPE(VC)
  /RANDOM INTERCEPT | SUBJECT(schoolid) COVTYPE(VC) .
```

There is no BY keyword in this syntax, meaning that the random factors in this model are not identified initially as categorical factors.

The first `RANDOM` subcommand declares that there is a random effect associated with the `INTERCEPT` for each subject identified by a combination of the classroom ID and school ID variables (CLASSID*SCHOOLID). Even though this syntax appears to be setting up a crossed effect for classroom by school, it is equivalent to the nested specification for these two random factors that was seen in the previous syntax (`CLASSID(SCHOOLID)`). The asterisk is necessary because nested `SUBJECT` variables cannot currently be specified when using the `MIXED` command. The second `RANDOM` subcommand specifies that there is a random effect in the model associated with the `INTERCEPT` for each subject identified by the SCHOOLID variable.

Software Note: Fitting Model 4.1 using the less efficient SPSS syntax takes a relatively long time compared to the time required to fit the equivalent model using the other packages; and it also takes longer compared to the more efficient version of the SPSS syntax. We use the more computationally efficient syntax for the remainder of the analysis in SPSS.

We now carry out a likelihood ratio test of Hypothesis 4.1 to decide if we wish to omit the nested random effects associated with classrooms from Model 4.1. To do this, we fit Model 4.1A by removing the first `RANDOM` subcommand from the more efficient syntax for Model 4.1:

```
* Model 4.1A .
MIXED
mathgain BY classid schoolid
/CRITERIA = CIN(95) MXITER(100) MXSTEP(5) SCORING(1)
SINGULAR(0.000000000001) HCONVERGE(0, ABSOLUTE) LCONVERGE
(0, ABSOLUTE) PCONVERGE(0.000001, ABSOLUTE)
/FIXED = | SSTYPE(3)
/METHOD = REML
/PRINT = SOLUTION
/RANDOM schoolid | COVTYPE(VC) .
```

To calculate the test statistic for Hypothesis 4.1, we subtract the –2 REML log-likelihood value for the reference model, Model 4.1, from the corresponding value for Model 4.1A (the nested model). We refer the resulting test statistic to a mixture of χ^2 distributions, with 0 and 1 degrees of freedom, and equal weight of 0.5 (see Subsection 4.5.1 for a discussion of this test). Based on the significant result of this test ($p = 0.002$), we decide to retain the nested random classroom effects in Model 4.1 and all future models. We also retain the random effects associated with schools in all models without testing their significance, to reflect the hierarchical structure of the data in the model specification.

Step 2: Build the Level 1 Model by adding Level 1 Covariates (Model 4.1 vs. Model 4.2).

Model 4.2 adds the fixed effects of the student-level covariates MATHKIND, SEX, MINORITY, and SES to Model 4.1, using the following syntax:

```
* Model 4.2 .
MIXED
 mathgain WITH mathkind sex minority ses BY classid schoolid
 /CRITERIA = CIN (95) MXITER(100) MXSTER(5) SCORING(1)
```

```
    SINGULAR (0.000000000001) HCONVERGE (0,ABSOLUTE) LCONVERGE
    (0,ABSOLUTER) PCONVERGE (0.000001,ABSOLUTE)
    /FIXED = mathkind sex minority ses | SSTYPE(3)
    /METHOD = REML
    /PRINT = SOLUTION
    /RANDOM classid(schoolid) | COVTYPE(VC)
    /RANDOM schoolid | COVTYPE(VC) .
```

Note that the four student-level covariates have been identified as continuous by listing them after the WITH keyword. This is an acceptable approach for SEX and MINORITY, even though they are categorical, because they are both indicator variables having values 0 and 1. All four student-level covariates are also listed in the FIXED = subcommand, so that fixed effect parameters associated with each covariate will be added to the model.

Software Note: If we had included SEX and MINORITY as categorical factors by listing them after the BY keyword, SPSS would have used the *highest* levels of each of these variables as the reference categories (i.e., SEX = 1, girls; and MINORITY = 1, minority students). The resulting parameter estimates would have given us the estimated fixed effect of being a boy, and of being a nonminority student, respectively, on math achievement score. These parameter estimates would have had the opposite signs of the estimates resulting from the syntax that we used, in which MINORITY and SEX were listed after the WITH keyword.

We test Hypothesis 4.2 to decide whether all fixed effects associated with the Level 1 (student-level) covariates are equal to zero, by performing a likelihood ratio test based on ML estimation (see Subsection 2.6.2.1 for a discussion of likelihood ratio tests for fixed effects). The test statistic is calculated by subtracting the –2 ML log-likelihood for Model 4.2 (the reference model) from the corresponding value for Model 4.1 (the nested model, excluding all fixed effects associated with the Level 1 covariates). We use the /METHOD = ML subcommand to request Maximum Likelihood estimation of both models:

```
* Model 4.1 (ML Estimation).
MIXED
mathgain BY classid schoolid
/CRITERIA = CIN(95) MXITER(100) MXSTEP(5) SCORING(1)
SINGULAR(0.000000000001) HCONVERGE(0, ABSOLUTE) LCONVERGE
(0, ABSOLUTE) PCONVERGE(0.000001, ABSOLUTE)
/FIXED = | SSTYPE(3)
/METHOD = ML
/PRINT = SOLUTION
/RANDOM classid(schoolid) | COVTYPE(VC)
/RANDOM schoolid | COVTYPE(VC) .

* Model 4.2 (ML Estimation).
MIXED
 mathgain WITH mathkind sex minority ses BY classid schoolid
 /CRITERIA = CIN(95) MXITER(100) MXSTEP(5) SCORING(1)
  SINGULAR(0.000000000001) HCONVERGE(0, ABSOLUTE)
  LCONVERGE (0, ABSOLUTE) PCONVERGE(0.000001, ABSOLUTE)
 /FIXED = mathkind sex minority ses | SSTYPE(3)
 /METHOD = ML
```

```
/PRINT = SOLUTION
/RANDOM classid(schoolid) | COVTYPE(VC)
/RANDOM schoolid | COVTYPE(VC) .
```

We reject the null hypothesis ($p < 0.001$) and decide to retain all fixed effects associated with the Level 1 covariates. This result suggests that the Level 1 fixed effects explain at least some of the variation at Level 1 (the student level) of the data, which we had previously attributed to residual variance in Model 4.1.

Step 3: Build the Level 2 Model by adding Level 2 Covariates (Model 4.3).

We fit Model 4.3 using the default REML estimation method by adding the Level 2 (classroom-level) covariates YEARSTEA, MATHPREP, and MATHKNOW to the SPSS syntax. We add these covariates to the WITH subcommand, so that they will be treated as continuous covariates, and we also add them to the FIXED subcommand:

```
* Model 4.3 .
MIXED
 mathgain WITH mathkind sex minority ses yearstea mathprep mathknow
    BY classid schoolid
 /CRITERIA = CIN(95) MXITER(100) MXSTEP(5) SCORING(1)
  SINGULAR(0.000000000001) HCONVERGE(0, ABSOLUTE) LCONVERGE
  (0, ABSOLUTE) PCONVERGE(0.000001, ABSOLUTE)
 /FIXED = mathkind sex minority ses
    yearstea mathprep mathknow | SSTYPE(3)
 /METHOD = REML
 /PRINT = SOLUTION
 /RANDOM classid(schoolid) | COVTYPE(VC)
 /RANDOM schoolid | COVTYPE(VC) .
```

We cannot perform a likelihood ratio test of Hypotheses 4.3 through 4.5, owing to the presence of missing data on some of the classroom-level covariates; we instead use t-tests for the fixed effects of each of the Level 2 covariates added at this step. The results of these t-tests are reported in the SPSS output for Model 4.3. Because none of these t-tests are significant, we do not add these fixed effects to the model, and select Model 4.2 as our preferred model at this stage of the analysis.

Step 4: Build the Level 3 Model by adding the Level 3 Covariate (Model 4.4).

We fit Model 4.4 using REML estimation, by updating the syntax used to fit Model 4.2. We add the Level 3 (school-level) covariate, HOUSEPOV, to the WITH subcommand and to the FIXED subcommand, as shown in the following syntax:

```
* Model 4.4.
MIXED
 mathgain WITH mathkind sex minority ses housepov BY classid schoolid
 /CRITERIA = CIN(95) MXITER(100) MXSTEP(5) SCORING(1)
  SINGULAR(0.000000000001) HCONVERGE(0, ABSOLUTE) LCONVERGE
  (0, ABSOLUTE) PCONVERGE(0.000001, ABSOLUTE)
 /FIXED = mathkind sex minority ses housepov | SSTYPE(3)
 /METHOD = REML
 /PRINT = SOLUTION
 /RANDOM classid(schoolid) | COVTYPE(VC)
 /RANDOM schoolid | COVTYPE(VC) .
```

We use a *t*-test for Hypothesis 4.6, to decide whether we wish to add the fixed effect of HOUSEPOV to the model. The result of this *t*-test is reported in the SPSS output, and indicates that the fixed effect of HOUSEPOV is not significant ($p = 0.25$). We therefore choose Model 4.2 as our final model.

4.4.3   R

We begin by reading the comma-delimited raw data file, having the structure described in Table 4.2 and with variable names in the first row, into a data frame object named `class`. The `h = T` option instructs R to read a header record containing variable names from the first row of the raw data:

```
> class <- read.csv("c:\\temp\\classroom.csv", h = T)
```

4.4.3.1   Analysis Using the `lme()` Function

We first load the `nlme` package, so that the `lme()` function can be used in the analysis:

```
> library(nlme)
```

We now proceed with the analysis steps.

Step 1: Fit the initial "unconditional" (variance components) model (Model 4.1), and decide whether to omit the random classroom effects (Model 4.1 vs. Model 4.1A).

We fit Model 4.1 to the Classroom data using the `lme()` function as follows:

```
> # Model 4.1.
> model4.1.fit <- lme(mathgain ~ 1, random = ~ 1 | schoolid/classid,
    class, method = "REML")
```

We describe the syntax used in the `lme()` function below:

- `model4.1.fit` is the name of the object that will contain the results of the fitted model.

- The first argument of the function, `mathgain ~ 1`, defines the response variable, MATH-GAIN, and the single fixed effect in this model, which is associated with the intercept (denoted by a 1 after the ~).

- The second argument, `random = ~ 1 | schoolid/classid`, indicates the nesting structure of the random effects in the model. The `random = ~ 1` portion of the argument indicates that the random effects are to be associated with the intercept. The first variable listed after the vertical bar (|) is the random factor at the highest level (Level 3) of the data (i.e., SCHOOLID). The next variable after a forward slash (/) indicates the random factor (i.e., CLASSID) with levels nested within levels of the first random factor. This notation for the nesting structure is known as the **Wilkinson–Rogers notation**.

- The third argument, `class`, indicates the name of the data frame object to be used in the analysis.

- The final argument of the function, `method = "REML"`, tells R that REML estimation should be used for the variance components in the model. REML is the default estimation method in the `lme()` function.

Estimates saved in the model fit object can be obtained by applying the `summary()` function:

```
> summary(mode14.1.fit)
```

> **Software Note:** The `getVarCov()` function, which can be used to display blocks of the estimated marginal V matrix for a two-level model, currently does not have the capability of displaying blocks of the estimated V matrix for the models considered in this example, due to the multiple levels of nesting in the Classroom data set.

The EBLUPs of the random school effects and the nested random classroom effects in the model can be obtained by using the `random.effects()` function:

```
> random.effects(model4.1.fit)
```

At this point we perform a likelihood ratio test of Hypothesis 4.1, to decide if we need the nested random classroom effects in the model. We first fit a nested model, Model 4.1A, omitting the random effects associated with the classrooms. We do this by excluding the CLASSID variable from the nesting structure for the random effects in the `lme()` function:

```
> # Model 4.1A.
> model4.1A.fit <- lme(mathgain ~ 1, random = ~1 | schoolid,
  data = class, method = "REML")
```

The `anova()` function can now be used to carry out a likelihood ratio test for Hypothesis 4.1, to decide if we wish to retain the nested random effects associated with classrooms.

```
> anova(model4.1.fit, model4.1A.fit)
```

The `anova()` function subtracts the –2 REML log-likelihood value for Model 4.1 (the reference model) from that for Model 4.1A (the nested model), and refers the resulting test statistic to a χ^2 distribution with 1 degree of freedom. However, because the appropriate null distribution for the likelihood ratio test statistic for Hypothesis 4.1 is a mixture of two χ^2 distributions, with 0 and 1 degrees of freedom and equal weights of 0.5, we multiply the p-value provided by the `anova()` function by 0.5 to obtain the correct p-value. Based on the significant result of this test ($p < 0.01$), we retain the nested random classroom effects in Model 4.1 and in all future models. We also retain the random school effects as well, to reflect the hierarchical structure of the data in the model specification.

Step 2: Build the Level 1 Model by adding Level 1 Covariates (Model 4.1 vs. Model 4.2).

After obtaining the estimates of the fixed intercept and the variance components in Model 4.1, we modify the syntax to fit Model 4.2, which includes the fixed effects of the four Level 1 (student-level) covariates MATHKIND, SEX, MINORITY, and SES. Note that these covariates are added on the right-hand side of the ~ in the first argument of the `lme()` function:

```
> # Model 4.2.
> model4.2.fit <- lme(mathgain ~ mathkind + sex + minority + ses,
  random = ~1 | schoolid/classid, class,
  na.action = "na.omit", method = "REML")
```

Because some of the students might have missing data on these covariates (which is actually not the case for the Level 1 covariates in the Classroom data set), we include the argument `na.action = "na.omit"`, to tell the two functions to drop cases with missing data from the analysis.

> **Software Note:** Without the `na.action = "na.omit"` specification, the `lme()` function will not run if there are missing data on any of the variables input to the function.

The 1 that was used to identify the intercept in the fixed part of Model 4.1 does not need to be specified in the syntax for Model 4.2, because the intercept is automatically included in any model with at least one fixed effect.

We assess the results of fitting Model 4.2 using the `summary()` function:

```
> summary(model4.2.fit)
```

We now test Hypothesis 4.2, to decide whether the fixed effects associated with all Level 1 (student-level) covariates in Model 4.2 are equal to zero, by carrying out a likelihood ratio test using the `anova()` function. To do this we refit the nested model, Model 4.1, and the reference model, Model 4.2, using ML estimation. The test statistic is calculated by the `anova()` function by subtracting the –2 ML log-likelihood for Model 4.2 (the reference model) from that for Model 4.1 (the nested model), and referring the test statistic to a χ^2 distribution with 4 degrees of freedom.

```
> # Model 4.1: ML estimation with lme().
> model4.1.ml.fit <- lme(mathgain ~ 1,
    random = ~1 | schoolid/classid, class, method = "ML")
> # Model 4.2: ML estimation with lme().
> model4.2.ml.fit <- lme(mathgain ~ mathkind + sex + minority + ses,
    random = ~1 | schoolid/classid, class,
    na.action = "na.omit", method = "ML")
> anova(model4.1.ml.fit, model4.2.ml.fit)
```

We see that at least one of the fixed effects associated with the Level 1 covariates is significant, based on the result of this test ($p < 0.001$); Subsection 4.5.2 presents details on testing Hypothesis 4.2. We therefore proceed with Model 4.2 as our preferred model.

Step 3: Build the Level 2 Model by adding Level 2 Covariates (Model 4.3).

We fit Model 4.3 by adding the fixed effects of the Level 2 (classroom-level) covariates, YEARSTEA, MATHPREP, and MATHKNOW, to Model 4.2:

```
> # Model 4.3.
> model4.3.fit <- update(model4.2.fit,
fixed = ~ mathkind + sex + minority + ses + yearstea + mathprep + mathknow)
```

We investigate the resulting parameter estimates and standard errors for the estimated fixed effects by applying the `summary()` function to the model fit object:

```
> summary(model4.3.fit)
```

We cannot consider a likelihood ratio test for the fixed effects added to Model 4.2, because some classrooms have missing data on the MATHKNOW variable, and Models 4.2

and 4.3 are fitted using different observations as a result. Instead, we test the fixed effects associated with the classroom-level covariates (Hypotheses 4.3 through 4.5) using t-tests. None of these fixed effects are significant based on the results of these t-tests (provided by the `summary()` function), so we choose Model 4.2 as the preferred model at this stage of the analysis.

Step 4: Build the Level 3 Model by adding the Level 3 Covariate (Model 4.4).

Model 4.4 can be fitted by adding the Level 3 (school-level) covariate to the formula for the fixed-effects portion of the model in the `lme()` function. We add the fixed effect of the HOUSEPOV covariate to the model by updating the `fixed =` argument for Model 4.2:

```
> # Model 4.4.
> model4.4.fit <- update(model4.2.fit, fixed = ~ mathkind + sex + minority
                                        + ses + housepov)
```

We apply the `summary()` function to the model fit object to obtain the resulting parameter estimates and t-tests for the fixed effects (in the case of `model4.4.fit`):

```
> summary(model4.4.fit)
```

The t-test for the fixed effect of HOUSEPOV is not significant, so we choose Model 4.2 as our final model for the Classroom data set.

4.4.3.2 Analysis Using the `lmer()` Function

We begin by loading the `lme4` package, so that the `lmer()` function can be used in the analysis:

```
> library(lme4)
```

We now proceed with the analysis steps.

Step 1: Fit the initial "unconditional" (variance components) model (Model 4.1), and decide whether to omit the random classroom effects (Model 4.1 vs. Model 4.1A).

We fit Model 4.1 to the Classroom data using the `lmer()` function as follows:

```
> # Model 4.1.
> model4.1.fit.lmer <- lmer(mathgain ~ 1 + (1|schoolid) + (1|classid),
  class, REML = T)
```

We describe the syntax used in the `lmer()` function below:

- `model4.1.fit.lmer` is the name of the object that will contain the results of the fitted model.

- Like the `lme()` function, the first argument of the function, `mathgain ~ 1`, defines the response variable, MATHGAIN, and the single fixed effect in this model, which is associated with the intercept (denoted by a 1 after the ~).

- Next, random intercepts associated with each level of SCHOOLID and CLASSID are added to the model formula (using + notation), using the syntax `(1|schoolid)` and `(1|classid)`. Note that a specific nesting structure does not need to be indicated here.

- The third argument, `class`, once again indicates the name of the data frame object to be used in the analysis.

- The final argument of the function, `REML = T`, tells R that REML estimation should be used for the variance components in the model. REML is also the default estimation method in the `lmer()` function.

Estimates from the model fit can be obtained using the `summary()` function:

```
> summary(mode14.1.fit.lmer)
```

The EBLUPs of the random school effects and the nested random classroom effects in the model can be obtained using the `ranef()` function:

```
> ranef(model4.1.fit.lmer)
```

At this point we perform a likelihood ratio test of Hypothesis 4.1, to decide if we need the nested random classroom effects in the model. We first fit a nested model, Model 4.1A, omitting the random effects associated with the classrooms. We do this by excluding (1|CLASSID) from the model formula for the `lmer()` function:

```
> # Model 4.1A.
> model4.1A.fit.lmer <- lmer(mathgain ~ 1 + (1|schoolid),
    class, REML = T)
```

The `anova()` function can now be used to carry out a likelihood ratio test for Hypothesis 4.1, to decide if we wish to retain the nested random effects associated with classrooms.

```
> anova(model4.1.fit.lmer, model4.1A.fit.lmer)
```

The `anova()` function subtracts the -2 REML log-likelihood value for Model 4.1 (the reference model) from that for Model 4.1A (the nested model), and refers the resulting test statistic to a χ^2 distribution with 1 degree of freedom. However, because the appropriate null distribution for the likelihood ratio test statistic for Hypothesis 4.1 is a mixture of two χ^2 distributions, with 0 and 1 degrees of freedom and equal weights of 0.5, we multiply the p-value provided by the `anova()` function by 0.5 to obtain the correct p-value. Based on the significant result of this test ($p < 0.01$), we retain the nested random classroom effects in Model 4.1 and in all future models. We also retain the random school effects as well, to reflect the hierarchical structure of the data in the model specification.

Step 2: Build the Level 1 Model by adding Level 1 Covariates (Model 4.1 vs. Model 4.2).

After obtaining the estimates of the fixed intercept and the variance components in Model 4.1, we modify the syntax to fit Model 4.2, which includes the fixed effects of the four Level 1 (student-level) covariates MATHKIND, SEX, MINORITY, and SES. Note that these covariates are added on the right-hand side of the ~ in the first argument of the `lmer()` function:

```
> # Model 4.2.
> model4.2.fit.lmer <- lmer(mathgain ~ mathkind + sex + minority + ses
    + (1|schoolid) + (1|classid),
    class, na.action = "na.omit", REML = T)
```

Because some of the students might have missing data on these covariates (which is actually not the case for the Level 1 covariates in the Classroom data set), we include the argument `na.action = "na.omit"`, to tell the two functions to drop cases with missing data from the analysis.

Software Note: Without the `na.action = "na.omit"` specification, the `lmer()` function will not run if there are missing data on any of the variables input to the function.

Software Note: The version of the `lmer()` function in the `lme4` package does not automatically compute p-values for the t-statistics that are generated by dividing the fixed-effect parameter estimates by their standard errors (for testing the hypothesis that a given fixed-effect parameter is equal to zero). This is primarily due to the lack of agreement in the literature over an appropriate distribution for this test statistic under the null hypothesis. Instead, approximate tests available in the `lmerTest` package can be used; see Chapter 3 for an example using two-level models. When possible, we use likelihood ratio tests for the fixed-effect parameters in this chapter.

The 1 that was used to identify the intercept in the fixed part of Model 4.1 does not need to be specified in the syntax for Model 4.2, because the intercept is automatically included by the two functions in any model with at least one fixed effect.

We assess the results of fitting Model 4.2 using the `summary()` function:

```
> summary(model4.2.fit.lmer)
```

We now test Hypothesis 4.2, to decide whether the fixed effects associated with all Level 1 (student-level) covariates in Model 4.2 are equal to zero, by carrying out a likelihood ratio test using the `anova()` function. To do this, we refit the nested model, Model 4.1, and the reference model, Model 4.2, using ML estimation (note the `REML = F` arguments below). The test statistic is calculated by the `anova()` function by subtracting the –2 ML log-likelihood for Model 4.2 (the reference model) from that for Model 4.1 (the nested model), and referring the test statistic to a χ^2 distribution with 4 degrees of freedom.

```
> # Model 4.1: ML estimation with lmer().
> model4.1.lmer.ml.fit <- lmer(mathgain ~ 1 + (1|schoolid) + (1|classid),
    class, REML = F)
> # Model 4.2: ML estimation with lmer().
> model4.2.lmer.ml.fit <- lmer(mathgain ~ mathkind + sex + minority + ses
    + (1|schoolid) + (1|classid),
    class, REML = F)
> anova(model4.1.lmer.ml.fit, model4.2.lmer.ml.fit)
```

We see that at least one of the fixed effects associated with the Level 1 covariates is significant, based on the result of this test ($p < 0.001$); Subsection 4.5.2 presents details on testing Hypothesis 4.2. We therefore proceed with Model 4.2 as our preferred model.

Step 3: Build the Level 2 Model by adding Level 2 Covariates (Model 4.3).

We fit Model 4.3 by adding the fixed effects of the Level 2 (classroom-level) covariates, YEARSTEA, MATHPREP, and MATHKNOW, to Model 4.2:

```
> # Model 4.3.
> model4.3.fit.lmer <- lmer(mathgain ~ mathkind + sex + minority + ses
    + yearstea + mathprep + mathknow
    + (1|schoolid) + (1|classid),
    class, na.action = "na.omit", REML = T)
```

We investigate the resulting parameter estimates and standard errors for the estimated fixed effects by applying the **summary()** function to the model fit object:

```
> summary(model4.3.fit.lmer)
```

We cannot consider a likelihood ratio test for the fixed effects added to Model 4.2, because some classrooms have missing data on the MATHKNOW variable, and Models 4.2 and 4.3 are fitted using different observations as a result. Instead, we can refer the test statistics provided by the **summary()** function to standard normal distributions to make approximate inferences about the importance of the effects (where a test statistic larger than 1.96 in absolute value would suggest a significant fixed effect at the 0.05 significance level, under asymptotic assumptions). These tests would suggest that none of these fixed effects are significant, so we would not retain them in this model.

Step 4: Build the Level 3 Model by adding the Level 3 Covariate (Model 4.4).

Model 4.4 can be fitted by adding the Level 3 (school-level) covariate HOUSEPOV to the formula for the fixed-effects portion of Model 4.2:

```
> # Model 4.4.
> model4.4.fit.lmer <- lmer(mathgain ~ mathkind + sex + minority + ses
    + housepov + (1|schoolid) + (1|classid),
    class, na.action = "na.omit", REML = T)
```

We apply the **summary()** function to the model fit object to obtain the resulting parameter estimates and standard errors:

```
> summary(mode14.4.fit.lmer)
```

Based on the test statistic for the fixed effect of HOUSEPOV (-1.151), we once again do not have enough evidence to say that this effect is different from 0 (at the 0.05 level), so we choose Model 4.2 as our final model for the Classroom data set.

4.4.4 Stata

First, we read the raw comma-delimited data into Stata using the **insheet** command:

```
. insheet using "C:\temp\classroom.csv", comma clear
```

Users of web-aware Stata can also import the data directly from the book's web page:

```
. insheet using http://www-personal.umich.edu/~bwest/classroom.csv
```

The **mixed** command can then be used to fit three-level hierarchical models with nested random effects.

Step 1: Fit the initial "unconditional" (variance components) model (Model 4.1), and decide whether to omit the random classroom effects (Model 4.1 vs. Model 4.1A).

We first specify the `mixed` syntax to fit Model 4.1, including the random effects of schools and of classrooms nested within schools:

```
. * Model 4.1.
. mixed mathgain || schoolid: || classid:, variance reml
```

The first variable listed after invoking `mixed` is the continuous dependent variable, MATHGAIN. No covariates are specified after the dependent variable, because the only fixed effect in Model 4.1 is the intercept, which is included by default.

After the first clustering indicator (||), we list the random factor identifying clusters at Level 3 of the data set, SCHOOLID, followed by a colon (:). We then list the nested random factor, CLASSID, after a second clustering indicator. This factor identifies clusters at Level 2 of the data set, and is again followed by a colon.

> **Software Note:** If a multilevel data set is organized by a series of nested groups, such as classrooms nested within schools as in this example, the random effects structure of the mixed model is specified in `mixed` by listing the random factors defining the structure, separated by two vertical bars (||). The nesting structure reads left to right; e.g., SCHOOLID is the highest level of clustering, with levels of CLASSID nested within each school.
>
> If no variables are specified after the colon at a given level of the nesting structure, the model will only include a single random effect (associated with the intercept) for each level of the random factor. Additional covariates with random effects at a given level of the nesting structure can be specified after the colon.

Finally, the `variance` and `reml` options specified after a comma request that the estimated variances of the random school and classroom effects, rather than their estimated standard deviations, should be displayed in the output, and that REML estimation should be used to fit this model (ML estimation is the default in Stata 12+).

Information criteria, including the REML log-likelihood, can be obtained by using the `estat ic` command after submitting the `mixed` command:

```
. estat ic
```

In the output associated with the fit of Model 4.1, Stata automatically reports a likelihood ratio test, calculated by subtracting the –2 REML log-likelihood of Model 4.1 (including the random school effects and nested random classroom effects) from the –2 REML log-likelihood of a simple linear regression model without the random effects. Stata reports the following note along with the test:

> Note: LR test is conservative and provided only for reference

Stata performs a classical likelihood ratio test here, where the distribution of the test statistic (under the null hypothesis that both variance components are equal to zero) is asymptotically χ^2_2 (where the 2 degrees of freedom correspond to the two variance components in Model 4.1). Appropriate theory for testing a model with multiple random effects (e.g., Model 4.1) vs. a model without any random effects has yet to be developed, and Stata discusses this issue in detail if users click on the LR test is conservative note. The *p*-value for this test statistic is known to be larger than it should be (making it conservative).

We recommend testing the need for the random effects by using individual likelihood ratio tests, based on REML estimation of nested models. To test Hypothesis 4.1, and decide whether we want to retain the nested random effects associated with classrooms in Model 4.1, we fit a nested model, Model 4.1A, again using REML estimation:

```
. * Model 4.1A.
. mixed mathgain || schoolid:, variance reml
```

The test statistic for Hypothesis 4.1 can be calculated by subtracting the –2 REML log-likelihood for Model 4.1 (the reference model) from that of Model 4.1A (the nested model). The *p*-value for the test statistic (7.9) is based on a mixture of χ^2 distributions with 0 and 1 degrees of freedom, and equal weight 0.5. Because of the significant result of this test ($p = 0.002$), we retain the nested random classroom effects in Model 4.1 and in all future models (see Subsection 4.5.1 for a discussion of this test). We also retain the random effects associated with schools, to reflect the hierarchical structure of the data set in the model.

Step 2: Build the Level 1 Model by adding Level 1 Covariates (Model 4.1 vs. Model 4.2).

We fit Model 4.2 by adding the fixed effects of the four student-level covariates, MATHKIND, SEX, MINORITY, and SES, using the following syntax:

```
. * Model 4.2.
. mixed mathgain mathkind sex minority ses || schoolid: || classid:,
    variance reml
```

Information criteria associated with the fit of this model can be obtained by using the estat ic command after the mixed command has finished running.

We test Hypothesis 4.2 to decide whether the fixed effects that were added to Model 4.1 to form Model 4.2 are all equal to zero, using a likelihood ratio test. We first refit the nested model, Model 4.1, and the reference model, Model 4.2, using ML estimation. We specify the mle option to request maximum likelihood estimation for each model. The est store command is then used to store the results of each model fit in new objects.

```
. * Model 4.1: ML Estimation.
. mixed mathgain || schoolid: || classid:, variance mle
. est store model4_1_ml_fit
```

```
. * Model 4.2: ML Estimation.
. mixed mathgain mathkind sex minority ses || schoolid: || classid:,
    variance mle
. est store model4_2_ml_fit
```

We use the lrtest command to perform the likelihood ratio test. The likelihood ratio test statistic is calculated by subtracting the –2 ML log-likelihood for Model 4.2 from that for Model 4.1, and referring the difference to a χ^2 distribution with 4 degrees of freedom. The likelihood ratio test requires that both models are fitted using the same cases.

```
. lrtest model4_1_ml_fit model4_2_ml_fit
```

Based on the significant result ($p < 0.001$) of this test, we choose Model 4.2 as our preferred model at this stage of the analysis. We discuss the likelihood ratio test for Hypothesis 4.2 in more detail in Subsection 4.5.2.

Step 3: Build the Level 2 Model by adding Level 2 Covariates (Model 4.3).

To fit Model 4.3, we modify the `mixed` command used to fit Model 4.2 by adding the fixed effects of the classroom-level covariates, YEARSTEA, MATHPREP, and MATHKNOW, to the fixed portion of the command. We again use the default REML estimation for this model, and obtain the model information criteria by using the post-estimation command `estat ic`.

```
. * Model 4.3.
. mixed mathgain mathkind sex minority ses yearstea mathprep mathknow
   || schoolid: || classid:, variance reml
. estat ic
```

We do not consider a likelihood ratio test for the fixed effects added to Model 4.2 to form Model 4.3, because Model 4.3 was fitted using different cases, owing to the presence of missing data on some of the classroom-level covariates. Instead, we consider the z-tests reported by Stata for Hypotheses 4.3 through 4.5. None of the z-tests reported for the fixed effects of the classroom-level covariates are significant. Therefore, we do not retain these fixed effects in Model 4.3, and choose Model 4.2 as our preferred model at this stage of the analysis.

Step 4: Build the Level 3 Model by adding the Level 3 Covariate (Model 4.4).

To fit Model 4.4, we add the fixed effect of the school-level covariate, HOUSEPOV, to the model, by updating the `mixed` command that was used to fit Model 4.2. We again use the default REML estimation method, and use the `estat ic` post-estimation command to obtain information criteria for Model 4.4.

```
. * Model 4.4.
. mixed mathgain mathkind sex minority ses housepov
   || schoolid: || classid:, variance reml
. estat ic
```

To test Hypothesis 4.6, we use the z-test reported by the `mixed` command for the fixed effect of HOUSEPOV. Because of the nonsignificant test result ($p = 0.25$), we do not retain this fixed effect, and choose Model 4.2 as our final model for the analysis of the Classroom data.

4.4.5   HLM

We assume that the first MDM file discussed in the initial data summary (Subsection 4.2.2) has been generated using HLM3, and proceed to the model-building window.

Step 1: Fit the initial "unconditional" (variance components) model (Model 4.1), and decide whether to omit the random classroom effects (Model 4.1 vs. Model 4.1A).

We begin by specifying the Level 1 (student-level) model. In the model-building window, click on MATHGAIN, and identify it as the **Outcome variable**. Go to the **Basic Settings** menu and identify the outcome variable as a **Normal** (Continuous) variable. Choose a title for this analysis (such as "Classroom Data: Model 4.1"), and choose a location and name for the output (.html) file that will contain the results of the model fit. Click **OK** to return to the model-building window. Under the **File** menu, click **Preferences**, and then click **Use level subscripts** to display subscripts in the model-building window.

Three models will now be displayed. The **Level 1 model** describes the "means-only" model at the student level. We show the Level 1 model below as it is displayed in the HLM model-building window:

Model 4.1: Level 1 Model

$$\text{MATHGAIN}_{ijk} = \pi_{0jk} + e_{ijk}$$

The value of MATHGAIN for an individual student i, within classroom j nested in school k, depends on the intercept for classroom j within school k, π_{0jk}, plus a residual, e_{ijk}, associated with the student.

The **Level 2 model** describes the classroom-specific intercept in **Model 4.1** at the classroom level of the data set:

Model 4.1: Level 2 Model

$$\pi_{0jk} = \beta_{00k} + r_{0jk}$$

The classroom-specific intercept, π_{0jk}, depends on the school-specific intercept, β_{00k}, and a random effect, r_{0jk}, associated with the j-th classroom within school k.

The **Level 3 model** describes the school-specific intercept in Model 4.1:

Model 4.1: Level 3 Model

$$\beta_{00k} = \gamma_{000} + u_{00k}$$

The school-specific intercept, β_{00k}, depends on the overall (grand) mean, γ_{000}, plus a random effect, u_{00k} associated with the school.

The overall "means-only" mixed model derived from the preceding Level 1, Level 2, and Level 3 models can be displayed by clicking on the **Mixed** button:

Model 4.1: Overall Mixed Model

$$\text{MATHGAIN}_{ijk} = \gamma_{000} + r_{0jk} + u_{00k} + e_{ijk}$$

An individual student's MATHGAIN depends on an overall fixed intercept, γ_{000} (which represents the overall mean of MATHGAIN across all students), a random effect associated with the student's classroom, r_{0jk}, a random effect associated with the student's school, u_{00k}, and a residual, e_{ijk}. Table 4.3 shows the correspondence of this HLM notation with the general notation used in (4.1).

To fit Model 4.1, click **Run Analysis**, and select **Save as and Run** to save the .hlm command file. You will be prompted to supply a name and location for this .him file. After the estimation has finished, click on **File**, and select **View Output** to see the resulting parameter estimates and fit statistics.

At this point, we test the significance of the random effects associated with classrooms nested within schools (Hypothesis 4.1). However, because the HLM3 procedure does not allow users to remove all random effects from a given level of a hierarchical model (in this example, the classroom level, or the school level), we cannot perform a likelihood ratio test of Hypothesis 4.1, as was done in the other software procedures. Instead, HLM provides chi-square tests that are calculated using methodology described in Raudenbush & Bryk (2002). The following output is generated by HLM3 after fitting Model 4.1:

```
Final estimation of level-1 and level-2 variance components:
---------------------------------------------------------------
Random Effect        Standard   Variance   df   Chi-    P-value
                     Deviation  Component        square
---------------------------------------------------------------
INTRCPT1,       r0   10.02212    100.44281  205  301.95331 <0.001
level-1,        e    32.05828   1027.73315
```

```
  ╭─────────────────────────────────────────────────────────────────────╮
  │  Final estimation of level-3 variance components:                     │
  │  -------------------------------------------------------------------  │
  │  Random Effect          Standard   Variance   df    Chi-    P-value    │
  │                         Deviation  Component         square           │
  │  -------------------------------------------------------------------  │
  │  INTRCPT1/       u00     8.66240    75.03712  106  165.74813  <0.001   │
  ╰──   INTRCPT2,                                                          │
      ╰────────────────────────────────────────────────────────────────╯
```

The chi-square test statistic for the variance of the nested random classroom effects (301.95) is significant ($p < 0.001$), so we reject the null hypothesis for Hypothesis 4.1 and retain the random effects associated with both classrooms and schools in Model 4.1 and all future models. We now proceed to fit Model 4.2.

Step 2: Build the Level 1 Model by adding Level 1 Covariates (Model 4.1 vs. Model 4.2).

We specify the **Level 1 Model** for Model 4.2 by clicking on **Level 1** to add fixed effects associated with the student-level covariates to the model. We first select the variable MATH-KIND, choose **add variable uncentered**, and then repeat this process for the variables SEX, MINORITY, and SES. Notice that as each covariate is added to the Level 1 model, the Level 2 and Level 3 models are also updated. The new Level 1 model is as follows:

Model 4.2: Level 1 Model

$$\text{MATHGAIN}_{ijk} = \pi_{0jk} + \pi_{1jk}(\text{SEX}_{ijk}) + \pi_{2jk}(\text{MINORITY}_{ijk})$$
$$+ \pi_{3jk}(\text{MATHKIND}_{ijk}) + \pi_{4jk}(\text{SES}_{ijk}) + e_{ijk}$$

This updated Level 1 model shows that a student's MATHGAIN now depends on the intercept specific to classroom j, π_{0jk}, the classroom-specific effects ($\pi_{1jk}, \pi_{2jk}, \pi_{3jk}$ and π_{4jk}) of each of the student-level covariates, and a residual, e_{ijk}.

The **Level 2** portion of the model-building window displays the classroom-level equations for the student-level intercept (π_{0jk}) and for each of the student-level effects (π_{1jk} through π_{4jk}) defined in this model. The equation for each effect from HLM is as follows:

Model 4.2: Level 2 Model

$$\pi_{0jk} = \beta_{00k} + r_{0jk}$$
$$\pi_{1jk} = \beta_{10k}$$
$$\pi_{2jk} = \beta_{20k}$$
$$\pi_{3jk} = \beta_{30k}$$
$$\pi_{4jk} = \beta_{40k}$$

The equation for the student-level intercept (π_{0jk}) has the same form as in Model 4.1. It includes an intercept specific to school k, β_{00k}, plus a random effect, r_{0jk}, associated with each classroom in school k. Thus, the student-level intercepts are allowed to vary randomly from classroom to classroom within the same school.

The equations for each of the effects associated with the four student-level covariates (π_{1jk} through π_{4jk}) are all constant at the classroom level. This means that the effects of being female, being a minority student, kindergarten math achievement, and student-level SES are assumed to be the same for students within all classrooms (i.e., these coefficients do not vary across classrooms within a given school).

The **Level 3** portion of the model-building window shows the school-level equations for the school-specific intercept, β_{00k}, and for each of the school-specific effects in the classroom-level model, β_{10k} through β_{40k}:

Model 4.2: Level 3 Model

$$\beta_{00k} = \gamma_{000} + u_{00k}$$
$$\beta_{10k} = \gamma_{100}$$
$$\beta_{20k} = \gamma_{200}$$
$$\beta_{30k} = \gamma_{300}$$
$$\beta_{40k} = \gamma_{400}$$

The equation for the school-specific intercept includes a parameter for an overall fixed intercept, γ_{000}, plus a random effect, u_{00k}, associated with the school. Thus, the intercepts are allowed to vary randomly from school to school, as in Model 4.1. However, the effects (β_{10k} through β_{40k}) associated with each of the covariates measured at the student level are not allowed to vary from school to school. This means that the effects of being female, being a minority student, of kindergarten math achievement, and of student-level SES (socioeconomic status) are assumed to be the same across all schools.

Click the **Mixed** button to view the overall linear mixed model specified for Model 4.2:

Model 4.2: Overall Mixed Model

$$\begin{aligned}
\text{MATHGAIN}_{ijk} = \;\; & \gamma_{000} + \gamma_{100} * \text{SEX}_{ijk} + \gamma_{200} * \text{MINORITY}_{ijk} \\
& + \gamma_{300} * \text{MATHKIND}_{ijk} + \gamma_{400} * \text{SES}_{ijk} + r_{0jk} + u_{00k} + e_{ijk}
\end{aligned}$$

The HLM specification of the model at each level results in the same overall linear mixed model (Model 4.2) that is fitted in the other software procedures. Table 4.3 shows the correspondence of the HLM notation with the general model notation used in (4.1).

At this point we wish to test Hypothesis 4.2, to decide whether the fixed effects associated with the Level 1 (student-level) covariates should be added to Model 4.1. We set up the likelihood ratio test for Hypothesis 4.2 in HLM before running the analysis for Model 4.2.

To set up a likelihood ratio test of Hypothesis 4.2, click on **Other Settings** and select **Hypothesis Testing**. Enter the **Deviance** (or –2 ML log-likelihood)[3] displayed in the output for Model 4.1 (deviance = 11771.33) and the **Number of Parameters** from Model 4.1 (number of parameters = 4: the fixed intercept, and the three variance components) in the Hypothesis Testing window. After fitting Model 4.2, HLM calculates the appropriate likelihood ratio test statistic and corresponding p-value for Hypothesis 4.2 by subtracting the deviance statistic for Model 4.2 (the reference model) from that for Model 4.1 (the nested model).

After setting up the analysis for Model 4.2, click **Basic Settings**, and enter a new title for this analysis, in addition to a new file name for the saved output. Finally, click **Run Analysis**, and choose **Save as and Run** to save a new .hlm command file for this model. After the analysis has finished running, click **File** and **View Output** to see the results.

Based on the significant ($p < 0.001$) result of the likelihood ratio test for the student-level fixed effects, we reject the null for Hypothesis 4.2 and conclude that the fixed effects at Level 1 should be retained in the model. The results of the test of Hypothesis 4.2 are discussed in more detail in Subsection 4.5.2.

The significant test result for Hypothesis 4.2 also indicates that the fixed effects at Level 1 help to explain residual variation at the student level of the data. A comparison of the estimated residual variance for Model 4.2 vs. that for Model 4.1, both calculated using ML estimation in HLM3, provides evidence that the residual variance at Level 1 is in fact substantially reduced in Model 4.2 (as discussed in Subsection 4.7.2). We retain the fixed effects of the Level 1 covariates in Model 4.2 and proceed to consider Model 4.3.

---

[3]HLM reports the value of the –2 ML log-likelihood for a given model as the model deviance.

Step 3: Build the Level 2 Model by adding Level 2 covariates (Model 4.3).

Before fitting Model 4.3, we need to add the MATHPREP and MATHKNOW variables to the MDM file (as discussed in Subsection 4.2.2). We then need to recreate Model 4.2 in the model-building window.

We obtain Model 4.3 by adding the Level 2 (classroom-level) covariates to Model 4.2. To do this, first click on **Level 2**, then click on the Level 2 model for the intercept term (π_{0jk}); include the nested random classroom effects, r_{0jk}, and add the uncentered versions of the classroom-level variables, YEARSTEA, MATHPREP, and MATHKNOW, to the Level 2 model for the intercept. This results in the following Level 2 model for the classroom-specific intercepts:

Model 4.3: Level 2 Model for Classroom-Specific Intercepts

$$\pi_{0jk} = \beta_{00k} + \beta_{01k}(\text{YEARSTEA}_{jk}) + \beta_{02k}(\text{MATHKNOW}_{jk}) + \beta_{03k}(\text{MATHPREP}_{jk}) + r_{0jk}$$

We see that adding the classroom-level covariates to the model implies that the randomly varying intercepts at Level 1 (the values of π_{0jk}) depend on the school-specific intercept (β_{00k}), the classroom-level covariates, and the random effect associated with each classroom (i.e., the value of r_{0jk}).

The effects of the student-level covariates (π_{1jk} through π_{4jk}) have the same expressions as in Model 4.2 (they are again assumed to remain constant from classroom to classroom).

Adding the classroom-level covariates to the Level 2 model for the intercept causes HLM to include additional Level 3 equations for the effects of the classroom-level covariates in the model-building window, as follows:

Model 4.3: Level 3 Model (Additional Equations)

$$\beta_{01k} = \gamma_{010}$$
$$\beta_{02k} = \gamma_{020}$$
$$\beta_{03k} = \gamma_{030}$$

These equations show that the effects of the Level 2 (classroom-level) covariates are constant at the school level. That is, the classroom-level covariates are not allowed to have effects that vary randomly at the school level, although we could set up the model to allow this.

Click the **Mixed** button in the HLM model-building window to view the overall mixed model for Model 4.3:

Model 4.3: Overall Mixed Model

$$\text{MATHGAIN}_{ijk} = \gamma_{000} + \gamma_{010} * \text{YEARSTEA}_{jk} + \gamma_{020} * \text{MATHKNOW}_{jk} + \gamma_{030} * \text{MATHPREP}_{jk} + \gamma_{100} * \text{SEX}_{ijk} + \gamma_{200} * \text{MINORITY}_{ijk} + \gamma_{300} * \text{MATHKIND}_{ijk} + \gamma_{400} * \text{SES}_{ijk} + r_{0jk} + u_{00k} + e_{ijk}$$

We see that the linear mixed model specified here is the same model that is being fit using the other software procedures. Table 4.3 shows the correspondence of the HLM model parameters with the parameters that we use in (4.1).

After setting up Model 4.3, click **Basic Settings** to enter a new name for this analysis and a new name for the .html output file. Click **OK**, and then click **Run Analysis**, and choose **Save as and Run** to save a new .html command file for this model before fitting the model. After the analysis has finished running, click **File** and **View Output** to see the results.

We use t-tests for Hypotheses 4.3 through 4.5 to decide if we want to keep the fixed effects associated with the Level 2 covariates in Model 4.3 (a likelihood ratio test based on the deviance statistics for Models 4.2 and 4.3 is not appropriate, due to the missing data on the classroom-level covariates). Based on the nonsignificant t-tests for each of the classroom-level fixed effects displayed in the HLM output, we choose Model 4.2 as our preferred model at this stage of the analysis.

Step 4: Build the Level 3 Model by adding the Level 3 covariate (Model 4.4).

In this step, we add the school-level covariate to Model 4.2 to obtain Model 4.4. We first open the .hlm file corresponding to Model 4.2 from the model-building window by clicking **File**, and then **Edit/Run old command file**. After locating the .hlm file saved for Model 4.2, open the file, and click the **Level 3** button. Click on the first Level 3 equation for the intercept that includes the random school effects (u_{00k}). Add the uncentered version of the school-level covariate, HOUSEPOV, to this model for the intercept. The resulting Level 3 model is as follows:

Model 4.4: Level 3 Model for School-Specific Intercepts

$$\beta_{00k} = \gamma_{000} + \gamma_{001}(\text{HOUSEPOV}_k) + u_{00k}$$

The school specific intercepts, β_{00k}, in this model now depend on the overall fixed intercept, γ_{000}, the fixed effect, γ_{001}, of HOUSEPOV, and the random effect, u_{00k}, associated with school k.

After setting up Model 4.4, click **Basic Settings** to enter a new name for this analysis and a new name for the .html output file. Click **OK**, and then click **Run Analysis**, and choose **Save as and Run** to save a new .html command file before fitting the model. After the analysis has finished running, click **File** and **View Output** to see the results.

We test Hypothesis 4.6 using a t-test for the fixed effect associated with HOUSEPOV in Model 4.4. Based on the nonsignificant result of this t-test ($p = 0.25$), we do not retain the fixed effect of HOUSEPOV, and choose Model 4.2 as our final model in the analysis of the Classroom data set.

We now generate residual files to be used in checking model diagnostics (discussed in Section 4.10) for Model 4.2. First, open the .hlm file for Model 4.2, and click **Basic Settings**. In this window, specify names and file types (we choose to save SPSS-format data files in this example) for the Level 1, Level 2, and Level 3 "Residual" files (click on the buttons for each of the three files). The Level 1 file will contain the Level 1 residuals in a variable named **l1resid**, and the conditional predicted values of the dependent variable in a variable named **fitval**. The Level 2 residual file will include a variable named **ebintrcp**, and the Level 3 residual file will include a variable named **eb00**; these variables will contain the Empirical Bayes (EB) predicted values (i.e., the EBLUPs) of the random classroom and school effects, respectively. These three files can be used for exploration of the distributions of the EBLUPs and the Level 1 residuals.

Covariates measured at the three levels of the Classroom data set can also be included in the three files, although we do not use that option here. Rerun the analysis for Model 4.2 to generate the residual files, which will be saved in the same folder where the .html output file was saved. We apply SPSS syntax to the resulting residual files in Section 4.10, to check the diagnostics for Model 4.2.

4.5 Results of Hypothesis Tests

4.5.1 Likelihood Ratio Tests for Random Effects

When the "step-up" approach to model building is used for three-level random intercept models, as for the Classroom data, random effects are usually retained in the model, regardless of the results of significance tests for the associated covariance parameters. However, when tests of significance for random effects are desired, we recommend using likelihood ratio tests, which require fitting a nested model (in which the random effects in question are omitted) and a reference model (in which the random effects are included). Both the nested and reference models should be fitted using REML estimation.

Likelihood ratio tests for the random effects in a three-level random intercept model are not possible when using the HLM3 procedure, because (1) HLM3 uses ML rather than REML estimation, and (2) HLM in general will not allow models to be specified that do not include random effects at each level of the data. Instead, HLM implements alternative chi-square tests for the variance of random effects, which are discussed in more detail in Raudenbush & Bryk (2002).

In this section, we present the results of a likelihood ratio test for the random effects in Model 4.1, based on fitting the reference and nested models using SAS.

Hypothesis 4.1. The random effects associated with classrooms nested within schools can be omitted from Model 4.1.

We calculate the likelihood ratio test statistic for Hypothesis 4.1 by subtracting the value of the –2 REML log-likelihood for Model 4.1 (the reference model) from the value for Model 4.1A (the nested model excluding the random classroom effects). The resulting test statistic is equal to 7.9 (see Table 4.5). Because a variance cannot be less than zero, the null hypothesis value of $\sigma^2_{int:classroom} = 0$ is at the boundary of the parameter space, and the null distribution of the likelihood ratio test statistic is a mixture of χ^2_0 and χ^2_1 distributions, each having equal weight 0.5 (Verbeke & Molenberghs, 2000). The calculation of the p-value for the likelihood ratio test statistic is as follows:

$$p\text{-value} = 0.5 \times p(\chi^2_0 > 7.9) + 0.5 \times p(\chi^2_1 > 7.9) < 0.01$$

Based on the result of this test, we conclude that there is significant variance in the MATHGAIN means between classrooms nested within schools, and we retain the random effects associated with classrooms in Model 4.1 and in all subsequent models. We also retain the random school effects, without testing them, to reflect the hierarchical structure of the data in the model specification.

4.5.2 Likelihood Ratio Tests and *t*-Tests for Fixed Effects

Hypothesis 4.2. The fixed effects, β_1, β_2, β_3, and β_4, associated with the four student-level covariates, MATHKIND, SEX, MINORITY, and SES, should be added to Model 4.1.

We test Hypothesis 4.2 using a likelihood ratio test, based on ML estimation. We calculate the likelihood ratio test statistic by subtracting the –2 ML log-likelihood for Model 4.2 (the reference model including the four student-level fixed effects) from the corresponding value for Model 4.1 (the nested model excluding the student-level fixed effects). The distribution of the test statistic, under the null hypothesis that the four fixed-effect parameters are all equal to zero, is asymptotically a χ^2 with 4 degrees of freedom. Because the p-value

TABLE 4.5: Summary of Hypothesis Test Results for the Classroom Analysis

Hypo-thesis Label	Test	Models Compared (Nested vs. Reference)[a]	Estima-tion Method[b]	Test Statistic Values (Calculation)	p-value
4.1	LRT	4.1A vs. 4.1	REML	$\chi^2(0:1) = 7.9$ (11776.7–11768.8)	< 0.01
4.2	LRT	4.1 vs. 4.2	ML	$\chi^2(4) = 380.4$ (11771.3–11390.9)	< 0.01
4.3	t-test	4.3	REML ML	$t(792) = 0.34$ $t(177) = 0.35$	0.73 0.72
4.4	t-test	4.3	REML ML	$t(792) = 0.97$ $t(177) = 0.97$	0.34 0.34
4.5	t-test	4.3	REML ML	$t(792) = 1.67$ $t(177) = 1.67$	0.10 0.10
4.6	t-test	4.4	REML ML	$t(873) = -1.15$ $t(105) = -1.15$	0.25 0.25

Note: See Table 4.4 for null and alternative hypotheses, and distributions of test statistics under H_0.
[a]Nested models are not necessary for the t-test of Hypotheses 4.4 through 4.6.
[b]The HLM3 procedure uses ML estimation only; we also report results based on REML estimation from SAS `proc mixed`.

is significant ($p < 0.001$), we add the fixed effects associated with the four student-level covariates to the model and choose Model 4.2 as our preferred model at this stage of the analysis.

Recall that likelihood ratio tests are only valid if both the reference and nested models are fitted using the same observations, and the fits of these two models are based on all 1190 cases in the data set.

Hypotheses 4.3, 4.4, and 4.5. The fixed effects, β_5, β_6, and β_7, associated with the classroom-level covariates, YEARSTEA, MATHKNOW, and MATHPREP, should be retained in Model 4.3.

We are unable to use a likelihood ratio test for the fixed effects of all the Level 2 (classroom-level) covariates, because cases are lost due to missing data on the MATHKNOW variable. Instead, we consider individual t-tests for the fixed effects of the classroom-level covariates in Model 4.3.

To illustrate testing Hypothesis 4.3, we consider the t-test reported by HLM3 for the fixed effect, β_5, of YEARSTEA in Model 4.3. Note that HLM used ML estimation for all models fitted in this chapter, so the estimates of the three variance components will be biased and, consequently, the t-tests calculated by HLM will also be biased (see Subsection 2.4.1). However, under the null hypothesis that $\beta_5 = 0$, the test statistic reported by HLM approximately follows a t-distribution with 177 degrees of freedom (see Subsec-

tion 4.11.3 for a discussion of the calculation of degrees of freedom in HLM3). Because the
t-test for Hypothesis 4.3 is not significant ($p = 0.724$), we decide not to include the fixed effect associated with YEARSTEA in the model and conclude that there is not a relationship
between the MATHGAIN score of the student and the years of experience of their teacher.

Similarly, we use t-statistics to test Hypotheses 4.4 and 4.5. Because neither of these
tests is significant (see Table 4.5), we conclude that there is not a relationship between the
MATHGAIN score of the student and the math knowledge or math preparation of their
teacher, as measured for this study. Because the results of hypothesis tests 4.3 through 4.5
were not significant, we do not add the fixed effects associated with any classroom-level
covariates to the model, and proceed with Model 4.2 as our preferred model at this stage
of the analysis.

Hypothesis 4.6. The fixed effect, β_8, associated with the school-level covariate, HOUSE-
POV, should be retained in Model 4.4.

We consider a t-test for Hypothesis 4.6. Under the null hypothesis, the t-statistic reported by HLM3 for the fixed effect of HOUSEPOV in Model 4.4 approximately follows a
t-distribution with 105 degrees of freedom (see Table 4.5). Because this test is not significant
($p = 0.253$), we do not add the fixed effect associated with HOUSEPOV to the model, and
conclude that the MATHGAIN score of a student is not related to the poverty level of the
households in the neighborhood of their school. We choose Model 4.2 as our final model for
the Classroom data analysis.

4.6 Comparing Results across the Software Procedures

In Tables 4.6 to 4.9, we present comparisons of selected results generated by the five software
procedures after fitting Models 4.1, 4.2, 4.3, and 4.4, respectively, to the Classroom data.

4.6.1 Comparing Model 4.1 Results

The initial model fitted to the Classroom data, Model 4.1, is variously described as an
unconditional, variance components, or "means-only" model. It has a single fixed-effect
parameter, the intercept, which represents the mean value of MATHGAIN for all students.
Despite the fact that HLM3 uses ML estimation, and the other five software procedures
use REML estimation for this model, all six procedures produce the same estimates for the
intercept and its standard error.

The REML estimates of the variance components and their standard errors are very
similar across the procedures in SAS, SPSS, R, and Stata, whereas the ML estimates
from HLM are somewhat different. Looking at the REML estimates, the estimated variance of the random school effects ($\sigma^2_{int:school}$) is 77.5, the estimated variance of the
nested random classroom effects ($\sigma^2_{int:classroom}$) is 99.2, and the estimated residual variance (σ^2) is approximately 1028.2; the largest estimated variance component is the residual
variance.

Table 4.6 also shows that the –2 REML log-likelihood values calculated for Model 4.1
agree across the procedures in SAS, SPSS, R, and Stata. The AIC and BIC information
criteria based on the –2 REML log-likelihood values disagree across the procedures that
compute them, owing to the different formulas that are used to calculate them (as discussed
in Subsection 3.6.1). The HLM3 procedure does not calculate these information criteria.

TABLE 4.6: Comparison of Results for Model 4.1

	SAS: proc mixed	SPSS: MIXED	R: lme() function	R: lmer() function	Stata: mixed	HLM3
Estimation Method	REML	REML	REML	REML	REML	ML
Fixed-Effect Parameter	*Estimate (SE)*	*Estimate (SE)*	*Estimate (SE)*	*Estimate (SE)*	*Estimate (SE)*	*Estimate (SE)*
β_0(intercept)	57.43(1.44)	57.43(1.44)	57.43(1.44)	57.43(1.44)	57.43(1.44)	57.43(1.44)[a]
Covariance Parameter	*Estimate (SE)*	*Estimate (SE)*	*Estimate (SE)[b]*	*Estimate (n.c.)*	*Estimate (SE)*	*Estimate (SE)*
$\sigma^2_{int:school}$	77.44(32.61)	77.49(32.62)	77.49[c]	77.49	77.50(32.62)	75.04(31.70)
$\sigma^2_{int:classroom}$	99.19(41.80)	99.23(41.81)	99.23	99.23	99.23(41.81)	100.44(38.45)
σ^2(residual variance)	1028.28(49.04)	1028.23(49.04)	1028.23	1028.23	1028.23(49.04)	1027.73(48.06)
Model Information Criteria						
−2 RE/ML log-likelihood	11768.8	11768.8	11768.8	11768	11768.8	11771.3
AIC	11774.8	11774.8	11776.8	11777	11776.8	n.c.
BIC	11782.8	11790.0	11797.1	11797	11797.1	n.c.

Note: (n.c.) = not computed

Note: 1190 Students at Level 1; 312 Classrooms at Level 2; 107 Schools at Level 3.

[a] Model-based standard errors are presented for the fixed-effect parameter estimates in HLM; robust (sandwich-type) standard errors are also produced in HLM by default.

[b] Standard errors for the estimated covariance parameters are not reported in the output generated by the summary() function in R; 95% confidence intervals for the parameter estimates can be generated by applying the intervals() function in the nlme package to the object containing the results of an lme() fit (e.g., intervals(model4.1.fit)).

[c] These are squared values of the estimated standard deviations reported by the lme() function in R.

4.6.2 Comparing Model 4.2 Results

Model 4.2 includes four additional parameters, representing the fixed effects of the four student-level covariates. As Table 4.7 shows, the estimates of these fixed-effect parameters and their standard errors are very similar across the six procedures. The estimates produced using ML estimation in HLM3 are only slightly different from the estimates produced by the other four procedures.

The estimated variance components in Table 4.7 are all smaller than the estimates in Table 4.6, across the six procedures, owing to the inclusion of the fixed effects of the Level 1 (student-level) covariates in Model 4.2. The estimate of the variance between schools was the least affected, while the residual variance was the most affected (as expected).

Table 4.7 also shows that the –2 REML log-likelihood values agree across the procedures that use REML estimation, as was noted for Model 4.1.

4.6.3 Comparing Model 4.3 Results

Table 4.8 shows that the estimates of the fixed-effect parameters in Model 4.3 (and their standard errors) are once again nearly identical across the five procedures that use REML estimation of the variance components (SAS, SPSS, R, and Stata). The parameter estimates are slightly different when using HLM3, due to the use of ML estimation (rather than REML) by this procedure.

The five procedures that use REML estimation agree quite well (with small differences likely due to rounding error) on the values of the estimated variance components. The variance component estimates from HLM3 are somewhat smaller.

The –2 REML log-likelihood values agree across the procedures in SAS, SPSS, R, and Stata. The AIC and BIC model fit criteria calculated using the –2 REML log-likelihood values for each program differ, due to the different calculation formulas used for the information criteria across the software procedures.

We have included the t-tests and z-tests reported by five of the six procedures for the fixed-effect parameters associated with the classroom-level covariates in Table 4.8, to illustrate the differences in the degrees of freedom computed by the different procedures for the approximate t-statistics. Despite the different methods used to calculate the approximate degrees of freedom for the t-tests (see Subsections 3.11.6 or 4.11.3), the results are nearly identical across the procedures. Note that the z-statistics calculated by Stata do not involve degrees of freedom; Stata refers these test statistics to a standard normal distribution to calculate p-values, and this methodology yields very similar test results. We remind readers that p-values for the t-statistics are not computed when using the `lme4` package version of the `lmer()` function in R to fit the model, because of the different approaches that exist for defining the reference distribution.

4.6.4 Comparing Model 4.4 Results

The comparison of the results produced by the software procedures in Table 4.9 is similar to the comparisons in the other three tables. Test statistics calculated for the fixed effect of HOUSEPOV again show that the procedures that compute the test statistics agree in terms of the results of the tests, despite the different degrees of freedom calculated for the approximate t-statistics.

TABLE 4.7: Comparison of Results for Model 4.2

Estimation Method	SAS: proc mixed REML	SPSS: MIXED REML	R: lme() function REML	R: lmer() function REML	Stata: mixed REML	HLM3 ML
Fixed-Effect Parameter	*Estimate (SE)*	*Estimate (SE)*	*Estimate (SE)*	*Estimate (SE)*	*Estimate (SE)*	*Estimate (SE)*
β_0(Intercept)	282.79(10.85)	282.79(10.85)	282.79(10.85)	282.79(10.85)	282.79(10.85)	282.73(10.83)
β_1(MATHKIND)	−0.47(0.02)	−0.47(0.02)	−0.47(0.02)	−0.47(0.02)	−0.47(0.02)	−0.47(0.02)
β_2(SEX)	−1.25(1.66)	−1.25(1.66)	−1.25(1.66)	−1.25(1.66)	−1.25(1.66)	−1.25(1.65)
β_3(MINORITY)	−8.26(2.34)	−8.26(2.34)	−8.26(2.34)	−8.26(2.34)	−8.26(2.34)	−8.25(2.33)
β_4(SES)	5.35(1.24)	5.35(1.24)	5.35(1.24)	5.35(1.24)	5.35(1.24)	5.35(1.24)
Covariance Parameter	*Estimate (SE)*	*Estimate (SE)*	*Estimate (n.c.)*	*Estimate (n.c.)*	*Estimate (SE)*	*Estimate (SE)*
$\sigma^2_{int:school}$	75.22(25.92)	75.20(25.92)	75.20	75.20	75.20(25.92)	72.88(26.10)
$\sigma^2_{int:classroom}$	83.24(29.37)	83.28(29.38)	83.28	83.29	83.28(29.38)	82.98(28.82)
σ^2(residual variance)	734.59(34.70)	734.57(34.70)	734.57	734.57	734.57(34.70)	732.22(34.30)
Model Information Criteria						
−2 RE/ML log-likelihood	11385.8	11385.8	11385.8	11386	11385.8	11390.9
AIC	11391.8	11391.8	11401.8	11402	11401.8	n.c.
BIC	11399.8	11407.0	11442.4	11442	11442.5	n.c.

Note: (n.c.) = not computed
Note: 1190 Students at Level 1; 312 Classrooms at Level 2; 107 Schools at Level 3.

TABLE 4.8: Comparison of Results for Model 4.3

	SAS: proc mixed	SPSS: MIXED	R: lme() function	R: lmer() function	Stata: mixed	HLM3
Estimation Method	REML	REML	REML	REML	REML	ML
Fixed-Effect Parameter	*Estimate (SE)*	*Estimate (SE)*	*Estimate (SE)*	*Estimate (SE)*	*Estimate (SE)*	*Estimate (SE)*
β_0(Intercept)	282.02(11.70)	282.02(11.70)	282.02(11.70)	282.02(11.70)	282.02(11.70)	281.90(11.65)
β_1(MATHKIND)	−0.48(0.02)	−0.48(0.02)	−0.48(0.02)	−0.48(0.02)	−0.48(0.02)	−0.47(0.02)
β_2(SEX)	−1.34(1.72)	−1.34(1.72)	−1.34(1.72)	−1.34(1.72)	−1.34(1.72)	−1.34(1.71)
β_3(MINORITY)	−7.87(2.42)	−7.87(2.42)	−7.87(2.42)	−7.87(2.42)	−7.87(2.42)	−7.83(2.40)
β_4(SES)	5.42(1.28)	5.42(1.28)	5.42(1.28)	5.42(1.28)	5.42(1.28)	5.43(1.27)
β_5(YEARSTEA)	0.04(0.12)	0.04(0.12)	0.04(0.12)	0.04(0.12)	0.04(0.12)	0.04(0.12)
β_6(MATHPREP)	1.09(1.15)	1.09(1.15)	1.09(1.15)	1.09(1.15)	1.09(1.15)	1.10(1.14)
β_7(MATHKNOW)	1.91(1.15)	1.91(1.15)	1.91(1.15)	1.91(1.15)	1.91(1.15)	1.89(1.14)
Covariance Parameter	*Estimate (SE)*	*Estimate (SE)*	*Estimate (n.c.)*	*Estimate (n.c.)*	*Estimate (SE)*	*Estimate (SE)*
$\sigma^2_{int:school}$	75.24(27.35)	75.19(27.35)	75.19	75.19	75.19(27.35)	72.16(27.44)
$\sigma^2_{int:classroom}$	86.52(31.39)	86.68(31.43)	86.68	86.68	86.68(31.43)	82.69(30.32)
σ^2(residual variance)	713.91(35.47)	713.83(35.47)	713.83	713.83	713.83(35.47)	711.50(35.00)
Model Information Criteria						
−2 RE/ML log-likelihood	10313.0	10313.0	10313.0	10312.0	10313.0	10320.1
AIC	10319.0	10319.0	10335.0	10335.0	10335.0	n.c.
BIC	10327.0	10333.9	10389.8	10390.0	10389.8	n.c.
Tests for Fixed Effects	*t-tests*	*t-tests*	*t-tests*	N/A	*z-tests*	*t-tests*
β_5(YEARSTEA)	$t(792.0) = 0.34$, $p = 0.73$	$t(227.7) = 0.34$, $p = 0.74$	$t(177.0) = 0.34$, $p = 0.73$		$Z = 0.34$, $p = 0.73$	$t(177.0) = 0.35$, $p = 0.72$
β_6(MATHPREP)	$t(792.0) = 0.95$, $p = 0.34$	$t(206.2) = 0.95$, $p = 0.34$	$t(177.0) = 0.95$, $p = 0.34$		$Z = 0.95$, $p = 0.34$	$t(177.0) = 0.97$, $p = 0.34$
β_7(MATHKNOW)	$t(792.0) = 1.67$, $p = 0.10$	$t(232.3) = 1.67$, $p = 0.10$	$t(177.0) = 1.67$, $p = 0.10$		$Z = 1.67$, $p = 0.10$	$t(177.0 = 1.67,)$ $p = 0.10$

Note: (n.c.) = not computed

Note: 1081 Students at Level 1; 285 Classrooms at Level 2; 105 Schools at Level 3.

TABLE 4.9: Comparison of Results for Model 4.4

Estimation Method	SAS: proc mixed REML	SPSS: MIXED REML	R: lme() function REML	R: lmer() function REML	Stata: mixed REML	HLM3 ML
Fixed-Effect Parameter	*Estimate (SE)*	*Estimate (SE)*	*Estimate (SE)*	*Estimate (SE)*	*Estimate (SE)*	*Estimate (SE)*
β_0 (Intercept)	285.06(11.02)	285.06(11.02)	285.06(11.02)	285.06(11.02)	285.06(11.02)	284.92(10.99)
β_1 (MATHKIND)	−0.47(0.02)	−0.47(0.02)	−0.47(0.02)	−0.47(0.02)	−0.47(0.02)	−0.47(0.02)
β_2 (SEX)	−1.23(1.66)	−1.23(1.66)	−1.23(1.66)	−1.23(1.66)	−1.23(1.66)	−1.23(1.65)
β_3 (MINORITY)	−7.76(2.39)	−7.76(2.38)	−7.76(2.38)	−7.76(2.38)	−7.76(2.38)	−7.74(2.37)
β_4 (SES)	5.24(1.25)	5.24(1.25)	5.24(1.25)	5.24(1.25)	5.24(1.25)	5.24(1.24)
β_8 (HOUSEPOV)	−11.44(9.94)	−11.44(9.94)	−11.44(9.94)	−11.44(9.94)	−11.44(9.94)	−11.30(9.83)
Covariance Parameter	*Estimate (SE)*	*Estimate (SE)*	*Estimate (n.c.)*	*Estimate (n.c.)*	*Estimate (SE)*	*Estimate (SE)*
$\sigma^2_{int:school}$	77.77(25.99)	77.76(25.99)	77.76	77.76	77.76(25.99)	74.14(26.16)
$\sigma^2_{int:classroom}$	81.52(29.07)	81.56(29.07)	81.56	81.56	81.56(29.07)	80.96(28.61)
σ^2 (residual variance)	734.44(34.67)	734.42(34.67)	734.42	734.42	734.42(34.67)	732.08(34.29)
Model Information Criteria						
−2 RE/ML log-likelihood	11378.1	11378.1	11378.1	11378	11378.1	11389.6
AIC	11384.1	11384.1	11396.1	11396	11396.1	n.c.
BIC	11392.1	11399.3	11441.8	11442.0	11441.8	n.c.
Tests for Fixed Effects	*t-tests*	*t-tests*	*t-tests*	N/A	*z-tests*	*t-tests*
β_8 (HOUSEPOV)	$t(873.0) = -1.15$, $p = 0.25$	$t(119.5) = -1.15$, $p = 0.25$	$t(105.0) = -1.15$, $p = 0.25$		$Z = -1.15$, $p = 0.25$	$t(105.0) = -1.15$, $p = 0.25$

Note: (n.c.) = not computed
Note: 1190 Students at Level 1; 312 Classrooms at Level 2; 107 Schools at Level 3.

4.7 Interpreting Parameter Estimates in the Final Model

We consider results generated by the HLM3 procedure in this section.

4.7.1 Fixed-Effect Parameter Estimates

Based on the results from Model 4.2, we see that gain in math score in the spring of first grade (MATHGAIN) is significantly related to math achievement score in the spring of kindergarten (MATHKIND), minority status (MINORITY), and student SES. The portion of the HLM3 output for Model 4.2 presented below shows that the individual tests for each of these fixed-effect parameters are significant ($p < 0.05$). The estimated fixed effect of SEX (females relative to males) is the only nonsignificant fixed effect in Model 4.2 ($p = 0.45$).

```
The outcome variable is MATHGAIN
Final estimation of fixed effects:
----------------------------------------------------------------------------
                                        Standard            Approx.
Fixed Effect            Coefficient      Error     T-ratio   d.f.   P-value
----------------------------------------------------------------------------
For INTRCPT1, P0
For INTRCPT2, B00
INTRCPT3, G000          282.726785     10.828453    26.110    106    0.000
For SEX slope, P1
For INTRCPT2, B10
INTRCPT3, G100           -1.251422      1.654663    -0.756    767    0.450
For MINORITY slope, P2
For INTRCPT2, B20
INTRCPT3, G200           -8.253782      2.331248    -3.540    767    0.001
For MATHKIND slope, P3
For INTRCPT2, B30
INTRCPT3, G300           -0.469668      0.022216   -21.141    767    0.000
For SES slope, P4
For INTRCPT2, B40
INTRCPT3, G400            5.348526      1.238400     4.319    767    0.000
----------------------------------------------------------------------------
```

The Greek letters for the fixed-effect parameters in the HLM version of Model 4.2 (see Table 4.3 and Subsection 4.4.5) are shown in the left-most column of the output, in their Latin form, along with the name of the variable whose fixed effect is included in the table. For example, G100 represents the overall fixed effect of SEX (γ_{100} in HLM notation). This fixed effect is actually the intercept in the Level 3 equation for the school-specific effect of SEX (hence, the INTRCPT3 notation). The column labeled "Coefficient" contains the fixed-effect parameter estimate for each of these covariates. The standard errors of the parameter estimates are also provided, along with the T-ratios (*t*-test statistics), approximate degrees of freedom (d.f.) for the T-ratios, and the *p*-value. We describe the HLM calculation of degrees of freedom for these approximate *t*-tests in Subsection 4.11.3.

The estimated fixed effect of kindergarten math score, MATHKIND, on math achievement score in first grade, MATHGAIN, is negative (-0.47), suggesting that students with higher math scores in the spring of their kindergarten year have a lower predicted gain in math achievement in the spring of first grade, after adjusting for the effects of other covariates (i.e., SEX, MINORITY, and SES). That is, students doing well in math in kindergarten will not improve as much over the next year as students doing poorly in kindergarten.

Minority students are predicted to have a mean MATHGAIN score that is 8.25 units lower than their nonminority counterparts, after adjusting for the effects of other covariates. In addition, students with higher SES are predicted to have higher math achievement gain than students with lower SES, controlling for the effects of the other covariates in the model.

4.7.2 Covariance Parameter Estimates

The HLM output below presents the estimated variance components for Model 4.2, based on the HLM3 fit of this model.

```
Final estimation of level-1 and level-2 variance components:
-----------------------------------------------------------------
Random Effect          Standard    Variance    df   Chi-square  P-value
                       Deviation   Component
-----------------------------------------------------------------
INTRCPT1,       r0     9.10959      82.98470   205   298.96800   0.000
Level-1,        e     27.05951     732.21715


Final estimation of level-3 variance components:
-----------------------------------------------------------------
Random Effect           Standard    Variance   df    Chi-square   P-value
                        Deviation   Component
-----------------------------------------------------------------
INTRCPT1/INTRCPT2, u00  8.53721      72.88397  106   183.59757    0.000
-----------------------------------------------------------------
```

The variance components in this three-level model are reported in two blocks of output. The first block of output contains the estimated standard deviation of the nested random effects associated with classrooms (labeled r0, and equal to 9.11), and the corresponding estimated variance component (equal to 82.98). In addition, a chi-square test (discussed in the following text) is reported for the significance of this variance component. The first block of output also contains the estimated standard deviation of the residuals (labeled e, and equal to 27.06), and the corresponding estimated variance component (equal to 732.22). No test of significance is reported for the residual variance.

The second block of output above contains the estimated standard deviation of the random effects associated with schools (labeled u00), and the corresponding estimated variance component (equal to 72.88). HLM also reports a chi-square test of significance for the variance component at the school level.

The addition of the fixed effects of the student-level covariates to Model 4.1 (to produce Model 4.2) reduced the estimated residual variance by roughly 29% (estimated residual variance = 1027.73 in Model 4.1, vs. 732.22 in Model 4.2). The estimates of the classroom- and school-level variance components were also reduced by the addition of the fixed effects associated with the student-level covariates, although not substantially (the estimated classroom-level variance was reduced by roughly 17.4%, and the estimated school-level variance was reduced by about 2.9%). This suggests that the four student-level covariates are effectively explaining some of the random variation in the response values at the different levels of the data set, especially at the student level (as expected).

The magnitude of the variance components in Model 4.2 (and the significant chi-square tests reported for the variance components by HLM3) suggests that there is still unexplained random variation in the response values at all three levels of this data set.

We see in the output above that HLM3 produces chi-square tests for the variance components in the output (see Raudenbush & Bryk (2002) for details on these tests). These tests suggest that the variances of the random effects at the school level (u00) and the classroom level (r0) in Model 4.2 are both significantly greater than zero, even after the inclusion of the fixed effects of the student-level covariates. These test results indicate that a significant amount of random variation in the response values at all three levels of this data set remains unexplained. At this point, fixed effects associated with additional covariates could be added to the model, to see if they help to explain random variation at the different levels of the data.

4.8 Estimating the Intraclass Correlation Coefficients (ICCs)

In the context of a three-level hierarchical model with random intercepts, the **intraclass correlation coefficient** (ICC) is a measure describing the similarity (or homogeneity) of observed responses within a given cluster. For each level of clustering (e.g., classroom or school), an ICC can be defined as a function of the variance components. For brevity in this section, we represent the variance of the random effects associated with schools as σ_s^2 (instead of $\sigma_{int:school}^2$), and the variance of the random effects associated with classrooms nested within schools as σ_c^2 (instead of $\sigma_{int:classroom}^2$).

The school-level ICC is defined as the proportion of the total random variation in the observed responses (the denominator in (4.5)) due to the variance of the random school effects (the numerator in (4.5)):

$$\text{ICC}_{school} = \frac{\sigma_s^2}{\sigma_s^2 + \sigma_c^2 + \sigma^2} \tag{4.5}$$

The value of ICC_{school} is high if the total random variation is dominated by the variance of the random school effects. In other words, the ICC_{school} is high if the MATHGAIN scores of students in the same school are relatively homogeneous, but the MATHGAIN scores across schools tend to vary widely.

Similarly, the classroom-level ICC is defined as the proportion of the total random variation (the denominator in (4.6)) due to random between-school and between-classroom variation (the numerator in (4.6)):

$$\text{ICC}_{classroom} = \frac{\sigma_s^2 + \sigma_c^2}{\sigma_s^2 + \sigma_c^2 + \sigma^2} \tag{4.6}$$

This ICC is high if there is little variation in the responses of students within the same classroom (σ^2 is low) compared to the total random variation.

The ICCs for classrooms and for schools are estimated by substituting the estimated variance components from a random intercept model into the preceding formulas. Because variance components are positive or zero by definition, the resulting ICCs are also positive or zero.

The software procedures discussed in this chapter provide clearly labeled variance component estimates in the computer output when fitting a random intercepts model, allowing for easy calculation of estimates of these ICCs. We can use the estimated variance components from Model 4.1 to compute estimates of the intraclass correlation coefficients (ICCs) defined in (4.5) and (4.6). We estimate the ICC of observations on students within the same school to be $77.5/(77.5 + 99.2 + 1028.2) = 0.064$, and we estimate the ICC of observations on students within the same classroom nested within a school to be $(77.5 + 99.2)/(77.5 +$

99.2 + 1028.2) = 0.147. Observations on students in the same school are modestly correlated, while observations on students within the same classroom have a somewhat higher correlation.

To further illustrate ICC calculations, we consider the marginal variance-covariance matrix V_k implied by Model 4.1 for a hypothetical school, k, having two classrooms, with the first classroom having two students, and the second having three students. The first two rows and columns of this matrix correspond to observations on the two students from the first classroom, and the last three rows and columns correspond to observations on the three students from the second classroom:

$$
V_k = \left(
\begin{array}{cc}
\left(
\begin{array}{cc}
\sigma_s^2 + \sigma_c^2 + \sigma^2 & \sigma_s^2 + \sigma_c^2 \\
\sigma_s^2 + \sigma_c^2 & \sigma_s^2 + \sigma_c^2 + \sigma^2 \\
\sigma_s^2 & \sigma_s^2 \\
\sigma_s^2 & \sigma_s^2 \\
\sigma_s^2 & \sigma_s^2
\end{array}
\right)
&
\begin{array}{ccc}
\sigma_s^2 & \sigma_s^2 & \sigma_s^2 \\
\sigma_s^2 & \sigma_s^2 & \sigma_s^2 \\
\left(
\begin{array}{ccc}
\sigma_s^2 + \sigma_c^2 + \sigma^2 & \sigma_s^2 + \sigma_c^2 & \sigma_s^2 + \sigma_c^2 \\
\sigma_s^2 + \sigma_c^2 & \sigma_s^2 + \sigma_c^2 + \sigma^2 & \sigma_s^2 + \sigma_c^2 \\
\sigma_s^2 + \sigma_c^2 & \sigma_s^2 + \sigma_c^2 & \sigma_s^2 + \sigma_c^2 + \sigma^2
\end{array}
\right)
\end{array}
\end{array}
\right)
$$

The corresponding marginal correlation matrix for these observations can be calculated by dividing all elements in the matrix above by the total variance of a given observation, $[var(y_{ijk}) = \sigma_s^2 + \sigma_c^2 + \sigma^2]$, as shown below. The ICCs defined in (4.5) and (4.6) can easily be identified in this implied correlation matrix:

$$
V_k(corr) = \left(
\begin{array}{cc}
\left(
\begin{array}{cc}
1 & \frac{\sigma_s^2 + \sigma_c^2}{\sigma_s^2 + \sigma_c^2 + \sigma^2} \\
\frac{\sigma_s^2 + \sigma_c^2}{\sigma_s^2 + \sigma_c^2 + \sigma^2} & 1 \\
\frac{\sigma_s^2}{\sigma_s^2 + \sigma_c^2 + \sigma^2} & \frac{\sigma_s^2}{\sigma_s^2 + \sigma_c^2 + \sigma^2} \\
\frac{\sigma_s^2}{\sigma_s^2 + \sigma_c^2 + \sigma^2} & \frac{\sigma_s^2}{\sigma_s^2 + \sigma_c^2 + \sigma^2} \\
\frac{\sigma_s^2}{\sigma_s^2 + \sigma_c^2 + \sigma^2} & \frac{\sigma_s^2}{\sigma_s^2 + \sigma_c^2 + \sigma^2}
\end{array}
\right)
&
\begin{array}{ccc}
\frac{\sigma_s^2}{\sigma_s^2 + \sigma_c^2 + \sigma^2} & \frac{\sigma_s^2}{\sigma_s^2 + \sigma_c^2 + \sigma^2} & \frac{\sigma_s^2}{\sigma_s^2 + \sigma_c^2 + \sigma^2} \\
\frac{\sigma_s^2}{\sigma_s^2 + \sigma_c^2 + \sigma^2} & \frac{\sigma_s^2}{\sigma_s^2 + \sigma_c^2 + \sigma^2} & \frac{\sigma_s^2}{\sigma_s^2 + \sigma_c^2 + \sigma^2} \\
\left(
\begin{array}{ccc}
1 & \frac{\sigma_s^2 + \sigma_c^2}{\sigma_s^2 + \sigma_c^2 + \sigma^2} & \frac{\sigma_s^2 + \sigma_c^2}{\sigma_s^2 + \sigma_c^2 + \sigma^2} \\
\frac{\sigma_s^2 + \sigma_c^2}{\sigma_s^2 + \sigma_c^2 + \sigma^2} & 1 & \frac{\sigma_s^2 + \sigma_c^2}{\sigma_s^2 + \sigma_c^2 + \sigma^2} \\
\frac{\sigma_s^2 + \sigma_c^2}{\sigma_s^2 + \sigma_c^2 + \sigma^2} & \frac{\sigma_s^2 + \sigma_c^2}{\sigma_s^2 + \sigma_c^2 + \sigma^2} & 1
\end{array}
\right)
\end{array}
\end{array}
\right)
$$

We obtain estimates of the ICCs from the marginal variance-covariance matrix for the MATHGAIN observations implied by Model 4.1 by using the v option in the **random** statement in SAS **proc mixed**. The estimated 11×11 V_1 matrix for the observations on the 11 students from school 1 is displayed as follows:

```
                      Estimated V Matrix for schoolid 1

Row   Col1      Col2      Col3      Col4      Col5     Col6      Col7      Col8      Col9      Col10     Col11
-------------------------------------------------------------------------------------------------------------
1   1204.910   176.630   176.630    77.442    77.442   77.442    77.442    77.442    77.442    77.442    77.442
2    176.630  1204.910   176.630    77.442    77.442   77.442    77.442    77.442    77.442    77.442    77.442
3    176.630   176.630  1204.910    77.442    77.442   77.442    77.442    77.442    77.442    77.442    77.442
4     77.442    77.442    77.442  1204.910   176.630  176.630   176.630   176.630   176.630   176.630   176.630
5     77.442    77.442    77.442   176.630  1204.910  176.630   176.630   176.630   176.630   176.630   176.630
6     77.442    77.442    77.442   176.630   176.630 1204.910   176.630   176.630   176.630   176.630   176.630
7     77.442    77.442    77.442   176.630   176.630  176.630  1204.910   176.630   176.630   176.630   176.630
8     77.442    77.442    77.442   176.630   176.630  176.630   176.630  1204.910   176.630   176.630   176.630
9     77.442    77.442    77.442   176.630   176.630  176.630   176.630   176.630  1204.910   176.630   176.630
10    77.442    77.442    77.442   176.630   176.630  176.630   176.630   176.630   176.630  1204.910   176.630
11    77.442    77.442    77.442   176.630   176.630  176.630   176.630   176.630   176.630   176.630  1204.910
```

The 3×3 submatrix in the upper-left corner of this matrix corresponds to the marginal variances and covariances of the observations for the three students in the first classroom, and the 8×8 submatrix in the lower-right corner represents the corresponding values for the eight students from the second classroom.

We note that the estimated covariance of observations collected on students in the same classroom is 176.63. This is the sum of the estimated variance of the nested random classroom effects, 99.19, and the estimated variance of the random school effects, 77.44. Observations collected on students attending the same school but having different teachers are estimated to have a common covariance of 77.44, which is the variance of the random school effects. Finally, all observations have a common estimated variance, 1204.91, which is equal to the sum of the three estimated variance components in the model (99.19 + 77.44 + 1028.28 = 1204.91), and is the value along the diagonal of this matrix.

The marginal variance-covariance matrices for observations on students within any given school would have the same structure, but would be of different dimensions, depending on the number of students within the school. Observations on students in different schools will have zero covariance, because they are assumed to be independent of each other.

The estimated marginal correlations of observations for students within school 1 implied by Model 4.1 can be derived by using the `vcorr` option in the `random` statement in SAS `proc mixed`. Note in the corresponding SAS output below that observations on different students within the same classroom in this school have an estimated marginal correlation of 0.1466, and observations on students in different classrooms within this school have an estimated correlation of 0.06427. These results match our initial ICC calculations based on the estimated variance components.

Covariates are not considered in the classical definitions of the ICC, either based on the random intercept model or the marginal model; however, covariates can easily be accommodated in the mixed model framework in either model setting. The ICC may be calculated from a model without fixed effects of other covariates (e.g., Model 4.1) or for a model including these fixed effects (e.g., Models 4.2 or 4.3). In either case, we can obtain the ICCs from the labeled variance component estimates or from the estimated marginal correlation matrix, as described earlier.

```
               Estimated V Correlation Matrix for schoolid 1

Row   Col1   Col2   Col3   Col4   Col5   Col6   Col7   Col8  Col9   Col10  Col11
---------------------------------------------------------------------------------
1    1.000  0.147  0.147  0.064  0.064  0.064  0.064  0.064  0.064  0.064  0.064
2    0.147  1.000  0.147  0.064  0.064  0.064  0.064  0.064  0.064  0.064  0.064
3    0.147  0.147  1.000  0.064  0.064  0.064  0.064  0.064  0.064  0.064  0.064
4    0.064  0.064  0.064  1.000  0.147  0.147  0.147  0.147  0.147  0.147  0.147
5    0.064  0.064  0.064  0.147  1.000  0.147  0.147  0.147  0.147  0.147  0.147
6    0.064  0.064  0.064  0.147  0.147  1.000  0.147  0.147  0.147  0.147  0.147
7    0.064  0.064  0.064  0.147  0.147  0.147  1.000  0.147  0.147  0.147  0.147
8    0.064  0.064  0.064  0.147  0.147  0.147  0.147  1.000  0.147  0.147  0.147
9    0.064  0.064  0.064  0.147  0.147  0.147  0.147  0.147  1.000  0.147  0.147
10   0.064  0.064  0.064  0.147  0.147  0.147  0.147  0.147  0.147  1.000  0.147
11   0.064  0.064  0.064  0.147  0.147  0.147  0.147  0.147  0.147  0.147  1.000
```

4.9   Calculating Predicted Values

4.9.1   Conditional and Marginal Predicted Values

In this section, we use the estimated fixed effects in Model 4.2, generated by the HLM3 procedure, to write formulas for calculating predicted values of MATHGAIN. Recall that three different sets of predicted values can be generated: conditional predicted values including the EBLUPs of the random school and classroom effects, and marginal predicted values based only on the estimated fixed effects. For example, considering the estimates for the fixed effects in Model 4.2, we can write a formula for the **conditional predicted values** of MATHGAIN for a student in a given classroom:

$$\widehat{\text{MATHGAIN}}_{ijk} = \ 282.73 - 0.47 \times \text{MATHKIND}_{ijk} - 1.25 \times \text{SEX}_{ijk}$$
$$- 8.25 \times \text{MINORITY}_{ijk} + 5.35 \times \text{SES}_{ijk} + \widehat{u}_k + \widehat{u}_{j|k} \qquad (4.7)$$

This formula includes the EBLUPs of the random effect for this student's school, u_k, and the random classroom effect for this student, $u_{j|k}$. Residuals calculated based on these conditional predicted values should be used to assess assumptions of normality and constant variance for the residuals (see Subsection 4.10.2). A formula similar to (4.7) that omits the EBLUPs of the random classroom effects $(u_{j|k})$ could be written for calculating a second set of conditional predicted values specific to schools (4.8):

$$\widehat{\text{MATHGAIN}}_{ijk} = \ 282.73 - 0.47 \times \text{MATHKIND}_{ijk} - 1.25 \times \text{SEX}_{ijk}$$
$$- 8.25 \times \text{MINORITY}_{ijk} + 5.35 \times \text{SES}_{ijk} + \widehat{u}_k \qquad (4.8)$$

A third set of **marginal predicted values**, based on the marginal distribution of MATHGAIN responses implied by Model 4.2, can be calculated based only on the estimated fixed effects:

$$\widehat{\text{MATHGAIN}}_{ijk} = \ 282.73 - 0.47 \times \text{MATHKIND}_{ijk} - 1.25 \times \text{SEX}_{ijk}$$
$$- 8.25 \times \text{MINORITY}_{ijk} + 5.35 \times \text{SES}_{ijk}$$

These predicted values represent average values of the MATHGAIN response (across schools and classrooms) for all students having given values on the covariates.

We discuss how to obtain both conditional and marginal predicted values based on the observed data using SAS, SPSS, R, and Stata in Chapter 3 and Chapters 5 through 7, respectively. Readers can refer to Subsection 4.4.5 for details on obtaining conditional predicted values in HLM.

4.9.2   Plotting Predicted Values Using HLM

The HLM software has several convenient graphical features that can be used to visualize the fit of a linear mixed model. For example, after fitting Model 4.2 in HLM, we can plot the marginal predicted values of MATHGAIN as a function of MATHKIND for each level of MINORITY, based on the estimated fixed effects in Model 4.2. In the model-building window of HLM, click **File**, **Graph Equations**, and then **Model graphs**. In the Equation Graphing window, we set the parameters of the plot. First, set the Level 1 **X focus** to be MATHKIND, which will set the horizontal axis of the graph. Next, set the first Level 1 **Z focus** to be MINORITY. Finally, click on **OK** in the main Equation Graphing window to generate the graph in Figure 4.4.

We can see the significant negative effect of MATHKIND on MATHGAIN in Figure 4.4, along with the gap in predicted MATHGAIN for students with different minority status. The fitted lines are parallel because we did not include an interaction between MATHKIND and MINORITY in Model 4.2. We also note that the values of SES and SEX are held fixed at their mean when calculating the marginal predicted values in Figure 4.4.

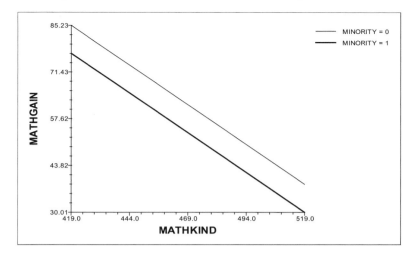

FIGURE 4.4: Marginal predicted values of MATHGAIN as a function of MATHKIND and MINORITY, based on the fit of Model 4.2 in HLM3.

We can also generate a graph displaying the fitted conditional MATHGAIN values as a function of MATHKIND for a sample of individual schools, based on both the estimated fixed effects and the predicted random school effects (i.e., EBLUPs) resulting from the fit of Model 4.2. In the HLM model-building window, click **File**, **Graph Equations**, and then **Level 1 equation graphing**. First, choose MATHKIND as the Level 1 **X focus**. For **Number of groups** (Level 2 units or Level 3 units), select **First ten groups**. Finally, set **Grouping** to be **Group at level 3**, and click OK. This plots the conditional predicted values of MATHGAIN as a function of MATHKIND for the first ten schools in the data set, in separate panels (not displayed here).

4.10   Diagnostics for the Final Model

In this section we consider diagnostics for our final model, Model 4.2, fitted using ML estimation in HLM.

4.10.1   Plots of the EBLUPs

Plots of the EBLUPs for the random classroom and school effects from Model 4.2 were generated by first saving the EBLUPs from the HLM3 procedure in SPSS data files (see Subsection 4.4.5), and then generating the plots in SPSS. Figure 4.5 below presents a normal Q–Q plot of the EBLUPs for the random classroom effects. This plot was created using the EBINTRCP variable saved in the Level 2 residual file by the HLM3 procedure:

```
PPLOT
/VARIABLES=ebintrcp
/NOLOG
/NOSTANDARDIZE
/TYPE=Q-Q
```

Normal Q-Q Plot of ebintrcp

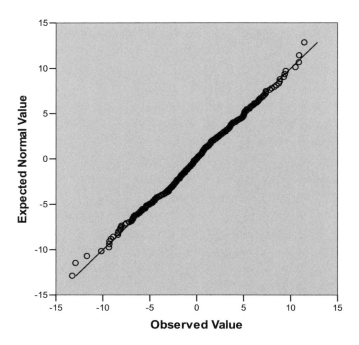

FIGURE 4.5: EBLUPs of the random classroom effects from Model 4.2, plotted using SPSS.

```
/FRACTION=BLOM
/TIES=MEAN
/DIST=NORMAL.
```

We do not see evidence of any outliers in the random classroom effects, and the distribution of the EBLUPs for the random classroom effects is approximately normal. In the next plot (Figure 4.6), we investigate the distribution of the EBLUPs for the random school effects, using the EB00 variable saved in the Level 3 residual file by the HLM3 procedure:

```
PPLOT
/VARIABLES=eb00
/NOLOG
/NOSTANDARDIZE
/TYPE=Q-Q
/FRACTION=BLOM
/TIES=MEAN
/DIST=NORMAL.
```

We do not see any evidence of a deviation from a normal distribution for the EBLUPs of the random school effects, and more importantly, we do not see any extreme outliers. Plots such as these can be used to identify EBLUPs that are potential outliers, and further investigate the clusters (e.g., schools or classrooms) associated with the extreme EBLUPs. Note that evidence of a normal distribution in these plots does not always imply that the distribution of the random effects is in fact normal (see Subsection 2.8.3).

Normal Q-Q Plot of eb00

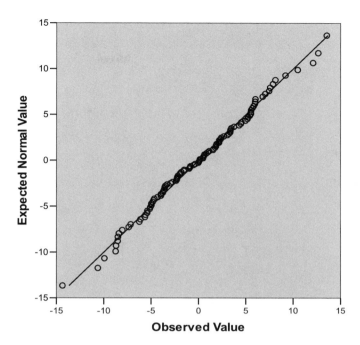

FIGURE 4.6: EBLUPs of the random school effects from Model 4.2, plotted using SPSS.

4.10.2 Residual Diagnostics

In this section, we investigate the assumptions of normality and constant variance for the residuals, based on the fit of Model 4.2. These plots were created in SPSS, using the Level 1 residual file generated by the HLM3 procedure. We first investigate a normal Q–Q plot for the residuals:

```
PPLOT
/VARIABLES=l1resid
/NOLOG
/NOSTANDARDIZE
/TYPE=Q-Q
/FRACTION=BLOM
/TIES=MEAN
/DIST=NORMAL.
```

If the residuals based on Model 4.2 followed an approximately normal distribution, all of the points in Figure 4.7 would lie on or near the straight line included in the figure. We see a deviation from this line at the tails of the distribution, which suggests a long-tailed distribution of the residuals (since only the points at the ends of the distribution deviate from normality). There appear to be small sets of extreme negative and positive residuals that may warrant further investigation. Transformations of the response variable (MATHGAIN) could also be performed, but the scale of the MATHGAIN variable (where some values are negative) needs to be considered; for example, a log transformation of the response would not be possible without first adding a constant to each response to produce a positive value.

Normal Q-Q Plot of l1resid

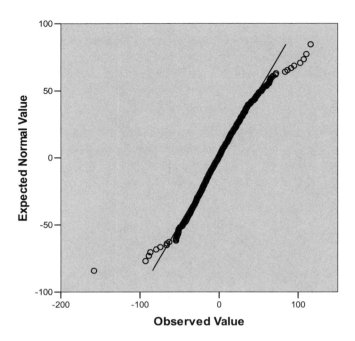

FIGURE 4.7: Normal quantile–quantile (Q–Q) plot of the residuals from Model 4.2, plotted using SPSS.

Next, we investigate a scatter plot of the conditional residuals vs. the fitted MATHGAIN values, which include the EBLUPs of the random school effects and the nested random classroom effects. These fitted values are saved by the HLM3 procedure in a variable named FITVAL in the Level 1 residual file. We investigate this plot to get a visual sense of whether or not the residuals have constant variance:

```
GRAPH
/SCATTERPLOT(BIVAR) = fitval WITH llresid
/MISSING = LISTWISE .
```

We have edited the scatter plot in SPSS (Figure 4.8) to include the fit of a smooth Loess curve, indicating the relationship of the fitted values with the residuals, in addition to a dashed reference line set at zero.

We see evidence of nonconstant variance in the residuals in Figure 4.8. We would expect there to be no relationship between the fitted values and the residuals (a line fitted to the points in this plot should look like the reference line, representing the zero mean of the residuals), but the Loess smoother shows that the residuals tend to get larger for larger predicted values of MATHGAIN.

This problem suggests that the model may be misspecified; there may be omitted covariates that would explain the large positive values and the low negative values of MATHGAIN that are not being well fitted. Scatter plots of the residuals against other covariates would be useful to investigate at this point, as there might be nonlinear relationships of the covariates with the MATHGAIN response that are not being captured by the strictly linear fixed effects in Model 4.2.

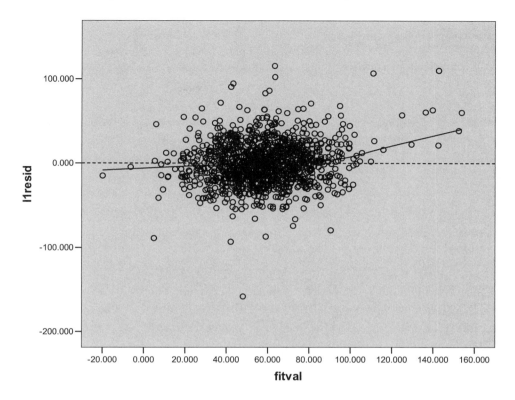

FIGURE 4.8: Residual vs. fitted plot from SPSS.

4.11 Software Notes

4.11.1 REML vs. ML Estimation

The procedures in SAS, SPSS, R, and Stata use restricted maximum likelihood (REML) estimation as the default estimation method for fitting models with nested random effects to three-level data sets. These four procedures estimate the variance and covariance parameters using REML (where ML estimation is also an option), and then use the estimated marginal V matrix to estimate the fixed-effect parameters in the models using generalized least squares (GLS). The procedure available in HLM (HLM3) utilizes ML estimation when fitting three-level models with nested random effects.

4.11.2 Setting up Three-Level Models in HLM

In the following text, we note some important differences in setting up three-level models using the HLM software as opposed to the other four packages:

- Three data sets, corresponding to the three levels of the data, are required to fit an LMM to a three-level data set. The other procedures require that all variables for each level of the data be included in a single data set, and that the data be arranged in the "long" format displayed in Table 4.2.

- Models in HLM are specified in multiple parts. For a three-level data set, Level 1, Level 2, and Level 3 models are identified. The Level 2 models are for the effects of covariates measured on the Level 1 units and specified in the Level 1 model; and the Level 3 models are for the effects of covariates measured on the Level 2 units and specified in the Level 2 models.

- In models for three-level data sets, the effects of any of the Level 1 predictors (including the intercept) are allowed to vary randomly across Level 2 and Level 3 units. Similarly, the effects of Level 2 predictors (including the intercept) are allowed to vary randomly across Level 3 units. In the models fitted in this chapter, we have allowed only the intercepts to vary randomly at different levels of the data.

4.11.3 Calculation of Degrees of Freedom for *t*-Tests in HLM

The degrees of freedom for the approximate *t*-statistics calculated by the HLM3 procedure, and reported for Hypotheses 4.3 through 4.6, are described in this subsection.

Level 1 Fixed Effects: df = number of Level 1 observations (i.e., number of students) − number of random effects at Level 2 − number of random effects at Level 3 − number of fixed-effect parameters associated with the covariates at Level 1.

For example, for the *t*-tests for the fixed effects associated with the Level 1 (student-level) covariates in Model 4.2, we have df = $1190 - 312 - 107 - 4 = 767$.

Level 2 Fixed Effects: df = number of random effects at Level 2 − number of random effects at Level 3 − number of fixed effects at Level 2.

For example, in Model 4.3, we have df = $285 - 105 - 3 = 177$ for the *t*-tests for the fixed effects associated with the Level 2 (classroom-level) covariates, as shown in Table 4.5. Note that there are three fixed effects at Level 2 (the classroom level) in Model 4.3.

Level 3 Fixed Effects: df = number of random effects at Level 3 − number of fixed effects at Level 3.

Therefore, in Model 4.4, we have df = $107 - 2 = 105$ for the *t*-test for the fixed effect associated with the Level 3 (school-level) covariate, as shown in Table 4.5. The fixed intercept is considered to be a Level 3 fixed effect, and there is one additional fixed effect at Level 3.

4.11.4 Analyzing Cases with Complete Data

We mention in the analysis of the Classroom data that likelihood ratio tests are not possible for Hypotheses 4.3 through 4.5, due to the presence of missing data for some of the Level 2 covariates.

An alternative way to approach the analyses in this chapter would be to begin with a data set having cases with complete data for all covariates. This would make either likelihood ratio tests or alternative tests (e.g., *t*-tests) appropriate for any of the hypotheses that we test. In the Classroom data set, MATHKNOW is the only classroom-level covariate with missing data. Taking that into consideration, we include the following syntax for each of the software packages that could be used to derive a data set where only cases with complete data on all covariates are included.

In SAS, the following data step could be used to create a new SAS data set, `classroom_nomiss`, which contains only observations with complete data:

```
data classroom_nomiss;
   set classroom;
   if mathknow ne .;
run;
```

In SPSS, the following syntax can be used to select cases that do not have missing data on MATHKNOW (the resulting data set should be saved under a different name):

```
FILTER OFF.
USE ALL.
SELECT IF (not MISSING (mathknow)).
EXECUTE.
```

In R, we could create a new data frame object excluding those cases with missing data on MATHKNOW:

```
> class.nomiss <- subset(class, !is.na(mathknow))
```

In Stata, the following command could be used to delete cases with missing data on MATHKNOW:

```
. keep if mathknow ! = .
```

Finally, in HLM, this can be accomplished by selecting **Delete data when** ... **making MDM** (rather than when running the analysis) when setting up the MDM file.

4.11.5   Miscellaneous Differences

Less critical differences between the five software procedures in terms of fitting three-level models are highlighted in the following text:

- Procedures in the HLM software package automatically generate both model-based standard errors and robust (or sandwich-type) standard errors for estimated fixed effects. The two different sets of standard errors are clearly distinguished in the HLM output. The robust standard errors are useful to report if one is unsure about whether the marginal variance-covariance matrix for the data has been correctly specified; if the robust standard errors differ substantially from the model-based standard errors, we would recommend reporting the robust standard errors (for more details see Raudenbush & Bryk (2002)). Robust standard errors can be obtained in SAS by using the `empirical` option when invoking `proc mixed`.

- EBLUPs for random effects cannot be calculated in SPSS when fitting models to data sets with multiple levels of clustering, such as the Classroom data.

- Fitting three-level random intercept models using the `MIXED` command in SPSS tends to be computationally intensive, and can take longer than in the other software procedures.

- The `mixed` command in Stata reports z-tests for the fixed effects, rather than the t-tests reported by the other four procedures. The z-tests are asymptotic, and thus require large sample sizes at all three levels of the data.

4.12 Recommendations

Three-level models for cross-sectional data introduce the possibility of extremely complex random-effects structures. In the example analyses presented in this chapter, we only considered models with random intercepts; we could have allowed the relationships of selected covariates at Level 1 (students) of the Classroom data to randomly vary across classrooms and schools, and the relationships of selected covariates at Level 2 (classrooms) to vary across schools. The decision to include many additional random effects in a three-level model will result in a much more complex implied covariance structure for the dependent variable, including several covariance parameters (especially if an unstructured D matrix is used, which is the default random-effects covariance structure in HLM and the two R functions). This may result in estimation difficulties, or the software appearing to "hang" or "freeze" when attempting to estimate a model. For this reason, we only recommend including a large number of random effects (above and beyond random intercepts) at higher levels if there is *explicit research interest* in empirically describing (and possibly attempting to explain) the variance in the relationships of selected covariates with the dependent variable across higher-level units. Including random intercepts at Level 2 and Level 3 of a given three-level data set will typically result in a reasonable implied covariance structure for a given continuous dependent variable in a cross-sectional three-level data set.

Because three-level models do introduce the possibility of allowing many relationships to vary across higher levels of the data hierarchy (e.g., the relationship of student-level SES with mathematical performance varying across schools), the ability to graphically explore variance in both the means of a dependent variable and the relationships of key independent variables with the dependent variable across higher-level units becomes very important when analyzing three-level data. For this reason, having good graphical tools "built in" to a given software package becomes very important. We find that the HLM software provides users with a useful set of "point-and-click" graphing procedures for exploring random coefficients without too much additional work (see Subsection 4.2.2.2). Creating similar graphs and figures in the other software tends to take more work and some additional programming, but is still possible.

5

Models for Repeated-Measures Data: The Rat Brain Example

5.1 Introduction

This chapter introduces the analysis of **repeated-measures data**, in which multiple measurements are made on the same subject (unit of analysis) under different conditions or across time. Repeated-measures data sets can be considered to be a type of two-level data, in which Level 2 represents the subjects and Level 1 represents the repeated measurements made on each subject. Covariates measured at Level 2 of the data (the subject level) describe between-subject variation and Level 1 covariates describe within-subject variation. The data that we analyze are from an experimental study of rats in which the dependent variable was measured in three brain regions for two levels of a drug treatment. Brain region and treatment are crossed **within-subject factors**; measurements were made for the same brain regions and the same treatments within each rat. **Between-subject factors** (e.g., sex or genotype) are not considered in this example.

Repeated-measures data typically arise in an experimental setting, and often involve measurements made on the same subject over time, although time is not a within-subject factor in this example. In Table 5.1, we present examples of repeated-measures data in different research settings.

In this chapter we highlight the SPSS software.

5.2 The Rat Brain Study

5.2.1 Study Description

The data used in this example were originally reported by Douglas et al. (2004).[1] The aim of their experiment was to examine nucleotide activation (guanine nucleotide bonding) in seven different brain nuclei (i.e., brain regions) among five adult male rats. The basal nucleotide activation, measured after treatment with saline solution, was compared to activation in the same region after treatment with the drug carbachol. Activation was measured as the mean optical density produced by autoradiography. We compare activation in a subset of three of the original seven brain regions studied by the authors: the bed nucleus of the stria terminalis (BST), the lateral septum (LS), and the diagonal band of Broca (VDB). The original data layout for this study is shown in Table 5.2.

---

[1]Data from this study are used with permission of the authors and represent a part of their larger study. Experiments were conducted in accordance with the National Institutes of Health Policy on Humane Care and Use of Laboratory Animals.

TABLE 5.1: Examples of Repeated-Measures Data in Different Research Settings

Level of Data		Research Setting		
		Linguistics	Medicine	Anesthesiology
Unit of analysis (Level 2)	Subject variable (random factor)	Person	Patient	Rat
	Subject-level covariates	Age, native language	Sex, severity score	Sex, genotype
Repeated measures (Level 1)	Within-subject factors	Word type, context	Time (minutes after administration of drug)	Brain region, treatment
	Dependent variable	Vowel duration (msec)	Pain relief (visual analog scale)	Nucleotide activation (optical density)

The following SPSS syntax can be used to read in the tab-delimited raw data from the original ratbrain.dat file, assumed to be in the `C:\temp` directory:

```
GET DATA /TYPE = TXT
/FILE = "C:\temp\ratbrain.dat"
/DELCASE = LINE
/DELIMITERS = "\t"
/ARRANGEMENT = DELIMITED
/FIRSTCASE = 2
/IMPORTCASE = ALL
/VARIABLES =
animal A7
Carb.BST F6.2
Carb.LS F6.2
Carb.VDB F6.2
Basal.BST F6.2
Basal.LS F6.2
Basal.VDB F6.2
.
CACHE.
EXECUTE.
```

Before we carry out an analysis of the Rat Brain data using SAS, SPSS, R, Stata, or HLM, we need to restructure the data set into the "long" format. The SPSS syntax to restructure the data is shown as follows. A portion of the restructured Rat Brain data is shown in Table 5.3.

TABLE 5.2: The Rat Brain Data in the Original "Wide" Data Layout. Treatments are "Carb" and "Basal"; brain regions are BST, LS, and VDB

Animal	Carb_BST	Carb_LS	Carb_VDB	Basal_BST	Basal_LS	Basal_VDB
R111097	371.71	302.02	449.70	366.19	199.31	187.11
R111397	492.58	355.74	459.58	375.58	204.85	179.38
R100797	664.72	587.10	726.96	458.16	245.04	237.42
R100997	515.29	437.56	604.29	479.81	261.19	195.51
R110597	589.25	493.93	621.07	462.79	278.33	262.05

```
VARSTOCASES
/MAKE activate FROM Basal_BST Basal_LS Basal_VDB Carb_BST
Carb_LS Carb_VDB
/INDEX = treatment(2) region(3)
/KEEP = animal
/NULL = KEEP.
VALUE LABELS treatment 1 'Basal' 2 'Carbachol'
/ region 1 'BST' 2 'LS' 3 'VDB'.
```

The following variables are included in the Rat Brain data set (note that there are no Level 2 covariates included in this study):

TABLE 5.3: Sample of the Rat Brain Data Set Rearranged in the "Long" Format

Rat (Level 2)	Repeated Measures (Level 1)		
Unit ID	Within-Subject Fixed Factors		Dependent Variable
ANIMAL	TREATMENT	REGION	ACTIVATE
R111097	1	1	366.19
R111097	1	2	199.31
R111097	1	3	187.11
R111097	2	1	371.71
R111097	2	2	302.02
R111097	2	3	449.70
R111397	1	1	375.58
R111397	1	2	204.85
R111397	1	3	179.38
R111397	2	1	492.58
R111397	2	2	355.74
R111397	2	3	459.58
R100797	1	1	458.16
R100797	1	2	245.04
R100797	1	3	237.42
...			

Note: "..." indicates portion of the data not displayed.

Rat (Level 2) Variable

- **ANIMAL** = Unique identifier for each rat

Repeated-Measures (Level 1) Variables

- **TREATMENT** = Level of drug treatment (1 = basal, 2 = carbachol)

- **REGION** = Brain nucleus (1 = BST, 2 = LS, 3 = VDB)

- **ACTIVATE** = Nucleotide activation (the dependent variable)

We recommend sorting the data in ascending order by ANIMAL and then by TREAT-MENT and REGION within each level of ANIMAL prior to running the analysis. Although this sorting is not necessary for the analysis, it makes the output displayed later (e.g., marginal variance-covariance matrices) easier to read.

5.2.2 Data Summary

The following SPSS syntax will generate descriptive statistics for the dependent variable, ACTIVATE, for each level of REGION by TREATMENT:

```
MEANS
TABLES = activate BY treatment BY region
/CELLS MEAN COUNT STDDEV MIN MAX.
```

The following table displays the SPSS output generated by submitting the syntax above.

Report

activate

treatment	region	Mean	N	Std. Deviation	Minimum	Maximum
Basal	BST	428.5060	5	53.31814	366.19	479.81
	LS	237.7440	5	34.67477	199.31	278.33
	VDB	212.2940	5	35.72899	179.38	262.05
	Total	292.8480	15	107.21452	179.38	479.81
Carbachol	BST	526.7100	5	109.86160	371.71	664.72
	LS	435.2700	5	112.44907	302.02	587.10
	VDB	572.3200	5	117.32236	449.70	726.96
	Total	511.4333	15	120.30398	302.02	726.96
Total	BST	477.6080	10	96.47086	366.19	664.72
	LS	336.5070	10	130.35415	199.31	587.10
	VDB	392.3070	10	206.61591	179.38	726.96
	Total	402.1407	30	157.77534	179.38	726.96

The mean activation level is generally higher for carbachol than for the basal treatment in each region. The mean activation also appears to differ by region, with BST having the highest mean activation in the basal condition and VDB having the highest mean activation in the carbachol condition. The standard deviations of activation appear to be much larger for the carbachol treatment than for the basal treatment.

We investigate the data by creating line graphs of the activation in the three brain regions for each animal for both the basal and carbachol treatments:

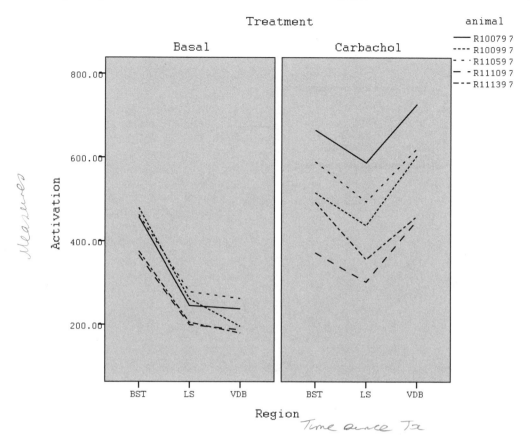

FIGURE 5.1: Line graphs of activation for each animal by region within levels of treatment for the Rat Brain data.

```
GRAPH
/LINE(MULTIPLE) MEAN(activate) BY region BY animal
/PANEL COLVAR = treatment COLOP = CROSS.
```

The effects of treatment and region on the activation within each animal are clear in Figure 5.1. Activation values are consistently higher in the carbachol treatment than in the basal treatment across all regions, and the LS region has a lower mean than the BST region for both carbachol and basal treatments. There also appears to be a greater effect of carbachol treatment in the VDB region than in either the BST or LS region. These observations suggest that the fixed effects of TREATMENT and REGION and the TREATMENT × REGION interaction are likely to be significant in a mixed model for the data.

In Figure 5.1, we also note characteristics of these data that will be useful in specifying the random effects in a mixed model. First, between-animal variation is apparent in the basal treatment, but is even greater in the carbachol treatment. To capture the between-animal variation in both treatments, we initially include in the model a random effect associated with the intercept for each rat. To address the greater between-animal variation in the carbachol treatment, we include a second random animal effect associated with TREATMENT (carbachol vs. basal). This results in each animal having two random intercepts, one for the basal condition and another for the carbachol treatment.

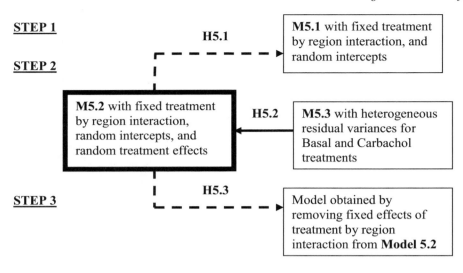

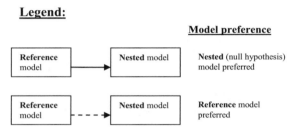

FIGURE 5.2: Model selection and related hypotheses for the analysis of the Rat Brain data.

5.3   Overview of the Rat Brain Data Analysis

We apply the "top-down" modeling strategy discussed in Chapter 2 (Subsection 2.7.1) to the analysis of the Rat Brain data. Subsection 5.3.1 outlines the analysis steps and informally introduces the related models and hypotheses to be tested. Subsection 5.3.2 presents a more formal specification of selected models, and Subsection 5.3.3 presents details about the hypotheses tested. The analysis steps outlined in this section are shown schematically in Figure 5.2. See Subsection 5.3.3 for details on the interpretation of Figure 5.2.

5.3.1   Analysis Steps

Step 1: Fit a model with a "loaded" mean structure (Model 5.1).

Fit a two-level model with a loaded mean structure and random animal-specific intercepts.

Model 5.1 includes fixed effects associated with region, treatment, and the interaction between region and treatment. This model also includes a random effect associated with the intercept for each animal and a residual associated with each observation. The residuals are assumed to be independent and to have the same variance across all levels of region and treatment. The assumption of homogeneous variance for the residuals, in conjunction with the single random effect associated with the intercept for each animal, implies that the six observations on each animal have the same marginal variance and that all pairs of observations have the same (positive) marginal covariance (i.e., a compound symmetry covariance structure).

Step 2: Select a structure for the random effects (Model 5.1 vs. Model 5.2).

Fit Model 5.2 by adding random animal-specific effects of treatment to Model 5.1, and decide whether to retain them in the model.

(*Note:* In this step we do not carry out a formal test of whether the random intercepts should be retained in Model 5.1, but assume that they should be kept, based on the study design.)

We observed in Figure 5.1 that the between-animal variation was greater for the carbachol treatment than for the basal treatment. To accommodate this difference in variation, we add a random animal-specific effect of treatment to Model 5.1 to obtain Model 5.2. The effect of treatment is fixed in Model 5.1 and therefore constant across all animals. The additional random effect associated with treatment included in Model 5.2 allows the implied marginal variance of observations for the carbachol treatment to differ from that for the basal treatment. Model 5.2 can also be interpreted as having two random intercepts per rat, i.e., one for the carbachol treatment and an additional one for the basal treatment.

We test Hypothesis 5.1 to decide if the random treatment effects should be kept in Model 5.2. Based on the test result, we retain the random treatment effects and keep Model 5.2 as the preferred model at this stage of the analysis.

Step 3: Select a covariance structure for the residuals (Model 5.2 vs. Model 5.3).

Fit Model 5.3 with heterogeneous residual variances for the basal and carbachol treatments and decide whether the model should have homogeneous residual variances (Model 5.2) or heterogeneous residual variances (Model 5.3)

We observed in Figure 5.1 that the variance of individual measurements appeared to be greater for the carbachol treatment than for basal treatment. In Step 2, we explored whether this difference can be attributed to between-subject variation by considering random treatment effects in addition to random intercepts. In this step, we investigate whether there is heterogeneity in the residual variances.

In Model 5.2, we assume that the residual variance is constant across all levels of region and treatment. Model 5.3 allows a more flexible specification of the residual variance; i.e., it allows the variance of the residuals to differ across levels of treatment. We test Hypothesis 5.2 to decide if the residual variance is equal for the carbachol and basal treatments. Based on the results of this test, we keep Model 5.2, with homogeneous residual variance, as our preferred model.

Step 4: Reduce the model by removing nonsignificant fixed effects and assess model diagnostics for the final model (Model 5.2).

Decide whether to keep the fixed effects of the region by treatment interaction in Model 5.2.

Based on the result of the test for Hypothesis 5.3, we conclude that the region by treatment interaction effects are significant and consider Model 5.2 to be our final model. We carry out diagnostics for the fit of Model 5.2 using informal graphical methods in SPSS in Section 5.9.

5.3.2 Model Specification

5.3.2.1 General Model Specification

We specify Models 5.2 and 5.3 in this subsection. We do not explicitly specify the simpler Model 5.1. All three models are summarized in Table 5.4. The general specification of Model 5.2 corresponds closely to the syntax used to fit this model in SAS, SPSS, Stata, and R.

The value of $ACTIVATE_{ti}$ for a given observation indexed by t ($t = 1, 2, ..., 6$) on the i-th animal ($i = 1, 2, ..., 5$) can be written as follows:

$$
\begin{aligned}
ACTIVATE_{ti} = \ & \left. \begin{array}{l} \beta_0 + \beta_1 \times REGION1_{ti} + \beta_2 \times REGION2_{ti} \\ + \beta_3 \times TREATMENT_{ti} + \beta_4 \times REGION1_{ti} \times TREATMENT_{ti} \\ + \beta_5 \times REGION2_{ti} \times TREATMENT_{ti} \end{array} \right\} \text{fixed} \\
& \quad \left. + u_{0i} + u_{3i} \times TREATMENT_{ti} + \varepsilon_{ti} \quad \right\} \text{random}
\end{aligned}
$$

$$(5.1)$$

In Model 5.2, we include two indicator variables for region, $REGION1_{ti}$ and $REGION2_{ti}$, which represent the BST and LS regions, respectively. $TREATMENT_{ti}$ is an indicator variable that indicates the carbachol treatment. In our parameterization of the model, we assume that fixed effects associated with REGION="VDB" and TREATMENT="Basal" are set to zero.

The fixed-effect parameters are represented by β_0 through β_5. The fixed intercept β_0 represents the expected value of $ACTIVATE_{ti}$ for the reference levels of region and treatment (i.e., the VDB region under the basal treatment). The parameters β_1 and β_2 represent the fixed effects of the BST and LS regions vs. the VDB region, respectively, for the *basal* treatment (given that the model includes an interaction between region and treatment). The parameter β_3 represents the fixed effect associated with $TREATMENT_{ti}$ (carbachol vs. basal) for the VDB region. The parameters β_4 and β_5 represent the fixed effects associated with the region by treatment interaction. These parameters can either be interpreted as *changes* in the carbachol effect for the BST and LS regions relative to the VDB region, or *changes* in the BST and LS region effects for the carbachol treatment relative to the basal treatment.

The u_{0i} term represents the random intercept associated with animal i, and u_{3i} represents the random effect associated with treatment (carbachol vs. basal) for animal i. We denote the random treatment effect as u_{3i} because this random effect is coupled with the fixed effect of TREATMENT (β_3). We assume that the distribution of the random effects associated with animal i, u_{0i} and u_{3i}, is bivariate normal:

$$
u_i = \begin{pmatrix} u_{0i} \\ u_{3i} \end{pmatrix} \sim \mathcal{N}(0, \boldsymbol{D})
$$

A set of three covariance parameters, $\sigma^2_{int}, \sigma^2_{treat}$, and $\sigma_{int,treat}$, defines the $\boldsymbol{D}$ matrix of variances and covariances for the two random effects in Model 5.2, as follows:

$$
\boldsymbol{D} = \begin{pmatrix} \sigma^2_{int} & \sigma_{int,treat} \\ \sigma_{int,treat} & \sigma^2_{treat} \end{pmatrix}
$$

In the $\boldsymbol{D}$ matrix, σ^2_{int} is the variance of the random intercepts, $\sigma_{int,treat}$ is the covariance of the two random effects, and σ^2_{treat} is the variance of the random treatment effects.

The residuals associated with the six observations on animal i are assumed to follow a multivariate normal distribution:

$$\varepsilon_i = \begin{pmatrix} \varepsilon_{1i} \\ \varepsilon_{2i} \\ \varepsilon_{3i} \\ \varepsilon_{4i} \\ \varepsilon_{5i} \\ \varepsilon_{6i} \end{pmatrix} \sim \mathcal{N}(0, \boldsymbol{R}_i)$$

where $\boldsymbol{R}_i$ is a 6×6 covariance matrix. We assume that the six components of the $\boldsymbol{\varepsilon_i}$ vector are ordered in the same manner as in Table 5.3: The first three residuals are associated with the basal treatment in the BST, LS, and VDB regions, and the next three are associated with the carbachol treatment in the same three regions.

In Model 5.2, the $\boldsymbol{R}_i$ matrix is simply $\boldsymbol{\sigma^2 I_6}$. The diagonal elements of the 6×6 matrix represent the variances of the residuals, all equal to σ^2 (i.e., the variances are homogeneous). The off-diagonal elements represent the covariances of the residuals, which are all zero:

$$\boldsymbol{R}_i = \begin{pmatrix} \sigma^2 & 0 & \cdots & 0 \\ 0 & \sigma^2 & \cdots & 0 \\ \vdots & \vdots & \ddots & \vdots \\ 0 & 0 & \cdots & \sigma^2 \end{pmatrix} = \sigma^2 I_6$$

In Model 5.3, we allow the residual variances to differ for each level of treatment, by including separate residual variances (σ_{basal}^2 and σ_{carb}^2) for the basal and carbachol treatments. This heterogeneous structure for the 6×6 $\boldsymbol{R}_i$ matrix is as follows:

$$\boldsymbol{R}_i = \begin{pmatrix} \sigma_{basal}^2 \boldsymbol{I}_3 & 0 \\ 0 & \sigma_{carb}^2 \boldsymbol{I}_3 \end{pmatrix}$$

The upper-left 3×3 submatrix corresponds to observations for the three regions in the basal treatment, and the lower right submatrix corresponds to the regions in the carbachol treatment. The treatment-specific residual variance is on the diagonal of the two 3×3 submatrices, and zeroes are off the diagonal. We assume that the residuals, ε_{ti} ($t = 1, ..., 6$), conditional on a given animal i, are independent in both Models 5.2 and 5.3.

5.3.2.2 Hierarchical Model Specification

We now present an equivalent hierarchical specification of Model 5.2, using the same notation as in Subsection 5.3.2.1. The correspondence between this notation and the HLM software notation is shown in Table 5.4. The hierarchical model has two components that reflect contributions from the two levels of the data: the repeated measures at Level 1 and the rats at Level 2. We write the **Level 1** component as:

Level 1 Model (Repeated Measures)

$$\begin{aligned} \text{ACTIVATE}_{ti} = \;\; & b_{0i} + b_{1i} \times \text{REGION1}_{ti} + b_{2i} \times \text{REGION2}_{ti} + b_{3i} \times \text{TREATMENT}_{ti} \\ & + b_{4i} \times \text{REGION1}_{ti} \times \text{TREATMENT}_{ti} \\ & + b_{5i} \times \text{REGION2}_{ti} \times \text{TREATMENT}_{ti} + \varepsilon_{ti} \end{aligned} \tag{5.2}$$

where the residuals ε_{ti} have the distribution defined in the general specification of Model 5.2 in Subsection 5.3.2.1, with constant variance and zero covariances in the $\boldsymbol{R}_i$ matrix.

TABLE 5.4: Summary of Selected Models for the Rat Brain Data

	Term/Variable	General Notation	HLM Notation	Model 5.1	Model 5.2	Model 5.3
Fixed effects	Intercept	β_0	β_{00}	✓	✓	✓
	REGION1 (BST vs. VDB)	β_1	β_{10}	✓	✓	✓
	REGION2 (LS vs. VDB)	β_2	β_{20}	✓	✓	✓
	TREATMENT (carbachol vs. basal)	β_3	β_{30}	✓	✓	✓
	REGION1 × TREATMENT	β_4	β_{40}	✓	✓	✓
	REGION2 × TREATMENT	β_5	β_{50}	✓	✓	✓
Random effects	Rat (i)					
	Intercept	u_{0i}	r_{0i}	✓	✓	✓
	TREATMENT (carbachol vs. basal)	u_{3i}	r_{3i}		✓	✓
Residuals	Measure (t) on rat (i)	ε_{ti}	e_{ti}	✓	✓	✓
Covariance parameters (θ_D) for D matrix	Rat level					
	Variance of intercepts	σ^2_{int}	$\tau[1,1]$	✓	✓	✓

TABLE 5.4: (Continued)

Term/Variable	General Notation	HLM Notation	Model 5.1	5.2	5.3
Covariance of intercepts and treatment effects	$\sigma_{int,treat}$	$\tau[2,1]$		✓	✓
Variance of treatment effects	σ_{treat}^2	$\tau[2,2]$		✓	✓
Covariance parameters ($\boldsymbol{\theta_R}$) for $\boldsymbol{R_i}$ matrix — **Repeated measures** Variances of residuals	$\sigma_{basal}^2, \sigma_{carb}^2$	σ^2	$\sigma_{basal}^2 = \sigma_{carb}^2 = \sigma^2$	$\sigma_{basal}^2 = \sigma_{carb}^2 = \sigma^2$	$\sigma_{basal}^2 \neq \sigma_{carb}^2$
Structure	R_i		$\sigma^2 I_6$	$\sigma^2 I_6$	Het[a]

[a]Heterogeneous residual variance across treatments (see Subsection 5.3.2.1).

In the **Level 1** model shown in (5.2), we assume that $ACTIVATE_{ti}$, the nucleotide activation for an individual observation t on rat i, follows a linear model, defined by the animal-specific intercept b_{0i}, the animal-specific effects b_{1i} and b_{2i} of $REGION1_{ti}$ and $REGION2_{ti}$ (BST and LS relative to the VDB region, respectively), the animal-specific effect b_{3i} of $TREATMENT_{ti}$ (carbachol vs. basal treatment), and the animal-specific interaction effects b_{4i} and b_{5i}.

The **Level 2** model describes variation between animals in terms of the animal-specific intercepts (b_{0i}) and the remaining animal-specific effects (b_{1i} through b_{5i}). Although the Level 2 model has six equations (which is consistent with the HLM software specification), there is a simple expression for each one:

Level 2 Model (Rat)

$$
\begin{aligned}
b_{0i} &= \beta_0 + u_{0i} \\
b_{1i} &= \beta_1 \\
b_{2i} &= \beta_2 \\
b_{3i} &= \beta_3 + u_{3i} \\
b_{4i} &= \beta_4 \\
b_{5i} &= \beta_5
\end{aligned}
\tag{5.3}
$$

where

$$
u_i = \begin{pmatrix} u_{0i} \\ u_{3i} \end{pmatrix} \sim \mathcal{N}(0, \boldsymbol{D})
$$

In this **Level 2** model, the intercept b_{0i} for rat i depends on the fixed intercept β_0 (i.e., the ACTIVATE mean for the VDB region in the basal treatment), and a random effect, u_{0i}, associated with rat i. The effect of treatment for rat i, b_{3i}, also depends on a fixed effect (β_3) and a random effect associated with rat i (u_{3i}). All remaining animal-specific effects (b_{1i}, b_{2i}, b_{4i}, and b_{5i}) are defined only by their respective fixed effects $\beta_1, \beta_2, \beta_4$, and β_5.

By substituting the expressions for b_{0i} through b_{5i} from the Level 2 model into the Level 1 model, we obtain the general linear mixed model (LMM) as specified in (5.1).

In Model 5.3, the residuals in the Level 1 model have the heterogeneous variance structure that was defined in Subsection 5.3.2.1.

5.3.3 Hypothesis Tests

Hypothesis tests considered in the analysis of the Rat Brain data are summarized in Table 5.5.

Hypothesis 5.1. The random effects (u_{3i}) associated with treatment for each animal can be omitted from Model 5.2.

Model 5.1 has a single random effect (u_{0i}) associated with the intercept for each animal, and Model 5.2 includes an additional random effect (u_{3i}) associated with treatment for each animal. We do not directly test the significance of the random animal-specific treatment effects in Model 5.2, but rather, we test a null version of the $\boldsymbol{D}$ matrix (for Model 5.1) vs. an alternative version for Model 5.2 (Verbeke & Molenberghs, 2000).

The null hypothesis $\boldsymbol{D}$ matrix has a single positive element, σ_{int}^2, which is the variance of the single random effect (u_{0i}). The alternative hypothesis $\boldsymbol{D}$ matrix is positive semidefinite and contains two additional parameters: $\sigma_{treat}^2 > 0$, which is the variance of the random treatment effects (u_{3i}), and $\sigma_{int,treat}$, which is the covariance of the two random effects, u_{0i} and u_{3i}, associated with each animal.

TABLE 5.5: Summary of Hypotheses Tested in the Analysis of the Rat Brain Data

| Label Null (H_0) | Alternative (H_A) | Test | Models Compared | | Est. Method | Test Stat. Dist. under H_0 |
			Nested Model (H_0)	Ref. Model (H_A)		
5.1 Drop u_{3i}	Retain u_{3i}	LRT	Model 5.1	Model 5.2	REML	$0.5\chi_1^2+0.5\chi_2^2$
5.2 Homogeneous residual variance ($\sigma^2_{carb} = \sigma^2_{basal}$)	Heterogeneous residual variances ($\sigma^2_{carb} \neq \sigma^2_{basal}$)	LRT	Model 5.2	Model 5.3	REML	χ_1^2
5.3 Drop TREATMENT × REGION effects ($\beta_4 = \beta_5 = 0$)	Retain TREATMENT × REGION effects ($\beta_4 \neq 0$, or $\beta_5 \neq 0$)	Type III F-test Wald χ^2test	N/A	Model 5.2	REML	$F(2, 16)^a$ χ_2^2

Note: N/A = Not applicable.
[a]We report the distribution of the F-statistic used by the MIXED command in SPSS, with the Satterthwaite approximation used to compute the denominator degrees of freedom.

$$H_0 \colon D = \begin{pmatrix} \sigma_{int}^2 & 0 \\ 0 & 0 \end{pmatrix}$$

$$H_A \colon D = \begin{pmatrix} \sigma_{int}^2 & \sigma_{int,treat} \\ \sigma_{int,treat} & \sigma_{treat}^2 \end{pmatrix}$$

We use a REML-based likelihood ratio test for Hypothesis 5.1. The test statistic is calculated by subtracting the –2 REML log-likelihood value for Model 5.2 (the reference model) from that for Model 5.1 (the nested model). The asymptotic null distribution of the test statistic is a mixture of χ_1^2 and χ_2^2 distributions, with equal weights of 0.5 (see Subsection 2.6.2.2).

Note that likelihood ratio tests, such as the one used for Hypothesis 5.1, rely on asymptotic (large-sample) theory, so we would not usually carry out this type of test for such a small data set (five rats). Rather, in practice, the random effects would probably be retained without testing, so that the appropriate marginal variance-covariance structure would be obtained for the data set. We present the calculation of this likelihood ratio test for the random effects (and those that follow in this chapter) strictly for illustrative purposes.

Hypothesis 5.2.   The variance of the residuals is constant across both treatments. The null and alternative hypotheses are

$$H_0 \colon \sigma_{basal}^2 = \sigma_{carb}^2$$
$$H_A \colon \sigma_{basal}^2 \neq \sigma_{carb}^2$$

We test Hypothesis 5.2 using a REML-based likelihood ratio test. The test statistic is calculated by subtracting the –2 REML log-likelihood value for Model 5.3 (the reference model with heterogeneous residual variances) from that for Model 5.2 (the nested model). The asymptotic distribution of the test statistic under the null hypothesis is a χ_1^2, and not a mixture of χ^2 distributions, as in the case of Hypothesis 5.1. This is because we are testing whether two variances are equal, which does not involve testing any parameter values that are on the boundary of a parameter space. The single degree of freedom corresponds to the one additional variance parameter in Model 5.3.

Hypothesis 5.3.   The fixed effects associated with the region by treatment interaction can be omitted from Model 5.2.

The null and alternative hypotheses are

$$H_0 \colon \beta_4 = \beta_5 = 0$$
$$H_A \colon \beta_4 \neq 0 \text{ or } \beta_5 \neq 0$$

We test Hypothesis 5.3 using F-tests in the software procedures that provide them (SAS, SPSS, and R), based on REML estimation of the parameters in Model 5.2. By default, the procedures in SAS and SPSS both calculate Type III F-tests, whereas R provides a Type I F-test only. In this case, because the interaction term is the last one added to the model, the Type III and Type I F-tests for the fixed interaction effects are comparable. We consider Wald chi-square tests for the fixed interaction effects in Stata and HLM.

For the results of these hypothesis tests, see Section 5.5.

5.4   Analysis Steps in the Software Procedures

The modeling results for all software procedures are presented and compared in Section 5.6.

5.4.1 SAS

We first read in the raw data from the tab-delimited file, which we assume is stored in the `C:\temp` folder. Note that the data actually begin on the second row of the file, so we use the `firstobs=2` option in the `infile` statement. We also use the `dlm="09"X` option to specify that the raw data file is tab-delimited. In addition, we create an indicator variable TREAT for the carbachol treatment:

```
data ratbrain;
infile "c:\temp\rat_brain.dat" firstobs=2 dlm="09"X;
input animal $ treatment region activate;
if treatment = 1 then treat = 0;
if treatment = 2 then treat = 1;
run;
```

We now proceed with the model-fitting steps.

Step 1: Fit a model with a "loaded" mean structure (Model 5.1).

The SAS syntax to fit Model 5.1 using `proc mixed` is as follows:

```
title "Model 5.1";
proc mixed data = ratbrain covtest;
class region animal;
model activate = region treat region*treat / solution;
random int / subject = animal type = vc solution v vcorr;
run;
```

We have specified the `covtest` option in the `proc mixed` statement to obtain the standard errors of the estimated variance components in the output for comparison with the other software procedures. This option also causes SAS to display a Wald test for the variance of the random effects associated with the animals, which we do not recommend for use in testing whether to include random effects in a model (see Subsection 2.6.3.2).

The `class` statement identifies the categorical variables that are required to specify the model. We include the fixed factor, REGION, and the random factor, ANIMAL, in the `class` statement. The variable TREAT is an indicator variable with a value of one for carbachol and zero for basal treatment, and therefore does not need to be included in the `class` statement.

The `model` statement sets up the fixed-effects portion of Model 5.1. We specify that the dependent variable, ACTIVATE, is a linear function of the fixed effects associated with the REGION factor, the TREAT indicator, and the REGION $\times$ TREAT interaction. The `solution` option requests that the estimate of each fixed-effect parameter be displayed in the output, along with its standard error and a t-test for the parameter.

Software Note: Because REGION is included in the `class` statement, SAS creates three indicator variables for REGION in the model: the first variable is for the BST region (REGION = 1), the second is for the LS region (REGION = 2), and the third is for the VDB region (REGION = 3). The `solution` option in the `model` statement sets the fixed-effect parameter for the highest level of REGION (the VDB region) to zero, and consequently, VDB becomes the reference category.

The `random` statement sets up the random-effects structure for the model. In this case, ANIMAL is identified as the `subject`, indicating that it is a random factor. By specifying `random int`, we include a random effect associated with the intercept for each animal.

We have included several options after the slash (/) in the `random` statement. The `type=` option defines the structure of the D covariance matrix for random effects. Because there is only one random effect specified in Model 5.1, the D matrix is 1×1, and defining its structure is not necessary. Although `proc mixed` uses a variance components structure by default, we specify it explicitly using the `type=vc` option. The `type` option becomes more important later when we specify models with two random effects, one associated with the intercept and another associated with the effect of treatment (Models 5.2 and 5.3).

We have also requested that the estimated marginal covariance matrix V_i and the estimated marginal correlation matrix be displayed in the output for the first animal, by specifying the `v` and `vcorr` options in the `random` statement.

The `solution` option in the `random` statement requests that the predicted random effect (i.e., the EBLUP) associated with each animal also be displayed in the output.

Step 2: Select a structure for the random effects (Model 5.1 vs. Model 5.2).

We now fit Model 5.2 by including an additional random effect associated with the treatment (carbachol vs. basal) for each animal.

```
title "Model 5.2";
proc mixed data = ratbrain covtest;
class animal region;
model activate = region treat region*treat / solution;
random int treat / subject = animal type = un solution v vcorr;
run;
```

The `random` statement from Model 5.1 has been updated to include TREAT. The `type =` option has also been changed to `type = un`, which specifies an "unstructured" covariance structure for the 2×2 D matrix.

The SAS syntax for an alternative random effects specification of Model 5.2 is as follows:

```
title "Alternative Random Effects Specification for Model 5.2";
proc mixed data = ratbrain covtest;
class region animal treatment;
model activate = region treatment region*treatment / solution;
random treatment / subject=animal type=un solution v vcorr;
lsmeans region*treatment / slice=region;
run;
```

Because we include the original TREATMENT variable in the class statement, SAS generates two dummy variables that will be used in the Z_i matrix, one for carbachol and the other for basal treatment. This has the effect of requesting two random intercepts, one for each level of treatment.

The first syntax for Model 5.2, in which a dummy variable, TREAT, was used for treatment, was specified to facilitate comparisons across software procedures. Specifically, the HLM2 procedure cannot directly accommodate categorical variables as predictors, unless they are specified as dummy variables.

We also include an `lsmeans` statement for the `region*treatment` interaction term to create a comparison of the effects of treatment within each region (`/slice=region`). We do not display these results generated by the `lsmeans` statement. Similar results obtained using SPSS are presented in Section 5.7.

To test Hypothesis 5.1, we use a REML-based likelihood ratio test. See the SAS section in Chapter 3 (Subsection 3.4.1) for two different versions of the syntax to carry out a likelihood ratio test for random effects, and Subsection 5.5.1 for the results of the test.

Step 3: Select a covariance structure for the residuals (Model 5.2 vs. Model 5.3).

To fit Model 5.3, which has heterogeneous residual variances for the two levels of TREAT-MENT, we add a `repeated` statement to the syntax used for Model 5.2 and include the original TREATMENT variable in the `class` statement:

```
title "Model 5.3";
proc mixed data = ratbrain;
class animal region treatment;
model activate = region treat region*treat / solution;
random int treat / subject = animal type = un solution v vcorr;
repeated region / subject = animal*treatment group = treatment;
run;
```

The `subject = animal*treatment` option in the `repeated` statement defines two 3×3 blocks on the diagonal of the 6×6 $\boldsymbol{R}_i$ matrix, with each block corresponding to a level of treatment (see Subsection 5.3.2.1). Because the `type=` option is omitted, the default "variance component" structure for each block is used. The `group = treatment` option specifies that the variance components defining the blocks of the $\boldsymbol{R}_i$ matrix be allowed to vary for each level of TREATMENT.

We carry out a likelihood ratio test of Hypothesis 5.2 to decide whether to retain the heterogeneous residual variance structure for the two treatments in Model 5.3. SAS syntax for this test is not shown here; refer to Subsection 3.4.1 for examples of relevant SAS syntax. Based on the nonsignificant result of this test ($p = 0.66$), we decide not to include the heterogeneous residual variances, and keep Model 5.2 as our preferred model at this stage of the analysis.

Step 4: Reduce the model by removing nonsignificant fixed effects (Model 5.2).

The result of the F-test for Hypothesis 5.3 in Model 5.2 is reported in Sections 5.5 and 5.6. The fixed effects associated with the REGION × TREAT interaction are significant ($p < 0.001$), so we retain these fixed effects and keep Model 5.2 as our final model.

5.4.2 SPSS

We assume that the data set created to carry out the data summary in Subsection 5.2.2 is currently open in SPSS. Prior to fitting any models we generate TREAT, an indicator variable for the carbachol treatment:

```
IF (treatment = 1) treat = 0.
IF (treatment = 2) treat = 1.
EXECUTE .
```

Step 1: Fit a model with a "loaded" mean structure (Model 5.1).

We begin the SPSS analysis by setting up the syntax to fit Model 5.1 using the `MIXED` command:

```
* Model 5.1.
MIXED
activate BY region WITH treat
/CRITERIA = CIN(95) MXITER(100) MXSTEP(5) SCORING(1)
SINGULAR(0.000000000001) HCONVERGE(0, ABSOLUTE)
LCONVERGE(0, ABSOLUTE) PCONVERGE(0.000001, ABSOLUTE)
/FIXED = region treat region*treat | SSTYPE(3)
/METHOD = REML
/PRINT = SOLUTION
/RANDOM INTERCEPT | SUBJECT(animal) COVTYPE(VC).
```

In this syntax, ACTIVATE is listed as the dependent variable. The variable REGION is specified as a categorical factor because it appears after the BY keyword. This causes SPSS to generate the appropriate indicator variables for the REGION factor and for any interactions involving REGION. The variable TREAT is specified as a covariate because it appears after the WITH keyword.

The /FIXED subcommand lists the variables and interactions that have associated fixed effects in Model 5.1. These terms include REGION, TREAT, and the REGION × TREAT interaction.

The /METHOD subcommand specifies the REML estimation method, which is the default.

The /PRINT subcommand requests that the SOLUTION for the estimated fixed-effect parameters in the model be displayed in the output.

The /RANDOM subcommand indicates that the only random effect for each level of the SUBJECT variable (ANIMAL) is associated with the INTERCEPT. The covariance structure for the random effects is specified as variance components (the default), using the COVTYPE(VC) syntax. Because there is only a single random effect for each animal included in Model 5.1, there is no need to use a different covariance structure (D is a 1×1 matrix).

Step 2: Select a structure for the random effects (Model 5.1 vs. Model 5.2).

We now fit Model 5.2, which adds a random effect associated with treatment for each animal. This allows the effect of the carbachol treatment (vs. basal) to vary from animal to animal.

```
* Model 5.2.
MIXED
activate BY region WITH treat
/CRITERIA = CIN(95) MXITER(100) MXSTEP(5) SCORING(1)
SINGULAR(0.000000000001) HCONVERGE(0, ABSOLUTE)
LCONVERGE(0, ABSOLUTE) PCONVERGE(0.000001, ABSOLUTE)
/FIXED = region treat region*treat | SSTYPE(3)
/METHOD = REML
/PRINT = SOLUTION G
/RANDOM INTERCEPT treat | SUBJECT(animal) COVTYPE(UN).
```

The /RANDOM subcommand has been updated by adding TREAT, which adds a random treatment effect to the model for each animal. The option COVTYPE(UN) specifies that the 2×2 D matrix defined in Subsection 5.3.2.1 for Model 5.2 is "unstructured." We request that the estimated D matrix be displayed in the output by including the G option in the /PRINT subcommand. Note that the G option requests that a single block of the estimated G matrix (defined in Subsection 2.2.3) be displayed in the output.

To test Hypothesis 5.1, and determine whether the random treatment effects can be omitted from Model 5.2, we perform a likelihood ratio test. We decide to retain the random treatment effects as a result of this significant test (see Subsection 5.5.1) and keep Model 5.2 as the preferred one at this stage of the analysis.

Step 3: Select a covariance structure for the residuals (Model 5.2 vs. Model 5.3).

In this step of the analysis we fit Model 5.3, in which the residual variances in the R_i matrix are allowed to vary for the different treatments.

```
* Model 5.3.
MIXED
activate BY region WITH treat
/CRITERIA = CIN(95) MXITER(100) MXSTEP(5) SCORING(1)
SINGULAR(0.000000000001) HCONVERGE(0, ABSOLUTE)
LCONVERGE(0, ABSOLUTE) PCONVERGE(0.000001, ABSOLUTE)
/FIXED = region treat region*treat | SSTYPE(3)
/METHOD = REML
/PRINT = SOLUTION
/RANDOM INTERCEPT treat | SUBJECT(animal) COVTYPE(UN)
/REPEATED = treatment | SUBJECT(animal*region) COVTYPE(DIAG).
```

The /REPEATED subcommand specifies that repeated measures, uniquely indexed by levels of the TREATMENT variable, are collected for each combination of levels of the ANIMAL and REGION variables, by including the SUBJECT(ANIMAL*REGION) option. Note that this specification of the /REPEATED subcommand is different from that used for proc mixed in SAS. This setup is required in SPSS to model heterogenous residual variances for each level of TREATMENT via the DIAG covariance structure.

A diagonal covariance structure for the R_i matrix is specified by COVTYPE(DIAG). The DIAG covariance type in SPSS means that there are heterogeneous variances for each level of TREATMENT on the diagonal of the 6×6 R_i matrix and that the residuals are not correlated (see Subsection 5.3.2.1). This specification allows observations at different levels of treatment on the same animal to have different residual variances.

After fitting Model 5.3, we test Hypothesis 5.2 using a likelihood ratio test to decide whether we should retain the heterogeneous residual variance structure. Based on the nonsignificant result of this test, we keep the simpler model (Model 5.2) at this stage of the analysis as our preferred model. See Subsection 5.5.2 for the result of this likelihood ratio test.

Step 4: Reduce the model by removing nonsignificant fixed effects (Model 5.2).

We use a Type III F-test to test Hypothesis 5.3 to decide if we wish to retain the fixed effects associated with the REGION $\times$ TREAT interaction in Model 5.2. Because this test (shown in the SPSS output for Model 5.2) is significant, we retain these fixed effects and select Model 5.2 is our final model (see Table 5.8).

We refit Model 5.2 in SPSS, and add syntax to generate pairwise comparisons of the means at each brain region for each treatment to aid interpretation of the significant region by treatment interaction:

```
* Model 5.2 (w/ interaction means).
MIXED
activate BY region WITH treat
/CRITERIA = CIN(95) MXITER(100) MXSTEP(5) SCORING(1)
SINGULAR(0.000000000001) HCONVERGE(0, ABSOLUTE)
```

```
LCONVERGE(0, ABSOLUTE) PCONVERGE(0.000001, ABSOLUTE)
/FIXED = region treat region*treat | SSTYPE(3)
/METHOD = REML
/PRINT = SOLUTION G
/RANDOM INTERCEPT treat | SUBJECT(animal) COVTYPE(UN)
/EMMEANS = TABLES(region) WITH(treat=1) COMPARE ADJ(BON)
/EMMEANS = TABLES(region) WITH(treat=0) COMPARE ADJ(BON).
```

Note that there are two instances of the /EMMEANS subcommand in the preceding syntax. These subcommands request pairwise comparisons of the estimated marginal activation means for each brain region, first for the carbachol treatment (treat=1) and then for the basal treatment (treat=0). A Bonferroni adjustment for multiple comparisons is requested with the ADJ(BON) option.

We consider diagnostics for Model 5.2 using SPSS in Section 5.9.

5.4.3   R

We begin the analysis of the Rat Brain data using R by reading the tab-delimited raw data set (having the "long" structure described in Table 5.3 and variable names in the first row of the raw data) into a data frame object. Recall that the h = T option tells R that the raw data file has a header (first row) containing variable names.

```
> rat.brain <- read.table("C:\\temp\\rat_brain.dat", h = T)
```

We then attach the vectors (or variables) in the rat.brain data frame object to R's working memory:

```
> attach(rat.brain)
```

Because R by default treats the lowest category (alphabetically or numerically) of a categorical fixed factor as the reference category in a model, we recode the REGION and TREATMENT variables to obtain results consistent with those in SAS, SPSS, and Stata. We first create a new factor REGION.F, which has VDB (REGION = 3) as the lowest value (equal to zero). We then create TREAT, which is an indicator variable for the carbachol treatment (TREAT = 1 for carbachol, TREAT = 0 for basal).

```
> region.f <- region
> region.f[region == 1] <- 1
> region.f[region == 2] <- 2
> region.f[region == 3] <- 0
> region.f <- factor(region.f)
> treat <- treatment
> treat[treatment == 1] <- 0
> treat[treatment == 2] <- 1
```

We add these new recoded variables to the rat.brain data frame object:

```
> rat.brain <- data.frame(rat.brain, region.f, treat)
```

Now that the appropriate variables in the Rat Brain data set have been attached to memory, we can analyze the data using the available functions in the nlme and lme4 packages.

5.4.3.1   Analysis Using the `lme()` Function

The `nlme` package first needs to be loaded, so that the `lme()` function can be used in the analysis:

```
> library(nlme)
```

Step 1: Fit a model with a "loaded" mean structure (Model 5.1).

Model 5.1 is fitted to the data using the `lme()` function:

```
> # Model 5.1.
> model5.1.fit <- lme(activate ~ region.f*treat,
random = ~1 | animal, method = "REML", data = rat.brain)
```

We describe each part of this specification of the `lme()` function:

- `model5.1.fit` is the name of the object that contains the results of the fitted linear mixed model.

- The first argument of the function, `activate ~ region.f*treat`, is the model formula, which defines the response variable (`activate`), and the terms with associated fixed effects in the model (`region.f` and `treat`). The asterisk (`*`) requests that the main effects associated with each variable (the REGION factor and the TREAT indicator) be included in the model, in addition to the fixed effects associated with the interaction between the variables.

- The second argument of the function, `random = ~1 | animal`, includes a random effect associated with the intercept (`~1`) for each level of the categorical random factor (`animal`).

- The third argument of the function, `method = "REML"`, tells R that REML estimation should be used for the desired covariance parameters in the model. This is the default estimation method for the `lme()` function.

- The final argument of the function, `data = rat.brain`, indicates the name of the data frame object to be used.

After the function is executed, estimates from the model fit can be obtained using the `summary()` function:

```
> summary(model5.1.fit)
```

Additional results of interest for this LMM fit can be obtained by using other functions in conjunction with the `model5.1.fit` object. For example, one can look at Type I (sequential) F-tests for the fixed effects in this model using the `anova()` function:

```
> anova(model5.1.fit)
```

Step 2: Select a structure for the random effects (Model 5.1 vs. Model 5.2).

We now fit Model 5.2 by updating Model 5.1 to include a second animal-specific random effect associated with the treatment (carbachol vs. basal):

```
> # Model 5.2.
> model5.2.fit <- update(model5.1.fit, random = ~ treat | animal)
```

Note that we use the **update()** function to specify Model 5.2, by modifying the random effects specification in Model 5.1 to include random TREAT effects associated with each animal. The random effects associated with the intercept for each animal will be included in the model by default, and an "unstructured" covariance structure for the D matrix will also be used by default (see Subsection 5.3.2.1 for specification of the D matrix for Models 5.2 and 5.3). We use the **summary()** function to display results generated by fitting Model 5.2:

```
> summary(model5.2.fit)
```

We also use the **anova()** function to obtain Type I F-tests of the fixed effects in Model 5.2:

```
> anova(model5.2.fit)
```

We test Hypothesis 5.1 to decide if we need the random treatment effects, using a likelihood ratio test. This would typically not be done with such a small sample of animals (given the asymptotic nature of likelihood ratio tests), but we perform this test for illustrative purposes. The test statistic is calculated by subtracting the –2 REML log-likelihood value for Model 5.2 (the reference model) from that for Model 5.1 (the value of the test statistic is $275.3 - 249.2 = 26.1$). The –2 REML log-likelihood values can be obtained from the output provided by the **summary()** function for each model. The test statistic has a null distribution that is a mixture of χ_1^2 and χ_2^2 distributions with equal weights of 0.5, so the **anova()** function cannot be used for the p-value. Instead, we calculate a p-value for the test statistic as follows:

```
> 0.5*(1 - pchisq(26.1,1)) + 0.5*(1 - pchisq(26.1,2))
[1] 1.237138e-06
```

See Subsection 5.5.1 for details. The test statistic is significant ($p < 0.001$), so we decide to reject the null hypothesis and retain the random treatment effects in the model; Model 5.2 is our preferred model at this stage of the analysis.

Step 3: Select a covariance structure for the residuals (Model 5.2 vs. Model 5.3).

We now fit Model 5.3, which has a heterogeneous variance structure for the R_i matrix (i.e., it allows the residual variances to differ for the two levels of TREAT). This is accomplished by using the **weights** argument, as shown in the following syntax:

```
> # Model 5.3.
> model5.3.fit <- lme(activate ~ region.f*treat,
random = ~ treat | animal,
weights = varIdent(form = ~ 1 | treat),
data = rat.brain)
```

We use the **summary()** function to obtain estimates of the parameters in Model 5.3:

```
> summary(model5.3.fit)
```

We now test Hypothesis 5.2 with a likelihood ratio test, to decide if we wish to retain the heterogeneous residual variance structure. To calculate the test statistic, we subtract the –2 REML log-likelihood of Model 5.3 (the reference model, with heterogeneous residual variances) from that for Model 5.2 (the nested model), using the **anova()** function:

```
> # Likelihood ratio test for Hypothesis 5.2.
> anova(model5.2.fit, model5.3.fit)
```

The result of this test is not significant ($p = 0.66$), so we keep Model 5.2 as our preferred model at this stage of the analysis (see Subsection 5.5.2).

Step 4: Reduce the model by removing nonsignificant fixed effects (Model 5.2).

The Type I F-tests reported for the fixed effects in Model 5.2 (see Section 5.6) indicate that the fixed effects associated with the REGION × TREAT interaction are significant ($p < 0.05$), so we reject the null hypothesis for Hypothesis 5.3. We therefore retain these fixed effects, and select Model 5.2 as our final model.

Software Note: The Type I F-test in R for the fixed effects associated with the REGION × TREAT interaction is comparable to the Type III F-tests performed in SAS and SPSS because the interaction is added last to the model, and the test is therefore conditional on the main fixed effects of REGION and TREAT (which is also the case for the Type III F-tests reported by SAS and SPSS).

5.4.3.2 Analysis Using the `lmer()` Function

The `lme4` package first needs to be loaded, so that the `lmer()` function can be used in the analysis:

```
> library(lme4)
```

Step 1: Fit a model with a "loaded" mean structure (Model 5.1).

Model 5.1 is fitted to the data using the `lmer()` function:

```
> # Model 5.1.
> model5.1.fit.lmer <- lmer(activate ~ region.f*treat + (1|animal),
data = rat.brain, REML = T)
```

We describe each part of this specification of the `lmer()` function:

- `model5.1.fit.lmer` is the name of the object that contains the results of the fitted linear mixed model.

- The first argument of the function, `activate ~ region.f*treat + (1|animal)`, is the model formula, which defines the response variable (`activate`), and the terms with associated fixed effects in the model (`region.f` and `treat`). The asterisk (`*`) requests that the main effects associated with each variable (the REGION factor and the TREAT indicator) be included in the model, in addition to the fixed effects associated with the interaction between the variables.

- The (`1|animal`) term in the model formula indicates that a random effect associated with the intercept should be included for each level of the categorical random factor `animal`.

- The third argument of the function, `data = rat.brain`, indicates the name of the data frame object to be used.

- The final argument of the function, `REML = T`, tells R that REML estimation should be used for the desired covariance parameters in the model. This is the default estimation method for the `lmer()` function.

After the function is executed, estimates from the model fit can be obtained using the `summary()` function:

```
> summary(model5.1.fit.lmer)
```

Additional results of interest for this LMM fit can be obtained by using other functions in conjunction with the `model5.1.fit.lmer` object. For example, one can look at Type I (sequential) *F*-tests for the fixed effects in this model using the `anova()` function:

```
> anova(model5.1.fit.lmer)
```

> **Software Note:** Applying the `summary()` and `anova()` functions to model fit objects produced by the `lme4` package version of the `lmer()` function does not result in *p*-values for the computed *t*-statistics and *F*-statistics. This is primarily due to the lack of consensus in the literature over appropriate degrees of freedom for these test statistics under the null hypothesis. In general, we recommend use of the `lmerTest` package in R for users interested in testing hypotheses about parameters estimated using the `lmer()` function. In this chapter, we illustrate likelihood ratio tests using selected functions available in the `lme4` package. See Chapter 3 for an example of using the `lmerTest` package, in the case of a random intercept model.

EBLUPs of the random effects for each animal can also be displayed:

```
> ranef(model5.1.fit.lmer)
```

Step 2: Select a structure for the random effects (Model 5.1 vs. Model 5.2).

We now fit Model 5.2 by updating Model 5.1 to include a second animal-specific random effect associated with the treatment (carbachol vs. basal):

```
> # Model 5.2.
> model5.2.fit.lmer <- lmer(activate ~ region.f*treat + (treat|animal),
data = rat.brain, REML = T)
```

Note that we modify the random-effects specification in Model 5.1 to include random TREAT effects associated with each animal. The random effects associated with the intercept for each animal will be included in the model by default, and an "unstructured" covariance structure for the D matrix will also be used by default (see Subsection 5.3.2.1 for specification of the D matrix for Models 5.2 and 5.3). We use the `summary()` function to display results generated by fitting Model 5.2:

```
> summary(model5.2.fit.lmer)
```

We next test Hypothesis 5.1 to decide if we need the random treatment effects, using a likelihood ratio test. This would typically not be done with such a small sample of animals (given the asymptotic nature of likelihood ratio tests), but we perform this test for illustrative purposes. The test statistic is calculated by subtracting the –2 REML log-likelihood value for Model 5.2 (the reference model) from that for Model 5.1 (the value of the test statistic is $275.3 - 249.2 = 26.1$). The –2 REML log-likelihood values can be obtained from the output provided by the `summary()` function for each model. The test statistic has a null distribution that is a mixture of χ_1^2 and χ_2^2 distributions with equal weights of 0.5, so the `anova()` function cannot be used for the *p*-value. Instead, we calculate a *p*-value for the test statistic as follows:

```
> 0.5*(1 - pchisq(26.1,1)) + 0.5*(1 - pchisq(26.1,2))
[1] 1.237138e-06
```

See Subsection 5.5.1 for details. The test statistic is significant ($p < 0.001$), so we decide to reject the null hypothesis and retain the random treatment effects in the model; Model 5.2 is our preferred model at this stage of the analysis.

Step 3: Select a covariance structure for the residuals (Model 5.2 vs. Model 5.3).

Unfortunately, we cannot fit models with heterogeneous residual variances when using the current implementation of the `lmer()` function. See Section 5.4.3.1 for an example of how this model can be fitted using the `lme()` function, and the results of the corresponding hypothesis tests.

Step 4: Reduce the model by removing nonsignificant fixed effects (Model 5.2).

The Type I F-tests reported for the fixed effects in Model 5.2 (see Section 5.6) suggest that the fixed effects associated with the REGION × TREAT interaction are significant (given the large F-statistics, despite the absence of computed p-values), so we reject the null hypothesis for Hypothesis 5.3. We remind readers that ML-based likelihood ratio tests could also be performed at this point to test fixed effects for significance (given larger sample sizes), by fitting a model without the interaction using ML estimation (using the `REML = F` argument in `lmer()`), and then using the `anova()` function to compare the two models (see, for example, Section 4.4.3.2). We therefore retain these fixed effects, and select Model 5.2 as our final model.

5.4.4   Stata

We begin the analysis in Stata by reading the tab-delimited raw data file (having the "long" structure) into Stata's working memory from the `C:\temp` folder:

```
. insheet using "C:\temp\rat_brain.dat", tab
```

Users of web-aware Stata can also import the data directly from the book's web page:

```
. insheet using http://www-personal.umich.edu/~bwest/rat_brain.dat
```

Next, we generate an indicator variable (TREAT) for the carbachol treatment:

```
. gen treat = 0 if treatment == 1
. replace treat = 1 if treatment == 2
```

We now proceed with the analysis steps.

Step 1: Fit a model with a "loaded" mean structure (Model 5.1).

We first fit Model 5.1 using the `mixed` command. Because Stata by default treats the lowest-valued level (alphabetically or numerically) of a categorical factor as the reference category (i.e., REGION = 1, or the BST region), we explicitly declare the value 3 for REGION to be the reference region, using the `ib3.` modifier as indicated below:

```
. * Model 5.1.
. mixed activate ib3.region treat ib3.region#c.treat
    || animal:, covariance(identity) variance reml
```

The `mixed` command has three parts. The first part specifies the fixed effects; the second part, the random effects; and the third part (after a comma), the covariance structure for the random effects, together with miscellaneous options. We describe each portion of the `mixed` syntax for Model 5.1 in the following paragraphs.

The first variable listed after the `mixed` command is the continuous dependent variable, ACTIVATE. In this particular command, we then include fixed effects associated with values 1 and 2 of the categorical factor REGION (using `ib3.region`), a fixed effect associated with value 1 of TREAT, and fixed effects associated with the interaction between REGION and TREAT (using `ib3.region#c.treat`). We note the use of `c.treat`, which more generally indicates that TREAT is a "continuous" variable in this specification of the interaction (we did not indicate factor variable coding for the TREAT effect included in the model). Although the indicator variable for TREAT is of course not strictly continuous, this simplifies the specification of the random effects in Model 5.2, which we will explain shortly.

After the dependent variable and the fixed effects have been identified, two vertical bars (||) precede the specification of the random effects in the model. We list the ANIMAL variable (`animal:`) as the variable that defines clusters of observations. Because no additional variables are listed after the colon, there will only be a single random effect in the model, associated with the intercept for each animal.

The covariance structure for the random effects is specified after the comma following the random effects. The `covariance(identity)` option tells Stata that an identity covariance structure is to be used for the single random effect associated with the intercept in Model 5.1 (this option is not necessary for models that only include a single random effect, because the D matrix will only have a single variance component).

Finally, the `variance` option requests that the estimated variances of the random animal effects and the residuals be displayed in the output (rather than the default estimated standard deviations), and the `reml` option requests REML estimation (as opposed to ML estimation, which is the default).

Information criteria associated with the model fit (i.e., the AIC and BIC statistics) can be obtained by using the following command:

```
. estat ic
```

Once a model has been fitted using the `mixed` command, EBLUPs of the random effects associated with the levels of the random factor (ANIMAL) can be saved in a new variable (named EBLUPS) using the following command:

```
. predict eblups, reffects
```

The saved EBLUPs can then be used to check for random animal effects that appear to be outliers.

Step 2: Select a structure for the random effects (Model 5.1 vs. Model 5.2).

We now fit Model 5.2, including a second random effect associated with each animal that allows the animals to have unique treatment effects (carbachol vs. basal):

```
. * Model 5.2.
. mixed activate ib3.region treat ib3.region#c.treat,
    || animal: treat, covariance(unstruct) variance reml
```

The random-effects portion of the model specified after the two vertical bars (||) has been changed. We have added the TREAT variable after the colon (`animal: treat`) to indicate that there will be an additional random effect in the model for each animal, associated with the effect of the carbachol treatment. Stata includes a random effect associated with the intercept for each level of ANIMAL by default.

The covariance structure of the random effects has also been changed to `covariance(unstruct)`. This will fit a model with an unstructured D matrix, as defined in Subsection 5.3.2.1. Alternatively, we could have used the following command to try and fit Model 5.2 (and also Model 5.1, minus the random treatment effects):

```
. * Model 5.2.
. mixed activate ib3.region ib0.treat ib3.region#ib0.treat,
   || animal: R.treat, covariance(unstruct) variance reml
```

Note in this command that we identify treat as a categorical factor, with 0 as the reference category, when including the fixed effect of TREAT in addition to the interaction term. Because we have specified TREAT as a categorical factor, and we wish to allow the TREAT effects to randomly vary across levels of ANIMAL, we need to use `R.treat` in the random effects specification (use of `ib0.treat` after the colon would produce an error message). Unfortunately, when random effects associated with the nonreference levels of a categorical factor are included in a model, Stata does not allow the user to specify an unstructured covariance structure (most likely to prevent the estimation of too many covariance parameters for the random effects of categorical factors with several levels). This is the reason that we decided to specify the TREAT indicator as if it were a "continuous" predictor, which is perfectly valid for binary predictor variables.

After fitting Model 5.2 (using the first command above), Stata automatically displays the following output, which is a likelihood ratio test for the two random effects in the model:

```
LR test vs. linear regression: chi2(3) = 42.07 Prob > chi2 = 0.0000

Note: LR test is conservative and provided only for reference
```

This likelihood ratio test is a conservative test of the need for both random effects associated with each animal in Model 5.2. The test statistic is calculated by subtracting the −2 REML log-likelihood value for Model 5.2 from that for a linear regression model with no random effects. The p-value for the test is obtained by referring the test statistic to a χ^2 distribution with 3 degrees of freedom, displayed as `chi2(3)`, because the reference model contains 3 more covariance parameters than the nested model.

Software Note: Stata uses a standard likelihood ratio test, which is a conservative approach. The appropriate theory for testing a model with multiple random effects (e.g., Model 5.2) vs. a nested model without any random effects (i.e., an ordinary linear regression model) has not yet been developed. Stata users can click on the link `LR test is conservative` in the Stata output for a detailed explanation, including references. We do not consider this test in any of the other software procedures.

We can now perform a likelihood ratio test of Hypothesis 5.1, to decide if we need the random treatment effects in Model 5.2, by subtracting the −2 REML log-likelihood for Model 5.2 (the reference model) from that for Model 5.1 (the nested model). The asymptotic null distribution of the test statistic is a mixture of χ_1^2 and χ_2^2 distributions with equal weights of 0.5. Because the test is significant at $\alpha = 0.05$, we choose Model 5.2 as our preferred model at this stage of the analysis (see Subsection 5.5.1 for a discussion of this test and its results).

Step 3: Select a covariance structure for the residuals (Model 5.2 vs. Model 5.3).

We now fit Model 5.3, allowing the variance of the residuals to vary depending on the level of the TREAT variable:

```
. mixed activate ib3.region treat ib3.region#c.treat,
    || animal: treat, covariance(unstruct)
    residuals(independent, by(treat)) variance reml
```

Note the addition of `residuals(independent, by(treat))` in the command above. This option indicates that the residuals remain independent, but that the residual variance should be allowed to vary across levels of the TREAT variable. In preparation for testing Hypothesis 5.2 and the need for heterogeneous residual variance, we save the results from this model fit in a new object:

```
. est store model53
```

We now fit Model 5.2 once again, save the results from that model fit in a separate object, and perform the likelihood ratio test using the `lrtest` command (where Model 5.2 is the nested model):

```
. mixed activate ib3.region treat ib3.region#c.treat,
    || animal: treat, covariance(unstruct) variance reml
. est store model52
. lrtest model53 model52
```

The resulting likelihood ratio test statistic suggests that the null hypothesis (i.e., the nested model with constant residual variance across the treatment groups) should not be rejected, and we therefore proceed with Model 5.2.

Step 4: Reduce the model by removing nonsignificant fixed effects (Model 5.2).

We now test Hypothesis 5.3 to decide whether we need to retain the fixed effects associated with the REGION × TREAT interaction in Model 5.2. To do this, we first refit Model 5.2, and then use the `test` command to generate a Wald χ^2 test:

```
. mixed activate ib3.region treat ib3.region#c.treat,
    || animal: treat, covariance(unstruct) variance reml
. test 1.region#c.treat 2.region#c.treat
```

The two terms following the `test` command are the indicator variables generated by Stata to represent the REGION × TREAT interaction (the products of the indicator variables for the nonreference categories of each factor). These product terms are listed in the output displayed for the estimated fixed effects, where the values of REGION associated with the coefficients for the interaction term (1 and 2) indicate the values that should be used in the specification of the `test` statement above. The result of this test is significant ($p < 0.05$), so we retain the fixed interaction effects and select Model 5.2 as our final model.

5.4.5   HLM

5.4.5.1   Data Set Preparation

To perform the analysis of the Rat Brain data using the HLM software, two separate data sets need to be prepared:

1. The **Level 1 (repeated-measures) data set:** Each row in this data set corresponds to an observation for a given level of REGION and TREATMENT on a given rat. This data set is similar in structure to the data set displayed in Table 5.3. The Level 1 data set includes ANIMAL, the REGION variable, the TREATMENT variable, and the response variable, ACTIVATE. This data set should be sorted by ANIMAL, TREATMENT, and REGION.

2. The **Level 2 (rat-level) data set:** This contains a single observation for each level of ANIMAL. The variables in this file represent measures that remain constant for a given rat. HLM requires that the Level 2 data set must include ANIMAL and at least one other variable measured at the rat level for the purpose of generating the MDM file. In this example, we do not have any covariates that remain constant for a given rat, so we include a variable (NOBS) representing the number of repeated measures for each rat (six). This data set should be sorted by ANIMAL.

To fit the LMMs used for this example in HLM, we need to create several new indicator variables and add them to the Level 1 data set. First, we need to create an indicator variable, TREAT, for the carbachol treatment. We also need to add two indicator variables representing the nonreference levels of REGION and two variables representing the interaction between REGION and TREAT. These new variables need to be created in the software package used to create the input data files prior to importing the data into HLM. We import the Level 1 and Level 2 data sets from SPSS to HLM for this example and show the SPSS syntax to create these five variables in the following Level 1 file:

```
IF (treatment = 1) treat = 0 .
IF (treatment = 2) treat = 1 .
EXECUTE .

COMPUTE region1 = (region = 1) .
COMPUTE region2 = (region = 2) .
EXECUTE .

COMPUTE reg1_tre = region1 * treat .
COMPUTE reg2_tre = region2 * treat .
EXECUTE .
```

After these two data sets have been created, we can proceed to create the multivariate data matrix (MDM), and fit Model 5.1.

5.4.5.2   Preparing the MDM File

In the main HLM menu, click **File**, **Make new MDM file** and then **Stat Package Input**. In the window that opens, select **HLM2** to fit a two-level hierarchical linear model, and click **OK**. Select the **Input File Type** as **SPSS/Windows**.

To prepare the MDM file for Model 5.1, locate the **Level 1 Specification** area, and **Browse** to the location of the Level 1 data set. Click **Open** after selecting the Level 1 SPSS file, click the **Choose Variables** button, and select the following variables: ANIMAL (click "ID" for the ANIMAL variable, because this variable identifies the Level 2 units), REGION1, REGION2, TREAT, REG1_TRT, and REG2_TRT (click "in MDM" for each of these variables). Check "in MDM" for the continuous response variable, ACTIVATE, as well. Click **OK** when finished.

Next, locate the **Level 2 Specification** area, and **Browse** to the location of the Level 2 SPSS data set. Click **Open** after selecting the file, and click the **Choose Variables** button to include ANIMAL (click on "ID") and the variable indicating the number of repeated measures on each animal, NOBS (click on "in MDM").

Software Note: Specifying at least one variable to be included in the MDM file at Level 2 (besides the ID variable) is not optional. This can be any variable, as long as it has nonmissing values for all levels of the ID variable (i.e., ANIMAL). This variable does not need to be included in the analysis and is not part of the analysis for this example.

After making these choices, check the **longitudinal (occasions within persons)** option for this repeated-measures data set (we use this option, although we actually have measures within rats; this selection will not affect the analysis, and only determines the notation that HLM uses to display the model). Select **No** for **Missing Data?** in the Level 1 data set, because we do not have any missing data in this analysis. In the upper-right corner of the MDM window, enter a name with a .mdm extension for the MDM file. Save the .mdmt template file under a new name (click **Save mdmt file**), and click **Make MDM**.

After HLM has processed the MDM file, click the **Check Stats** button to display descriptive statistics for the variables in the Level 1 and Level 2 files (this is not optional). Click **Done** to begin building Model 5.1.

Step 1: Fit a model with a "loaded" mean structure (Model 5.1).

In the model-building window, identify ACTIVATE as the **Outcome variable**. To add more informative subscripts to the models, click **File** and **Preferences** and then choose **Use level subscripts**.

To complete the specification of Model 5.1, we first add the effects of the uncentered Level 1 indicator variables TREAT, REGION1, and REGION2 to the model, in addition to the effects of the interaction variables REG1_TRE and REG2_TRE. The Level 1 model is displayed in HLM as follows:

Model 5.1: Level 1 Model

$$\text{ACTIVATE}_{ti} = \pi_{0i} + \pi_{1i}(\text{TREAT}_{ti}) + \pi_{2i}(\text{REGION1}_{ti}) + \pi_{3i}(\text{REGION2}_{ti})$$
$$+ \pi_{4i}(\text{REG1\_TRE}_{ti}) + \pi_{5i}(\text{REG2\_TRE}_{ti}) + e_{ti}$$

The Level 2 equation for the rat-specific intercept (π_{0i}) includes a constant fixed effect (β_{00}) and a random effect associated with the rat (r_{0i}), which allows the intercept to vary randomly from rat to rat. The Level 2 equations for the five rat-specific coefficients for TREAT, REGION1, REGION2, REG1_TRE, and REG2_TRE (π_{1i} through π_{5i} respectively) are simply defined as constant fixed effects (β_{10} through β_{50}):

Model 5.1: Level 2 Model

$$\pi_{0i} = \beta_{00} + r_{0i}$$
$$\pi_{1i} = \beta_{10}$$
$$\pi_{2i} = \beta_{20}$$
$$\cdots$$
$$\pi_{5i} = \beta_{50}$$

We display the overall LMM by clicking the **Mixed** button in the model building window:

Model 5.1: Overall Mixed Model

$$\text{ACTIVATE}_{ti} = \beta_{00} + \beta_{10} * \text{TREAT}_{ti} + \beta_{20} * \text{REGION1}_{ti} + \beta_{30} * \text{REGION2}_{ti}$$
$$+ \beta_{40} * \text{REG1\_TRE}_{ti} + \beta_{50} * \text{REG2\_TRE}_{ti} + r_{0i} + e_{ti}$$

This model is the same as the general specification of Model 5.1 introduced in Subsection 5.3.2.1, although the notation is somewhat different. The correspondence between the HLM notation and the notation used in Subsection 5.3.2.1 is shown in Table 5.4.

After specifying Model 5.1, click **Basic Settings** to enter a title for this analysis (such as "Rat Brain Data: Model 5.1") and a name for the output (.html) file that HLM generates when fitting this model. Note that the default outcome variable distribution is Normal (Continuous).

Click **OK** to return to the model-building window, and then click **Other Settings** and **Hypothesis Testing** to set up multivariate hypothesis tests for the fixed effects in Model 5.1 (see Subsection 3.4.5 for details). Set up the appropriate general linear hypothesis tests for the fixed effects associated with REGION, TREAT, and the REGION × TREAT interaction.

After setting up the desired tests for the fixed effects in the model, return to the model-building window, and click **File** and **Save As** to save this model specification in a new .hlm file. Finally, click **Run Analysis** to fit the model. After the estimation of the parameters in Model 5.1 has finished, click on **File** and **View Output** to see the resulting estimates.

Step 2: Select a structure for the random effects (Model 5.1 vs. Model 5.2).

We now fit Model 5.2, including a second random effect associated with each animal that allows for animal-specific treatment effects (carbachol vs. basal). We include this second random effect by clicking on the shaded r_{1i} term in the Level 2 equation for the rat-specific effect of TREAT.

Model 5.2: Level 2 Equation for Treatment Effects

$$\pi_{1i} = \beta_{10} + r_{1i}$$

The Level 2 equation for TREAT now implies that the rat-specific effect of TREAT (π_{1i}) depends on an overall fixed effect (β_{10}) and a random effect (r_{1i}) associated with rat i.

After adding the random effect to this equation, click **Basic Settings** to enter a different title for the analysis and change the name of the associated output file. HLM will again perform the same general linear hypothesis tests for the fixed effects in the model that were specified for Model 5.1. Save the .hlm file under a different name (**File, Save As ...**), and click **Run Analysis** to fit the model. Click **File** and **View Output** to see the resulting parameter estimates and significance tests when HLM has finished processing the model.

To test Hypothesis 5.1 (to decide if we need the random treatment effects), we subtract the –2 REML log-likelihood for Model 5.2 (reported as the **deviance** in the Model 5.2 output) from that for Model 5.1 and refer the difference to the appropriate mixture of χ^2 distributions (see Subsection 5.5.1 for more details).

Step 3: Select a covariance structure for the residuals (Model 5.2 vs. Model 5.3).

We now fit Model 5.3, which has the same fixed and random effects as Model 5.2 but allows different residual variances for the basal and carbachol treatments. To specify a model with heterogeneous residual variances at Level 1, click **Other Settings** in the menu of the model-building window, and then click **Estimation Settings**.

In the estimation settings window, click the **Heterogeneous sigma**^{2} button, which allows us to specify variables measured at Level 1 of the data set that define the Level 1 residual variance. In the window that opens, double-click on TREAT to identify it as a predictor of the Level 1 residual variance. Click **OK** to return to the estimation settings window, and then click **OK** to return to the model-building window.

Note that HLM has added another equation to the Level 1 portion of the model, shown as follows:

Model 5.3: Level 1 Model (Heterogeneous Residual Variance)

$$\text{Var}(e_{ti}) = \sigma_{ti}^2 \text{ and } \log(\sigma_{ti}^2) = \alpha_0 + \alpha_1(\text{TREAT}_{ti})$$

This equation defines the parameterization of the heterogeneous residual variance at Level 1: two parameters (α_0 and α_1) are estimated, which define the variance as a function of TREAT. Because HLM2 models a log-transformed version of the residual variance, both sides of the preceding equation need to be exponentiated to calculate the estimated residual variance for a given level of treatment. For example, once estimates of α_0 and α_1 have been computed and displayed in the HLM output, the estimated residual variance for the carbachol treatment can be calculated as $\exp(\alpha_0 + \alpha_1)$.

We enter a new title and a new output file name for Model 5.3 under **Basic Settings**, and then save the .hlm file under a different name. Click **Run Analysis** to fit the new model. HLM divides the resulting output into two sections: a section for Model 5.2 with homogeneous variance and a section for Model 5.3 with heterogeneous residual variance for observations from the carbachol and basal treatments.

Software Note: The current version of HLM only allows models with heterogeneous variances for residuals defined by groups at Level 1 of the data. In addition, REML estimation for models with heterogeneous residual variances is not available, so ML estimation is used instead.

In the output section for Model 5.3, HLM automatically performs a likelihood ratio test of Hypothesis 5.2. The test statistic is obtained by subtracting the –2 ML log-likelihood (i.e., the deviance statistic) of Model 5.3 (the reference model) from that of Model 5.2 (the nested model). This test is a helpful guide to ascertaining whether specifying heterogeneous residual variance improves the fit of the model; however, it is based on ML estimation, which results in biased estimates of variance parameters:

```
                    Summary of Model Fit
-----------------------------------------------------------------
Model                          Number of Parameters    Deviance
1. Homogeneous sigma squared         10                292.72297
2. Heterogeneous sigma squared       11                292.53284
-----------------------------------------------------------------
Model Comparison                   Chi-square       df   P-value
Model 1 vs Model 2                   0.19012          1   >.500
```

The resulting likelihood ratio test statistic is not significant when referred to a χ^2 distribution with 1 degree of freedom (corresponding to the extra covariance parameter, α_1, in Model 5.3). This nonsignificant result suggests that we should not include heterogeneous residual variances for the two treatments in the model. At this stage, we keep Model 5.2 as our preferred model. Although HLM uses an estimation method (ML) different from the procedures in SAS, SPSS, R, and Stata (see Subsection 5.5.2 for more details), the choice of the final model based on the likelihood ratio test is consistent with the other procedures.

Step 4: Reduce the model by removing nonsignificant fixed effects (Model 5.2).

We return to the output for Model 5.2 to investigate tests for the fixed effects of the region by treatment interaction (Hypothesis 5.3). Specifically, locate the "Results of General Linear Hypothesis Testing" in the HLM output. Investigation of the Wald chi-square tests reported for the REGION by TREATMENT interaction in Model 5.2 indicates that Hypothesis 5.3 should be rejected, because the fixed effects associated with the REGION × TREAT interaction are significant. See Sections 5.5 and 5.6 for more details.

5.5 Results of Hypothesis Tests

Table 5.6 presents results of the hypothesis tests carried out in the analysis of the Rat Brain data. The test results reported in this section were calculated based on the analysis in SPSS.

5.5.1 Likelihood Ratio Tests for Random Effects

Hypothesis 5.1. The random effects (u_{3i}) associated with treatment for each animal can be omitted from Model 5.2.

The likelihood ratio test statistic for Hypothesis 5.1 is calculated by subtracting the –2 REML log-likelihood for Model 5.2 (the reference model including the random treatment effects) from that for Model 5.1 (the nested model without the random treatment effects). This difference is calculated as $275.3 - 249.2 = 26.1$. Because the null hypothesis value for the variance of the random treatment effects is on the boundary of the parameter space (i.e., zero), the asymptotic null distribution of this test statistic is a mixture of χ_1^2 and χ_2^2 distributions, each with equal weights of 0.5 (Verbeke & Molenberghs, 2000). To evaluate the significance of the test, we calculate the p-value as follows:

$$p\text{-value} = 0.5 \times P(\chi_2^2 > 26.1) + 0.5 \times P(\chi_1^2 > 26.1) < 0.001$$

We reject the null hypothesis and retain the random effects associated with treatment in Model 5.2 and all subsequent models.

5.5.2 Likelihood Ratio Tests for Residual Variance

Hypothesis 5.2. The residual variance is homogeneous for both the carbachol and basal treatments.

We use a likelihood ratio test for Hypothesis 5.2. The test statistic is calculated by subtracting the –2 REML log-likelihood for Model 5.3, the reference model with heterogeneous residual variances, from that for Model 5.2, the nested model with homogeneous residual variance. The asymptotic null distribution of the test statistic is a χ^2 with one degree of freedom. The single degree of freedom is a consequence of the reference model having one additional covariance parameter (i.e., the additional residual variance for the carbachol treatment) in the $\boldsymbol{R}_i$ matrix. We do not reject the null hypothesis in this case ($p = 0.66$), so we conclude that Model 5.2, with homogeneous residual variance, is our preferred model.

TABLE 5.6: Summary of Hypothesis Test Results for the Rat Brain Analysis

Hypo-thesis Label	Test	Estima-tion Method	Models Compared (Nested vs. Reference)	Test Statistic Values (Calculation)	p-Value
5.1	LRT[a]	REML	5.1 vs. 5.2	$\chi^2(1:2) = 26.1$ $(275.3 - 249.2)$	$< .001$
5.2	LRT	REML	5.2 vs. 5.3	$\chi^2(1) = 0.2$ $(249.2 - 249.0)$	0.66
5.3	Type-III F-test	REML		$F(2, 16) = 81.0$	$< .001$
	Wald χ^2-test		5.2^b	$\chi^2(1) = 162.1$	$< .001$

Note: See Table 5.5 for null and alternative hypotheses and distributions of test statistics under H_0.
[a]Likelihood ratio test; the test statistic is calculated by subtracting the –2 REML log-likelihood for the reference model from that of the nested model.
[b]The use of an F-test (SAS, SPSS, or R) or a Wald χ^2-test (Stata or HLM) does not require fitting a nested model.

5.5.3 F-Tests for Fixed Effects

Hypothesis 5.3. The REGION by TREATMENT interaction effects can be omitted from Model 5.2.

To test Hypothesis 5.3, we use an F-test based on the results of the REML estimation of Model 5.2. We present the Type III F-test results based on the SPSS output in this section. This test is significant at $\alpha = 0.05$ ($p < .001$), which indicates that the fixed effect of the carbachol treatment on nucleotide activation differs by region, as we noted in our original data summary. We retain the fixed effects associated with the region by treatment interaction and select Model 5.2 as our final model.

5.6 Comparing Results across the Software Procedures

Table 5.7 shows a comparison of selected results obtained by using the five software procedures to fit Model 5.1 to the Rat Brain data. This model is "loaded" with fixed effects, has a random effect associated with the intercept for each animal, and has homogeneous residual variance across levels of TREATMENT and REGION.

Table 5.8 presents a comparison of selected results from the five procedures for Model 5.2, which has the same fixed and random effects as Model 5.1, and an additional random effect associated with treatment for each animal.

5.6.1 Comparing Model 5.1 Results

Table 5.7 shows that results for Model 5.1 agree across the software procedures in terms of the fixed-effect parameter estimates and their estimated standard errors. They also agree on the values of the estimated variances, σ^2_{int} and σ^2, and their standard errors, when reported.

The value of the –2 REML log-likelihood is the same across all five software procedures. However, there is some disagreement in the values of the information criteria (AIC and BIC) because of different calculation formulas that are used (see Subsection 3.6.1).

There are also differences in the types of tests reported for the fixed-effect parameters and, thus, in the results for these tests. SAS and SPSS report Type III F-tests, R reports Type I F-tests, and Stata and HLM report Wald χ^2 tests (see Subsection 2.6.3.1 for a detailed discussion of the differences in these tests of fixed effects). The values of the test statistics for the software procedures that report the same tests agree closely.

We note that the lmer() function in R does not compute denominator degrees of freedom for F-test statistics or degrees of freedom for t-statistics, primarily due to the variability in methods that are available for approximating these degrees of freedom and the absence of an accepted standard. As noted in the footnotes for Table 5.7, users of this function can use the lmerTest package to compute approximate degrees of freedom for these test statistics, along with p-values.

5.6.2 Comparing Model 5.2 Results

Table 5.8 shows that the estimated fixed-effect parameters and their respective standard errors for Model 5.2 agree across all five software procedures.

The estimated covariance parameters differ slightly, likely due to rounding differences. It is also noteworthy that R reports the estimated correlation of the two random effects in Model 5.2, as opposed to the covariances reported by the other four software procedures.

There are also differences in the types of the F-tests for fixed effects computed in SAS, SPSS, and R. These differences are discussed in general in Subsections 2.6.3.1 and 3.11.6.

5.7 Interpreting Parameter Estimates in the Final Model

The results that we present in this section were obtained by fitting the final model (Model 5.2) to the Rat Brain data, using REML estimation in SPSS.

5.7.1 Fixed-Effect Parameter Estimates

The fixed-effect parameter estimates, standard errors, significance tests (t-tests with Satterthwaite approximations for the degrees of freedom), and 95% confidence intervals obtained by fitting Model 5.2 to the Rat Brain data in SPSS are reported in the following SPSS output:

TABLE 5.7: Comparison of Results for Model 5.1

	SAS: proc mixed	SPSS: MIXED	R: lme() function	R: lmer() function	Stata: mixed	HLM: HLM2
Estimation Method	REML	REML	REML	REML	REML	REML
Fixed-Effect Parameter	*Estimate (SE)*	*Estimate (SE)*	*Estimate (SE)*	*Estimate (SE)*	*Estimate (SE)*	*Estimate (SE)*
β_0 (Intercept)	212.29(38.21)	212.29(38.21)	212.29(38.21)	212.29(38.21)	212.29(38.21)	212.29(38.21)
β_1 (BST vs. VDB)	216.21(31.31)	216.21(31.31)	216.21(31.31)	216.21(31.31)	216.21(31.31)	216.21(31.31)
β_2 (LS vs. VDB)	25.45(31.31)	25.45(31.31)	25.45(31.31)	25.45(31.31)	25.45(31.31)	25.45(31.31)
β_3 (TREATMENT)	360.03(31.31)	360.03(31.31)	360.03(31.31)	360.03(31.31)	360.03(31.31)	360.03(31.31)
β_4 (BST × TREATMENT)	−261.82(44.27)	−261.82(44.27)	−261.82(44.27)	−261.82(44.27)	−261.82(44.27)	−261.82(44.28)
β_5 (LS × TREATMENT)	−162.50(44.27)	−162.50(44.27)	−162.50(44.27)	−162.50(44.27)	−162.50(44.27)	−162.50(44.28)
Covariance Parameter	*Estimate (SE)*	*Estimate (SE)*	*Estimate (n.c.)*[a]	*Estimate (n.c.)*	*Estimate (SE)*	*Estimate (n.c.)*
σ^2_{int}	4849.81(3720.35)	4849.81(3720.35)	4849.81[b]	4849.8	4849.81(3720.35)	4849.74
σ^2 (Residual variance)	2450.29(774.85)	2450.30(774.85)	2450.30	2450.30	2450.30(774.85)	2450.38
Model Information Criteria						
−2 RE/ML log-likelihood	275.3	275.3	275.3	275.3	275.3	275.3
AIC	279.3	279.3	291.3	291.3	291.3	n.c.
BIC	278.5	281.6	300.7	302.5	302.5	n.c.
Tests for Fixed Effects	**Type III F-Tests**[c]	**Type III F-Tests**	**Type I F-Tests**	**Type I F-Tests**	**Wald χ^2-Tests**[c]	**Wald χ^2-Tests**[c]
Intercept	$t(4) = 5.6$, $p < .01$	$F(1, 4.7) = 75.7$, $p < .01$	$F(1, 20) = 153.8$, $p < .01$	$t = 5.6$ (no d.f.)[d]	$Z = 5.6$, $p < .01$	$t(4) = 5.6$, $p < .01$
REGION	$F(2, 20) = 28.5$, $p < .01$	$F(2, 20) = 28.5$, $p < .01$	$F(2, 20) = 20.6$, $p < .01$	$F(1,\text{no d.f.}) = 20.6$	$\chi^2(2) = 57.0$, $p < .01$	$\chi^2(2) = 57.0$, $p < .01$
TREATMENT	$F(1, 20) = 146.3$,	$F(1, 20) = 146.2$,	$F(1, 20) = 146.2$,	$F(1,\text{no d.f.}) = 146.2$	$\chi^2(1) = 132.3$,	$\chi^2(1) = 132.2$,

TABLE 5.7: (Continued)

Estimation Method	**SAS:** proc mixed	**SPSS:** MIXED	**R:** lme() function	**R:** lmer() function	**Stata:** mixed	**HLM:** HLM2
	REML	REML	REML	REML	REML	REML
REGION × TREATMENT	$p < .01$ $F_{(2,20)} = 17.8$, $p < .01$	$p < .01$ $F_{(2,20)} = 17.8$, $p < .01$	$p < .01$ $F_{(2,20)} = 17.8$, $p < .01$	$F_{(2,\text{no d.f.})} = 17.8$	$p < .01$ $\chi^2(2) = 35.7$, $p < .01$	$p < .01$ $\chi^2(2) = 35.7$, $p < .01$

Note: (n.c.) = not computed

Note: 30 Repeated Measures at Level 1; 5 Rats at Level 2

[a] The nlme version of the lme() function in R reports the estimated standard deviations of the random effects and the residuals; these estimates are squared to get the estimated variances.

[b] Users of the lme() function in R can use the function intervals(model5.1.fit) to obtain *approximate* (i.e., large-sample) 95% confidence intervals for covariance parameters.

[c] The test reported for the intercept differs from the tests for the other fixed effects in the model.

[d] Users of the lmer() function in R can use the lmerTest package to obtain approximate degrees of freedom for these test statistics based on a Satterthwaite approximation (which enables the computation of p-values for the test statistics).

TABLE 5.8: Comparison of Results for Model 5.2

Estimation Method	SAS: proc mixed	SPSS: MIXED	R: lme() function	R: lmer() function	Stata: mixed	HLM: HLM2
	REML	REML	REML	REML	REML	REML
Fixed-Effect Parameter	*Estimate (SE)*	*Estimate (SE)*	*Estimate (SE)*	*Estimate (SE)*	*Estimate (SE)*	*Estimate (SE)*
β_0 (Intercept)	212.29(19.10)	212.29(19.10)	212.29(19.10)	212.29(19.10)	212.29(19.10)	212.29(19.10)
β_1 (BST vs. VDB)	216.21(14.68)	216.21(14.68)	216.21(14.68)	216.21(14.68)	216.21(14.68)	216.21(14.68)
β_2 (LS vs. VDB)	25.45(14.68)	25.45(14.68)	25.45(14.68)	25.45(14.68)	25.45(14.68)	25.45(14.68)
β_3 (TREATMENT)	360.03(38.60)	360.03(38.60)	360.03(38.60)	360.03(38.60)	360.03(38.60)	360.03(38.60)
β_4 (BST × TREATMENT)	−261.82(20.76)	−261.82(20.76)	−261.82(20.76)	−261.82(20.76)	−261.82(20.76)	−261.82(20.76)
β_5 (LS × TREATMENT)	−162.50(20.76)	−162.50(20.76)	−162.50(20.76)	−162.50(20.76)	−162.50(20.76)	−162.50(20.76)
Covariance Parameter	*Estimate (SE)*	*Estimate (SE)*	*Estimate (n.c.)[a]*	*Estimate (n.c.)*	*Estimate (SE)*	*Estimate (n.c.)*
σ^2_{int}	1284.32(1037.12)	1284.32(1037.12)	1284.29	1284.30	1284.32(1037.12)	1284.30
$\sigma_{int,treat}$	2291.22(1892.63)	2291.22(1892.63)	0.80(corr.)	0.80(corr.)	2291.23(1892.63)	2291.25
σ^2_{treat}	6371.33(4760.94)	6371.33(4760.94)	6371.25	6371.30	6371.34(4760.95)	6371.29
σ^2	538.90(190.53)	538.90(190.53)	538.90	538.90	538.90(190.53)	538.90
Model Information Criteria						
−2 RE/ML log-likelihood	249.2	249.2	249.2	249.2	249.2	249.2
AIC	257.2	257.2	269.2	269.2	269.2	249.2
BIC	255.6	261.9	281.0	283.2	283.2	n.c.
Tests for Fixed Effects	**Type III F-Tests[a]**	**Type III F-Tests**	**Type I F-Tests**	**Type I F-Tests**	**Wald χ^2-Tests[a]**	**Wald χ^2-Tests[a]**
Intercept	$t(4) = 11.1$, $p<.01$	$F(1,4) = 292.9$, $p<.01$	$F(1,20) = 313.8$, $p<.01$	$t(\text{no d.f.}) = 11.1$	$Z = 11.1$, $p<.01$	$t(4) = 11.1$, $p<.01$
REGION	$F(2,16) = 129.6$,	$F(2,16) = 129.6$,	$F(2,20) = 93.7$,	$F(2,\text{no d.f.}) = 93.7$	$\chi^2(2) = 259.1$,	$\chi^2(2) = 259.1$,

TABLE 5.8: (Continued)

Estimation Method	SAS: proc mixed	SPSS: MIXED	R: lme() function	R: lmer() function	Stata: mixed	HLM: HLM2
	REML	REML	REML	REML	REML	REML
TREATMENT	$p < .01$, $F(1,4) = 35.5$, $p < .01$	$p < .01$, $F(1,4) = 35.5$, $p < .01$	$p < .01$, $F(1,20) = 35.5$, $p < .01$	$F(1,\text{no d.f.}) = 35.1$	$p < .01$, $\chi^2(1) = 87.0$, $p < .01$	$p < .01$, $\chi^2(1) = 87.0$, $p < .01$
REGION × TREATMENT	$F(2,16) = 81.1$, $p < .01$	$F(2,16) = 81.0$, $p < .01$	$F(2,20) = 81.0$, $p < .01$	$F(2,\text{no d.f.}) = 81.0$	$\chi^2(2) = 162.1$, $p < .01$	$\chi^2(2) = 162.1$, $p < .01$

Note: (n.c.) = not computed
Note: 30 Repeated Measures at Level 1; 5 Rats at Level 2
[a]The test used for the intercept differs from the tests for the other fixed effects in the model.

```
                        Estimates of Fixed Effects (a)
                                                           95% Confidence Interval
  Parameter           Estimate    Std. Error  df       t       Sig.   Lower Bound  Upper Bound
  Intercept           212.294000  19.095630   6.112   11.117   .000   165.775675   258.812325
  [region=1]          216.212000  14.681901   16      14.726   .000   185.087761   247.336239
  [region=2]           25.450000  14.681901   16       1.733   .102    -5.674239    56.574239
  [region=3]               0(b)        0        .         .       .          .            .
  treat               360.026000  38.598244   4.886    9.328   .000   260.103863   459.948137
  [region=1]*treat   -261.822000  20.763343   16     -12.610   .000  -305.838322  -217.805678
  [region=2]*treat   -162.500000  20.763343   16      -7.826   .000  -206.516322  -118.483678
  [region=3]*treat         0(b)        0        .         .       .          .            .

  a. Dependent Variable: activate.
  b. This parameter is set to zero because it is redundant.
```

Because of the presence of the REGION × TREAT interaction in the model, we need to be careful when interpreting the main effects associated with these variables. To aid interpretation of the results in the presence of the significant interaction, we investigate the estimated marginal means (EMMEANS) of activation for each region within each level of treatment, requested in the SPSS syntax for Model 5.2:

```
                            Estimates (a)

                                                 95% Confidence Interval
   region   Mean         Std. Error   df       Lower Bound  Upper Bound
   BST      526.710(b)   50.551       4.234    389.364      664.056
   LS       435.270(b)   50.551       4.234    297.924      572.616
   VDB      572.320(b)   50.551       4.234    434.974      709.666

   a. Dependent Variable: activate.
   b. Covariates appearing in the model are evaluated at the following
   values: treat = 1.00.
```

These are the estimated means for activation at the three regions when TREAT = 1 (carbachol). The estimated marginal means are calculated using the estimates of the fixed effects displayed earlier. SPSS also performs pairwise comparisons of the estimated marginal means for the three regions for carbachol treatment, requested with the COMPARE option in the /EMMEANS subcommand:

```
                           Pairwise Comparisons (a)
                          Mean
                          Difference                               95% Confidence Interval
  (I) region  (J) region  (I-J)       Std. Error   df    Sig.(c)  Lower Bound  Upper Bound
  BST         LS            91.440*    14.682       16    .000      52.195      130.685
              VDB          -45.610*    14.682       16    .020     -84.855       -6.365
  LS          BST          -91.440*    14.682       16    .000    -130.685      -52.195
              VDB         -137.050*    14.682       16    .000    -176.295      -97.805
  VDB         BST           45.610*    14.682       16    .020       6.365       84.855
              LS           137.050*    14.682       16    .000      97.805      176.295

  Based on estimated marginal means
  *. The mean difference is significant at the .05 level.
  a. Dependent Variable: activate.
  c. Adjustment for multiple comparisons: Bonferroni.
```

We see that all of the estimated marginal means are significantly different at $\alpha = 0.05$ after performing a Bonferroni adjustment for the multiple comparisons. We display similar tables for the basal treatment (TREAT = 0):

```
                          Estimates (a)

                                       95% Confidence Interval
        region  Mean       Std. Error  df    Lower Bound Upper Bound
        BST     428.506(b)   19.096    6.112   381.988    475.024
        LS      237.744(b)   19.096    6.112   191.226    284.262
        VDB     212.294(b)   19.096    6.112   165.776    258.812


        a. Dependent Variable: activate.
        b. Covariates appearing in the model are evaluated at the following
        values: treat = .00.
```

Note that the activation means at the LS and VDB regions are not significantly different for the basal treatment ($p = 0.307$):

```
                       Pairwise Comparisons (a)
                        Mean
                        Difference                       95% Confidence Interval
  (I) region  (J) region  (I-J)    Std. Error  df   Sig.(c) Lower Bound Upper Bound
  BST         LS          190.762*   14.682    16    .000    151.517    230.007
              VDB         216.212*   14.682    16    .000    176.967    255.457
  LS          BST        -190.762*   14.682    16    .000   -230.007   -151.517
              VDB          25.450    14.682    16    .307    -13.795     64.695
  VDB         BST        -216.212*   14.682    16    .000   -255.457   -176.967
              LS          -25.450*   14.682    16    .307    -64.695     13.795


  Based on estimated marginal means
  *. The mean difference is significant at the .05 level.
  a. Dependent Variable: activate.
  c. Adjustment for multiple comparisons: Bonferroni.
```

These results are in agreement with what we observed in our initial graph of the data (Figure 5.1), in which the LS and VDB regions had very similar activation means for the basal treatment, but different activation means for the carbachol treatment.

5.7.2   Covariance Parameter Estimates

The estimated covariance parameters obtained by fitting Model 5.2 to the Rat Brain data using the MIXED command in SPSS with REML estimation are reported in the following output:

```
              Estimates of Covariance Parameters (a)

      Parameter                           Estimate     Std. Error

      Residual                           538.895532    190.528343

      Intercept + treat    UN (1,1)     1284.319876   1037.116565
      [subject = animal]   UN (2,1)     2291.223258   1892.631626
                           UN (2,2)     6371.331082   4760.943896

      a. Dependent Variable: activate.
```

The first part of the output contains the estimated residual variance, which has a value of 538.9. The next part of the output, labeled "Intercept + treat [subject=animal]," lists the three elements of the estimated D covariance matrix for the two random effects in the model. These elements are labeled according to their position (row, column) in the D matrix. We specified the D matrix to be unstructured by using the COVTYPE(UN) option in the /RANDOM subcommand of the SPSS syntax.

The variance of the random intercepts, labeled UN(1,1) in this unstructured matrix, is estimated to be 1284.32, and the variance of the random treatment effects, labeled UN(2,2), is estimated to be 6371.33. The positive estimated covariance between the random intercepts and random treatment effects, denoted by UN(2,1) in the output, is 2291.22.

The estimated D matrix, referred to as the G matrix by SPSS, is shown as follows in matrix form. This output was requested by using the `/PRINT G` subcommand in the SPSS syntax for Model 5.2.

```
              Random Effect Covariance Structure (G) (a)

                          Intercept | animal   treat | animal
          Intercept | animal     1284.319876        2291.223258
          treat | animal         2291.223258        6371.331082

          Unstructured
          a. Dependent Variable: activate.
```

5.8   The Implied Marginal Variance-Covariance Matrix for the Final Model

The current version of the `MIXED` command in SPSS does not provide an option to display the estimated V_i covariance matrix for the marginal model implied by Model 5.2 in the output, so we use output from SAS and R in this section. The matrices of marginal covariances and marginal correlations for an individual subject can be obtained in SAS by including the v and vcorr options in the `random` statement in the `proc mixed` syntax for Model 5.2:

```
random int treat / subject = animal type = un v vcorr;
```

By default, SAS displays the marginal variance-covariance and corresponding correlation matrices for the first subject in the data file (in this case the matrices displayed correspond to animal R100797). Note that these matrices have the same structure for any given animal. Both the marginal V_i matrix and the marginal correlation matrix are of dimension 6×6, corresponding to the values of activation for each combination of region by treatment for a given rat i.

The "Estimated V Matrix for animal R100797" displays the estimated marginal variances of activation on the diagonal and the estimated marginal covariances off the diagonal. The 3×3 submatrix in the upper-left corner represents the marginal covariance matrix for observations on the BST, LS, and VDB regions in the basal treatment, and the 3×3 submatrix in the lower-right corner represents the marginal covariance matrix for observations on the three brain regions in the carbachol treatment. The remainder of the V_i matrix represents the marginal covariances of observations on the same rat across treatments.

```
              Estimated V Matrix for animal R100797

   Row   Col1     Col2     Col3     Col4     Col5      Col6
    1   1823.22  1284.32  1284.32  3575.54  3575.54  3575.54
    2   1284.32  1823.22  1284.32  3575.54  3575.54  3575.54
    3   1284.32  1284.32  1823.22  3575.54  3575.54  3575.54
    4   3575.54  3575.54  3575.54  12777    12238    12238
    5   3575.54  3575.54  3575.54  12238    12777    12238
    6   3575.54  3575.54  3575.54  12238    12238    12777
```

The inclusion of the random treatment effects in Model 5.2 implies that the marginal variances and covariances differ for the carbachol and basal treatments. We see in the estimated V_i matrix that observations for the carbachol treatment have a much larger estimated marginal variance (12777) than observations for the basal treatment (1823.22). This result is consistent with the initial data summary and with Figure 5.1, in which we noted that the between-rat variability in the carbachol treatment is greater than in the basal treatment.

The implied marginal covariances of observations within a given treatment are assumed to be constant, which might be viewed as a fairly restrictive assumption. We consider alternative models that allow these marginal covariances to vary in Section 5.11.

The 6×6 matrix of estimated marginal correlations implied by Model 5.2 (taken from the SAS output) is displayed below. The estimated marginal correlations of observations in the basal treatment and the carbachol treatment are both very high (.70 and .96, respectively). The estimated marginal correlation of observations for the same rat across the two treatments is also high (.74).

```
Estimated V Correlation Matrix for animal R100797

Row  Col1    Col2    Col3    Col4    Col5    Col6
1    1.0000  0.7044  0.7044  0.7408  0.7408  0.7408
2    0.7044  1.0000  0.7044  0.7408  0.7408  0.7408
3    0.7044  0.7044  1.0000  0.7408  0.7408  0.7408
4    0.7408  0.7408  0.7408  1.0000  0.9578  0.9578
5    0.7408  0.7408  0.7408  0.9578  1.0000  0.9578
6    0.7408  0.7408  0.7408  0.9578  0.9578  1.0000
```

In R, the estimated marginal variance-covariance matrix can be displayed by applying the `getVarCov()` function to a model fit object produced by the `lme()` function:

```
> getVarCov(model5.2.fit, individual = "R100797", type = "marginal")
```

These findings support our impressions in the initial data summary (Figure 5.1), in which we noted that observations on the same animal appeared to be very highly correlated (i.e., the level of activation for a given animal tended to "track" across regions and treatments).

We present a detailed example of the calculation of the implied marginal variance-covariance matrix for the simpler Model 5.1 in Appendix B.

5.9 Diagnostics for the Final Model

In this section we present an informal graphical assessment of the diagnostics for our final model (Model 5.2), fitted using REML estimation in SPSS.

The syntax in the following text was used to refit Model 5.2 with the `MIXED` command in SPSS, using REML estimation to get unbiased estimates of the covariance parameters. The `/SAVE` subcommand requests that the conditional predicted values, PRED, and the conditional residuals, RESID, be saved in the current working data set. The predicted values and the residuals are conditional on the random effects in the model and are saved in two new variables in the working SPSS data file. Optionally, marginal predicted values can be saved in the data set, using the `FIXPRED` option in the `/SAVE` subcommand. See Section 3.10 for a more general discussion of conditional predicted values and conditional residuals.

Software Note: The variable names used for the conditional predicted values and the conditional residuals saved by SPSS depend on how many previously saved versions of these variables already exist in the data file. If the current model is the first for which these variables have been saved, they will be named PRED_1 and RESID_1 by default. SPSS numbers successively saved versions of these variables as PRED_n and RESID_n, where n increments by one for each new set of conditional predicted and residual values.

```
* Model 5.2 (Diagnostics).
MIXED
activate BY region WITH treat
/CRITERIA = CIN(95) MXITER(100) MXSTEP(5) SCORING(1)
SINGULAR(0.000000000001) HCONVERGE(0, ABSOLUTE)
LCONVERGE(0, ABSOLUTE) PCONVERGE(0.000001, ABSOLUTE)
/FIXED = region treat region*treat | SSTYPE(3)
/METHOD = REML
/PRINT = SOLUTION G
/RANDOM INTERCEPT treat | SUBJECT(animal) COVTYPE(UN)
/SAVE = PRED RESID .
```

We include the following syntax to obtain a normal Q–Q plot of the conditional residuals:

```
PPLOT
/VARIABLES=RESID_1
/NOLOG
/NOSTANDARDIZE
/TYPE=Q-Q
/FRACTION=BLOM
/TIES=MEAN
/DIST=NORMAL.
```

The conditional residuals from this analysis appear to follow a normal distribution fairly well (see Figure 5.3). However, it is difficult to assess the distribution of the conditional residuals, because there are only 30 total observations (= 5 rats × 6 observations per rat). A Kolmogorov–Smirnov test for normality of the conditional residuals can be carried out using the following syntax:

```
NPAR TESTS
/K-S(NORMAL)= RESID_1
/MISSING ANALYSIS.
```

The result of the Kolmogorov–Smirnov test for normality[2] is not significant ($p = 0.95$). We consider normality of the residuals to be a reasonable assumption for this model.

We also investigate the assumption of equal residual variance in both treatments by examining a scatter plot of the conditional residuals vs. the conditional predicted values:

```
GRAPH
/SCATTERPLOT(BIVAR)=PRED_1 WITH RESID_1 BY treatment
/MISSING=LISTWISE .
```

---

[2]In general, the Shapiro–Wilk test for normality is more powerful than the Kolmogorov–Smirnov test when working with small sample sizes. Unfortunately, this test is not available in SPSS.

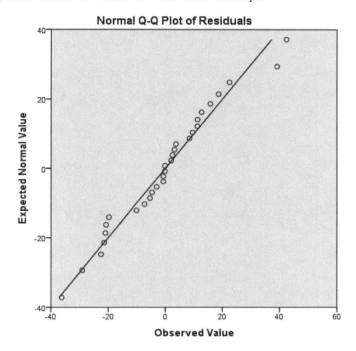

FIGURE 5.3: Distribution of conditional residuals from Model 5.2.

Figure 5.4 suggests that the residual variance is fairly constant across treatments (there is no pattern, and the residuals are symmetric). We formally tested the assumption of equal residual variances across treatments in Hypothesis 5.2, and found no significant difference in the residual variance for the carbachol treatment vs. the basal treatment in Model 5.2 (see Subsection 5.5.2).

The distributions of the EBLUPs of the random effects should also be investigated to check for possible outliers. Unfortunately, EBLUPs for the two random effects associated with each animal in Model 5.2 cannot be generated in the current version of the MIXED command in IBM SPSS Statistics (Version 21). Because we have a very small number of animals, we do not investigate diagnostics for the EBLUPs for this model.

5.10 Software Notes

5.10.1 Heterogeneous Residual Variances for Level 1 Groups

Recall that in Chapter 3 we used a heterogeneous residual variance structure for groups defined by a Level 2 variable (treatment) in Model 3.2B. The ability to fit such models is available only in proc mixed in SAS, the GENLINMIXED command in SPSS, the lme() function in R, and the mixed command in Stata. When we fit Model 5.3 to the Rat Brain data in this chapter, we defined a heterogeneous residual variance structure for different values of a Level 1 variable (TREATMENT). We were able to fit this model using all of the software procedures, with the exception of the lmer() function in R.

The HLM2 procedure only allows maximum likelihood estimation for models that are fitted with a heterogeneous residual variance structure. SAS, SPSS, the the lme() function

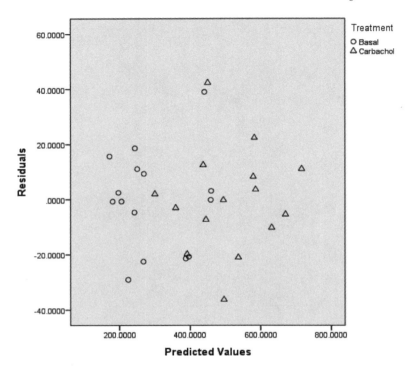

FIGURE 5.4: Scatter plot of conditional residuals vs. conditional predicted values based on the fit of Model 5.2.

in R, and the `mixed` command in Stata all allow ML or REML (default) estimation for these models. The parameterization of the heterogeneous residual variances in HLM2 employs a logarithmic transformation, so the parameter estimates for the variances from HLM2 need to be exponentiated before they can be compared with results from the other software procedures (see Subsection 5.4.5).

5.10.2 EBLUPs for Multiple Random Effects

Model 5.2 specified two random effects for each animal: one associated with the intercept and a second associated with treatment. The EBLUPs for multiple random effects per subject can be displayed using SAS, R, Stata, and HLM. However, it is not possible to obtain separate estimates of the EBLUPs for more than one random effect per subject when using the `MIXED` command in the current version of IBM SPSS Statistics (Version 21).

5.11 Other Analytic Approaches

5.11.1 Kronecker Product for More Flexible Residual Covariance Structures

Most residual covariance structures (e.g., AR(1) or compound symmetry) are designed for one within-subject factor (e.g., time). In the Rat Brain example, we have two within-subject factors: brain region and treatment. With such data, one can consider modeling a residual covariance structure using the **Kronecker Product** of the underlying within-subject factor-specific covariance matrices (Galecki, 1994). This method adds flexibility in building residual covariance structures and has an attractive interpretation in terms of independent within-subject factor-specific contributions to the overall within-subject covariance structure. Examples of this general methodology are implemented in SAS `proc mixed`.

The SAS syntax that implements an example of this methodology for the Rat Brain data is provided below:

```
title "Kronecker Product Covariance Structure";
proc mixed data = ratbrain;
class animal region treatment;
model activate = region treatment region*treatment / s;
random int / subject = animal type=vc v vcorr solution;
repeated region treatment / subject=animal type=un@un r rcorr;
run;
```

Note that both REGION and TREATMENT must be listed in the `class` statement. In the `random` statement we retain the random intercept but omit the random animal-specific treatment effects to avoid overparameterization of the model. The `repeated` statement includes the option `type=un@un`, which specifies the Kronecker product of the two matrices for the REGION and TREATMENT factors, with three and two levels, respectively. The syntax implies that REGION contributes an unstructured 3×3 matrix and TREATMENT contributes an unstructured 2×2 matrix to the overall 6×6 matrix $\boldsymbol{R}_i$. To ensure identifiability of the matrices, we assume that the upper-left element in the matrix contributed by TREATMENT is equal to 1 (which is automatically done by the software).

This syntax results in the following estimates of the elements of the underlying factor-specific matrices for both REGION and TREATMENT.

	Covariance Parameter Estimates		
	Cov Parm	Subject	Estimate
Intercept		animal	7637.3000
region	UN(1,1)	animal	2127.7400
	UN(2,1)	animal	1987.2900
	UN(2,2)	animal	2744.6600
	UN(3,1)	animal	1374.5100
	UN(3,2)	animal	2732.2200
	UN(3,3)	animal	3419.7000
treatment	UN(1,1)	animal	1.0000
	UN(2,1)	animal	-0.4284
	UN(2,2)	animal	0.6740

We can use these estimates to determine the unstructured variance-covariance matrices for the residuals contributed by the REGION and TREATMENT factors:

$$R_{REGION} = \begin{pmatrix} 2127.74 & 1987.29 & 1374.51 \\ 1987.29 & 2744.66 & 2732.22 \\ 1374.51 & 2732.22 & 3419.70 \end{pmatrix}$$

$$R_{TREATMENT} = \begin{pmatrix} 1.00 & -0.43 \\ -0.43 & 0.67 \end{pmatrix}$$

The Kronecker product of these two factor-specific residual variance-covariance matrices implies the following overall estimated R_i correlation matrix for a given rat:

```
     Estimated R Correlation Matrix for animal R100797

Row  Col1     Col2     Col3     Col4     Col5     Col6
 1   1.0000   0.8224   0.5096  -0.5219  -0.4292  -0.2659
 2   0.8224   1.0000   0.8918  -0.4292  -0.5219  -0.4654
 3   0.5096   0.8918   1.0000  -0.2659  -0.4654  -0.5219
 4  -0.5219  -0.4292  -0.2659   1.0000   0.8224   0.5096
 5  -0.4292  -0.5219  -0.4654   0.8224   1.0000   0.8918
 6  -0.2659  -0.4654  -0.5219   0.5096   0.8918   1.0000
```

The implied marginal correlation matrix of observations for a given rat based on this model is as follows. The structure of this marginal correlation matrix reveals a high level of correlation among observations on the same animal, as was observed for the implied marginal correlation matrix based on the fit of Model 5.2:

```
     Estimated V Correlation Matrix for animal R100797

Row  Col1     Col2     Col3     Col4     Col5     Col6
 1   1.0000   0.9559   0.8673   0.7146   0.7050   0.7153
 2   0.9559   1.0000   0.9678   0.6992   0.6511   0.6365
 3   0.8673   0.9678   1.0000   0.7038   0.6314   0.5887
 4   0.7146   0.6992   0.7038   1.0000   0.9676   0.9017
 5   0.7050   0.6511   0.6314   0.9676   1.0000   0.9760
 6   0.7153   0.6365   0.5887   0.9017   0.9760   1.0000
```

The AIC for this model is 258.0, which is very close to the value for Model 5.2 (AIC = 257.2) and better (i.e., smaller) than the value for Model 5.1 (AIC = 279.3).

Note that covariance structures based on Kronecker products can also be used for studies involving multiple dependent variables measured longitudinally (not considered in this book).

5.11.2 Fitting the Marginal Model

We can also take a strictly marginal approach (in which random animal effects are not considered) to modeling the Rat Brain data. However, there are only 5 animals and 30 observations; therefore, fitting a marginal model with an unstructured residual covariance matrix is not recommended, because the unstructured R_i matrix would require the estimation of 21 covariance parameters. When attempting to fit a marginal model with an unstructured covariance structure for the residuals using REML estimation in SPSS, the MIXED command issues a warning and does not converge to a valid solution.

We can consider marginal models with more restrictive residual covariance structures. For example, we can readily fit a model with a heterogeneous compound symmetry R_i matrix, which requires the estimation of seven parameters: six variances, i.e., one for each combination of treatment and region, and a constant correlation parameter (note that these are also a lot of parameters to estimate for this small data set). We use the following syntax in SPSS:

```
* Marginal model with heterogeneous compound symmetry R(i) matrix.
MIXED
activate BY region treatment
```

```
/CRITERIA = CIN(95) MXITER(100) MXSTEP(5) SCORING(1)
SINGULAR(0.000000000001) HCONVERGE(0, ABSOLUTE)
LCONVERGE(0, ABSOLUTE) PCONVERGE(0.000001, ABSOLUTE)
/FIXED = region treatment region*treatment | SSTYPE(3)
/METHOD = REML
/PRINT = SOLUTION R
/REPEATED Region Treatment | SUBJECT(animal) COVTYPE(CSH) .
```

In this syntax for the marginal model, note that the /RANDOM subcommand is not included. The structure of the R_i matrix is specified as CSH (compound symmetric heterogeneous), which means that the residual marginal variance is allowed to differ for each combination of REGION and TREATMENT, although the correlation between observations on the same rat is constant (estimated to be 0.81). The AIC for this model is 267.8, as compared to the value of 257.2 for Model 5.2. So, it appears that we have a better fit using the LMM with explicit random effects (Model 5.2).

5.11.3 Repeated-Measures ANOVA

A more traditional approach to repeated-measures ANOVA (Winer et al., 1991) starts with a data set in the wide format shown in Table 5.2. This type of analysis could be carried out, for example, using the GLM procedures in SPSS and SAS. However, if any missing values occur for a given subject (e.g., animal), that subject is dropped from the analysis altogether (complete case analysis). Refer to Subsection 2.9.4 for more details on the problems with this approach when working with missing data.

The correlation structure assumed in a traditional repeated-measures ANOVA is spherical (i.e., compound symmetry with homogeneous variance), with adjustments (Greenhouse–Geisser or Huynh–Feldt) made to the degrees of freedom used in the denominator for the F-tests of the within-subject effects when the assumption of sphericity is violated. There are no explicit random effects in this approach, but a separate mean square error is estimated for each within-subject factor, which is then used in the F-tests for that factor. Thus, in an analysis of the Rat Brain data, there would be a separate residual variance estimated for REGION, for TREATMENT, and for the REGION × TREATMENT interaction.

In general, the LMM approach to the analysis of repeated-measures data allows for much more flexible correlation structures than can be specified in a traditional repeated-measures ANOVA model.

5.12 Recommendations

This chapter has illustrated some important points with respect to how categorical fixed factors are handled in the specification of random-effects structures for linear mixed models in the different software procedures. If, for example, a categorical fixed factor has four levels, a standard reference (or "dummy variable") parameterization of the model is used (where fixed effects of indicators for the nonreference categories are included in the model), and there are multiple measurements at each level of the categorical factor within each of the higher-level units (as was the case for the two treatments in the Rat Brain example), then technically one could include random effects associated with each of the indicator variables. However, because some of the software procedures (e.g., R, HLM) use an *unstructured* covariance matrix for the random effects included in the model by default, analysts need to be wary of how many *covariance* parameters are being included when specifying random effects associated with categorical factors. Allowing the effects of the three nonreference levels of the four-category fixed factor to randomly vary across higher-level units, for example, not only introduces three additional variance components that need to be estimated, but several additional covariances as well (e.g., the covariance of the random effects associated with nonreference levels 1 and 2 of the categorical fixed factor). Analysts need to be careful when specifying these random effects, and the approach that we took in the Stata example in this chapter underscores this issue (where Stata does not allow for the estimation of covariances of random effects when the effects associated with a categorical fixed factor are allowed to randomly vary at higher levels). For this reason, we recommend "manual" indicator coding for fixed factors, enabling more control over exactly what random effects covariance structure is being estimated. We illustrated this approach with our handling of the TREATMENT variable in the Rat Brain example.

We also illustrated the ability of the different software procedures to allow the error variances to vary depending on the levels of categorical factors at Level 1 of the data hierarchy in a repeated measures design (i.e., within-subject factors). While allowing the error variances to vary across the two treatments did not improve the fits of the models to the Rat Brain data, we recommend that analysts always consider this possibility, beginning with graphical exploration of the data. It has been our experience that many real-world data sets have this feature, and the fits of models (and corresponding tests for various parameters) can be greatly improved by correctly modeling this aspect of the covariance structure for a given data set.

Finally, we illustrated various likelihood ratio tests in this chapter, and we remind readers that likelihood ratio tests rely on asymptotic theory (i.e., large sample sizes at each of the levels of the data hierarchy, and especially at Level 2 in repeated measures designs). The Rat Brain data set is certainly not large enough to make these tests valid in an asymptotic sense, and the likelihood ratio tests are only included for illustration purposes. For smaller data sets, analysts might consider Bayesian methods for hypothesis testing, as was discussed in Chapter 2.

6

Random Coefficient Models for Longitudinal Data: The Autism Example

6.1 Introduction

This chapter illustrates fitting **random coefficient models** to data arising from a **longitudinal study** of the social development of children with autism, whose socialization scores were observed at ages 2, 3, 5, 9 and 13 years. We consider models that allow the child-specific coefficients describing individual time trajectories to vary randomly. Random coefficient models are often used for the analysis of longitudinal data when the researcher is interested in modeling the effects of time and other time-varying covariates at Level 1 of the model on a continuous dependent variable, and also wishes to investigate the amount of between-subject variance in the effects of the covariates across Level 2 units (e.g., subjects in a longitudinal study). In the context of growth and development over time, random coefficient models are often referred to as **growth curve models**. Random coefficient models may also be employed in the analysis of clustered data, when the effects of Level 1 covariates, such as student's socioeconomic status, tend to vary across clusters (e.g., classrooms or schools). Table 6.1 illustrates some examples of longitudinal data that may be analyzed using linear mixed models with random coefficients.

We highlight the R software in this chapter.

6.2 The Autism Study

6.2.1 Study Description

The data used in this chapter were collected by researchers at the University of Michigan (Anderson et al., 2009) as part of a prospective longitudinal study of 214 children. The children were divided into three diagnostic groups when they were 2 years old: autism, pervasive developmental disorder (PDD), and nonspectrum children. We consider a subset of 158 autism spectrum disorder (ASD) children, including autistic and PDD children, for this example. The study was designed to collect information on each child at ages 2, 3, 5, 9, and 13 years, although not all children were measured at each age. One of the study objectives was to assess the relative influence of the initial diagnostic category (autism or PDD), language proficiency at age 2, and other covariates on the developmental trajectories of the socialization of these children.

Study participants were children who had had consecutive referrals to one of two autism clinics before the age of 3 years. Social development was assessed at each age using the Vineland Adaptive Behavior Interview survey form, a parent-reported measure of socialization. The dependent variable, VSAE (Vineland Socialization Age Equivalent), was a

TABLE 6.1: Examples of Longitudinal Data in Different Research Settings

Level of Data		Research Setting		
		Substance Abuse	Business	Autism Research
Subject (Level 2)	Subject variable (random factor)	College	Company	Child
	Covariates	Geographic region, public/private, rural/urban	Industry, geographic region	Gender, baseline language level
Time (Level 1)	Time variable	Year	Quarter	Age
	Dependent variable	Percent of students who use marijuana during each academic year	Stock value in each quarter	Socialization score at each age
	Covariates	School ranking, cost of tuition	Quarterly sales, workforce size	Amount of therapy received

combined score that included assessments of interpersonal relationships, play/leisure time activities, and coping skills. Initial language development was assessed using the Sequenced Inventory of Communication Development (SICD) scale; children were placed into one of three groups (SICDEGP) based on their initial SICD scores on the expressive language subscale at age 2.

Table 6.2 displays a sample of cases from the Autism data in the "long" form appropriate for analysis using the LMM procedures in SAS, SPSS, Stata, and R. The data have been sorted in ascending order by CHILDID and by AGE within each level of CHILDID. This sorting is helpful when interpreting analysis results, but is not required for the model-fitting procedures.

Note that the values of the subject-level variables, CHILDID and SICDEGP, are the same for each observation within a child, whereas the value of the dependent variable (VSAE) is different at each age. We do not consider any time-varying covariates other than AGE in this example. The variables that will be used in the analysis are defined below:

Subject (Level 2) Variables

- **CHILDID** = Unique child identifier

- **SICDEGP** = **S**equenced **I**nventory of **C**ommunication **D**evelopment **E**xpressive

---

TABLE 6.2: Sample of the Autism Data Set

Child (Level 2)		Longitudinal Measures (Level 1)	
Subject ID	**Covariate**	**Time Variable**	**Dependent Variable**
CHILDID	**SICDEGP**	**AGE**	**VSAE**
1	3	2	6
1	3	3	7
1	3	5	18
1	3	9	25
1	3	13	27
2	1	2	6
2	1	3	7
2	1	5	7
2	1	9	8
2	1	13	14
3	3	2	17
3	3	3	18
3	3	5	12
3	3	9	18
3	3	13	24
. . .			

Group: categorized expressive language score at age 2 years (1 = low, 2 = medium, 3 = high)

Time-Varying (Level 1) Variables

- **AGE** = Age in years (2, 3, 5, 9, 13); the time variable

- **VSAE** = **V**ineland **S**ocialization **A**ge **E**quivalent: parent-reported socialization, the dependent variable, measured at each age

6.2.2 Data Summary

The data summary for this example is carried out using the R software package. A link to the syntax and commands that can be used to perform similar analyses in the other software packages is included on the book's web page (see Appendix A).

We begin by reading the comma-separated raw data file (autism.csv) into R and "attaching" the data set to memory, so that shorter versions of the variable names can be used in R functions (e.g., `age` rather than `autism$age`).

```
> autism <- read.csv("c:\\temp\\autism.csv", h = T)
> attach(autism)
```

Alternatively, R users can import text data files directly from a web site:

```
> file <- "http://www-personal.umich.edu/~bwest/autism.csv"
> autism <- read.csv(file, h = T)
> attach(autism)
```

Next, we apply the `factor()` function to the numeric variables SICDEGP and AGE to create categorical versions of these variables (SICDEGP.F and AGE.F), and add the new variables to the data frame.

```
> sicdegp.f <- factor(sicdegp)
> age.f <- factor(age)
> # Add the new variables to the data frame object.
> autism.updated <- data.frame(autism, sicdegp.f, age.f)
```

After creating these factors, we request descriptive statistics for both the continuous and factor variables included in the analysis using the `summary()` function. Note that the `summary()` function produces different output for the continuous variable, VSAE, than for the factor variables.

```
> # Number of Observations at each level of AGE

> summary(age.f)
  2   3   5   9  13
156 150  91 120  95

> # Number of Observations at each level of AGE within each group
> # defined by the SICDEGP factor

> table(sicdegp.f, age.f)
         age.f
sicdegp.f  2  3  5  9 13
        1 50 48 29 37 28
        2 66 64 36 48 41
        3 40 38 26 35 26

> # Overall summary for VSAE

> summary(vsae)
 Min. 1st Qu. Median  Mean 3rd Qu.    Max. NA's
 1.00   10.00   4.00 26.41   27.00  198.00 2.00

> # VSAE means at each AGE

> tapply(vsae, age.f, mean, na.rm=TRUE)
        2        3        5        9       13
 9.089744 15.255034 21.483516 39.554622 60.600000

> # VSAE minimum values at each AGE
```

```
> tapply(vsae, age.f, min, na.rm=TRUE)
2 3 5 9 13
1 4 4 3  7

> # VSAE maximum values at each AGE

> tapply(vsae, age.f, max, na.rm=TRUE)
  2  3  5   9  13
20 63 77 171 198
```

The number of children examined at each age differs due to attrition over time. We also have fewer children at age 5 years because one of the clinics did not schedule children to be examined at that age. There were two children for whom VSAE scores were not obtained at age 9, although the children were examined at that age (missing values of VSAE are displayed as NAs in the output).

Overall, VSAE scores ranged from 1 to 198, with a mean of 26.41. The minimum values changed only slightly at each age, but the means and maximum values increased markedly at later ages.

We next generate graphs that show the observed VSAE scores as a function of age for each child within levels of SICDEGP (Figure 6.1). We also display the mean VSAE profiles by SICDEGP (Figure 6.2). The R syntax that can be used to generate these graphs is provided below:

```
> library(lattice) # Load the library for trellis graphics.
> trellis.device(color=F) # Color is turned off.

> # Load the nlme library, which is required for the
> # plots as well as for subsequent models.
> library(nlme)
```

We generate Figure 6.1 using the R code below. We use the model formula vsae ~ age | childid as an argument in the groupedData() function to create a "grouped" data frame object, autism.g1, in which VSAE is the y-axis variable, AGE is the x-axis variable, and CHILDID defines the grouping of the observations (one line for each child in the plot). The one-sided formula ~ sicdegp.f in the outer = argument defines the outer groups for the plot (i.e., requests one plot per level of the SICDEGP factor).

```
> autism.g1 <- groupedData(vsae ~ age | childid,
outer = ~ sicdegp.f, data = autism.updated)

> # Generate individual profiles in Figure 6.1.
> plot(autism.g1, display = "childid", outer = TRUE, aspect = 2,
key = F, xlab = "Age (Years)", ylab = "VSAE",
main = "Individual Data by SICD Group")
```

For Figure 6.2, we create a grouped data frame object, autism.g2, where VSAE and AGE remain the y-axis and x-axis variables, respectively. However, by replacing "| childid" with "| sicdegp.f", all children with the same value of SICDEGP are defined as a group and used to generate mean profiles. The argument order.groups = F preserves the numerical order of the SICDEGP levels.

Individual Data by SICD Group

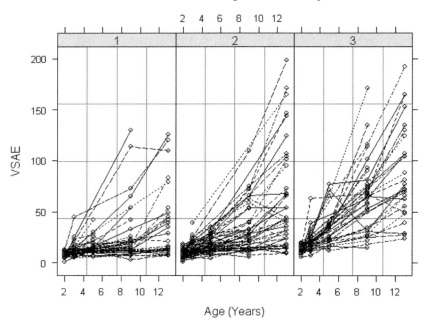

FIGURE 6.1: Observed VSAE values plotted against age for children in each SICD group.

```
> autism.g2 <- groupedData(vsae ~ age | sicdegp.f,
order.groups = F, data = autism.updated)

> # Generate mean profiles in Figure 6.2.
> plot(autism.g2, display = "sicdegp", aspect = 2, key = F,
xlab = "Age (Years)", ylab = "VSAE",
main = "Mean Profiles by SICD Group")
```

The plots of the observed VSAE values for individual children in Figure 6.1 show substantial variation from child to child within each level of SICD group; the VSAE scores of some children tend to increase as the children get older, whereas the scores for other children remain relatively constant. On the other hand, we do not see much variability in the initial values of VSAE at age 2 years for any of the levels of SICD group. Overall, we observe increasing between-child variability in the VSAE scores at each successive year of age. The random coefficient models fitted to the data account for this important feature, as we shall see later in this chapter.

The mean profiles displayed in Figure 6.2 show that the mean VSAE scores generally increase with age. There may also be a quadratic trend in VSAE scores, especially in SICD group two. This suggests that a model to predict VSAE should include both linear and quadratic fixed effects of age, and possibly interactions between the linear and quadratic effects of age and SICD group.

Mean Profiles by SICD Group

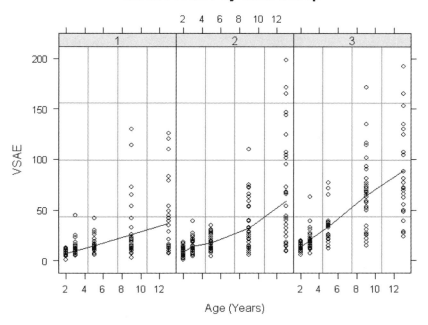

FIGURE 6.2: Mean profiles of VSAE values for children in each SICD group.

6.3   Overview of the Autism Data Analysis

For the analysis of the Autism data, we follow the "top-down" modeling strategy outlined in Subsection 2.7.1 of Chapter 2. In Subsection 6.3.1 we outline the analysis steps, and informally introduce related models and hypotheses to be tested. Subsection 6.3.2 presents a more formal specification of selected models that are fitted to the Autism data, and Subsection 6.3.3 details the hypotheses tested in the analysis. To follow the analysis steps outlined in this section, refer to the schematic diagram presented in Figure 6.3.

6.3.1   Analysis Steps

Step 1: Fit a model with a "loaded" mean structure (Model 6.1).

Fit an initial random coefficient model with a "loaded" mean structure and random child-specific coefficients for the intercept, age, and age-squared.

In Model 6.1, we fit a quadratic regression model for each child, which describes their VSAE as a function of age. This initial model includes the fixed effects of age[1], age-squared, SICD group, the SICD group by age interaction, and the SICD group by age-squared interaction. We also include three random effects associated with each child: a random intercept, a random age effect, and a random age-squared effect. This allows each child to have a unique parabolic trajectory, with coefficients that vary randomly around the fixed effects defining

---

[1]To simplify interpretation of the intercept, we subtract two from the value of AGE and create an auxiliary variable named AGE_2. The intercept can then be interpreted as the predicted VSAE score at age 2, rather than at age zero, which is outside the range of our data.

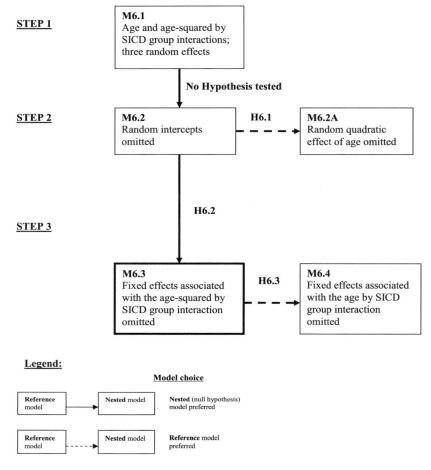

FIGURE 6.3: Model selection and related hypotheses for the analysis of the Autism data.

the mean growth curve for each SICD group. We use REML to estimate the variances and covariances of the three random effects. Model 6.1 also includes residuals associated with the VSAE observations, which conditional on a given child are assumed to be independent and identically distributed.

We encounter some estimation problems when we fit Model 6.1 using the software procedures. SAS `proc mixed` reports problems with estimating the covariance parameters for the random effects; the procedures in SPSS and R do not achieve convergence; the `mixed` command in Stata converges to a solution, but encounters difficulties in estimating the standard errors of the covariance parameters; and HLM2 requires more than 1000 iterations to converge to a solution. As a result, the estimates of the covariance parameters defined in Model 6.1 differ widely across the packages.

Step 2: Select a structure for the random effects (Model 6.2 vs. Model 6.2A).

Fit a model without the random child-specific intercepts (Model 6.2), and test whether to keep the remaining random effects in the model.

We noted in the initial data summary (Figures 6.1 and 6.2) that there was little variability in the VSAE scores at age 2, and therefore in Model 6.2 we attribute this variation entirely to random error rather than to between-subject variability. Compared to Model 6.1, we remove the random child-specific intercepts, while retaining the same fixed effects and the child-specific random effects of age and age-squared. Model 6.2 therefore implies that the children-specific predicted trajectories within a given level of SICD group have a common VSAE value at age 2 (i.e., there is no between-child variability in VSAE scores at age 2). We also assume for Model 6.2 that the age-related linear and quadratic random effects describe the between-subject variation.

We formally test the need for the child-specific quadratic effects of age (Hypothesis 6.1) by using a REML-based likelihood ratio test. To perform this test, we fit a nested model (Model 6.2A) that omits the random quadratic effects. Based on the significant result of this test, we decide to retain both the linear and quadratic child-specific random effects in all subsequent models.

Step 3: Reduce the model by removing nonsignificant fixed effects (Model 6.2 vs. Model 6.3), and check model diagnostics.

In this step, we test whether the fixed effects associated with the AGE-squared × SICDEGP interaction can be omitted from Model 6.2 (Hypothesis 6.2). We conclude that these fixed effects are not significant, and we remove them to form Model 6.3. We then test whether the fixed effects associated with the AGE × SICDEGP interaction can be omitted from Model 6.3 (Hypothesis 6.3), and find that these fixed effects are significant and should be retained in the model.

Finally, we refit Model 6.3 (our final model) using REML estimation to obtain unbiased estimates of the covariance parameters. We check the assumptions for Model 6.3 by examining the distribution of the residuals and of the EBLUPs for the random effects. We also investigate the agreement of the observed VSAE values with the conditional predicted values based on Model 6.3 using scatter plots. The diagnostic plots are generated using the R software package (Section 6.9).

In Figure 6.3 we summarize the model selection process and hypotheses considered in the analysis of the Autism data. Refer to Subsection 3.3.1 for details on the notation in this figure.

6.3.2 Model Specification

Selected models considered in the analysis of the Autism data are summarized in Table 6.3.

6.3.2.1 General Model Specification

The general form of Model 6.1 for an individual response, $VSAE_{ti}$, on child i at the t-th visit ($t = 1, 2, 3, 4, 5$, corresponding to ages 2, 3, 5, 9 and 13), is shown in (6.1). This specification corresponds closely to the syntax used to fit the model using the procedures in SAS, SPSS, R, and Stata.

$$
\begin{aligned}
VSAE_{ti} = \quad & \left. \beta_0 + \beta_1 \times AGE\_2_{ti} + \beta_2 \times AGE\_2SQ_{ti} + \beta_3 \times SICDEGP1_i \right. \\
& + \beta_4 \times SICDEGP2_i + \beta_5 \times AGE\_2_{ti} \times SICDEGP1_i \\
& + \beta_6 \times AGE\_2_{ti} \times SICDEGP2_i \times \beta_7 \times AGE\_2SQ_{ti} \times SICDEGP1_i \\
& + \beta_8 \times AGE\_2SQ_{ti} \times SICDEGP2_i \\
& \qquad + u_{0i} + u_{1i} \times AGE\_2_{ti} + u_{2i} \times AGE\_2SQ_{ti} + \varepsilon_{ti}
\end{aligned}
$$

$\left. \right\}$ fixed

$\}$ random

(6.1)

TABLE 6.3: Summary of Selected Models Considered for the Autism Data

	Term/Variable	General Notation	HLM Notation	Model 6.1	Model 6.2	Model 6.3
Fixed effects	Intercept	β_0	β_{00}	✓	✓	✓
	AGE_2	β_1	β_{10}	✓	✓	✓
	AGE_2SQ	β_2	β_{20}	✓	✓	✓
	SICDEGP1	β_3	β_{01}	✓	✓	✓
	SICDEGP2	β_4	β_{02}	✓	✓	✓
	AGE_2 × SICDEGP1	β_5	β_{11}	✓	✓	✓
	AGE_2 × SICDEGP2	β_6	β_{12}	✓	✓	✓
	AGE_2SQ × SICDEGP1	β_7	β_{21}	✓	✓	
	AGE_2SQ × SICDEGP2	β_8	β_{22}	✓	✓	
Random effects Child (i)	Intercept	u_{0i}	r_{0i}	✓	✓	✓
	AGE_2	u_{1i}	r_{1i}		✓	✓
	AGE_2SQ	u_{2i}	r_{2i}		✓	✓
Residuals Time (t)		ε_{ti}	e_{ti}	✓	✓	✓
Covariance Parameters (θ_D) for D Matrix Child level	Variance of intercepts	σ_{int}^2	$\tau[1,1]$	✓	✓	✓
	Covariance of intercepts, AGE_2 effects	$\sigma_{int,age}$	$\tau[1,2]$		✓	✓
	Covariance of intercepts, AGE_2SQ effects	$\sigma_{int,age\text{-}sq}$	$\tau[1,3]$		✓	✓
	Variance of AGE_2 effects	σ_{age}^2	$\tau[2,2]$		✓	✓
	Covariance of AGE_2 effects, AGE_2SQ effects	$\sigma_{age,age\text{-}sq}$	$\tau[2,3]$		✓	✓
	Variance of AGE_2SQ effects	$\sigma_{age\text{-}sq}$	$\tau[3,3]$		✓	✓
Covariance Parameters (θ_R) for R_i Matrix Time level	Residual variance	σ^2	σ^2	✓	✓	✓

In (6.1), the AGE_2 variable represents the value of AGE minus 2, and AGE_2SQ represents AGE_2 squared. We include two dummy variables, SICDEGP1 and SICDEGP2, to indicate the first two levels of the SICD group. Because we set the fixed effect for the third level of the SICD group to 0, we consider SICDEGP = 3 as the "reference category." There are two variables that represent the interaction between age and SICD group: AGE_2 × SICDEGP1 and AGE_2 × SICDEGP2. There are also two variables that represent the interaction between age-squared and SICD group: AGE_2SQ × SICDEGP1 and AGE_2SQ × SICDEGP2.

The parameters β_0 through β_8 represent the fixed effects associated with the intercept, the covariates, and the interaction terms in the model. Because the fixed intercept, β_0, corresponds to the predicted VSAE score when all covariates, including AGE_2, are equal to zero, the intercept can be interpreted as the mean predicted VSAE score for children at 2 years of age in the reference category of the SICD group (SICDEGP = 3).

The parameters β_1 and β_2 represent the fixed effects of age and age-squared for the reference category of the SICD group (SICDEGP = 3). The fixed effects β_3 and β_4 represent the difference in the intercept for the first two levels of the SICD group vs. the reference category.

The fixed effects β_5 and β_6 represent the differences in the linear effect of age between the first two levels of SICD group and the linear effect of age in the reference category of the SICD group. Similarly, the fixed effects β_7 and β_8 represent the differences in the quadratic effect of age between the first two levels of the SICD group and the quadratic effect of age in the reference category.

The terms u_{0i}, u_{1i}, and u_{2i} in (6.1) represent the random effects associated with the intercept, linear effect of age, and quadratic effect of age for child i. The distribution of the vector of the three random effects, $\boldsymbol{u}_i$, associated with child i is assumed to be multivariate normal:

$$\boldsymbol{u_i} = \begin{pmatrix} u_{0i} \\ u_{1i} \\ u_{2i} \end{pmatrix} \sim \mathcal{N}(0, \boldsymbol{D})$$

Each of the three random effects has a mean of 0, and the variance-covariance matrix, $\boldsymbol{D}$, for the random effects is:

$$\boldsymbol{D} = \begin{pmatrix} \sigma^2_{int} & \sigma_{int,age} & \sigma_{int,age\text{-}squared} \\ \sigma_{int,age} & \sigma^2_{age} & \sigma_{age,age\text{-}squared} \\ \sigma_{int,age\text{-}squared} & \sigma_{age,age\text{-}squared} & \sigma^2_{age\text{-}squared} \end{pmatrix}$$

The term ε_{ti} in (6.1) represents the residual associated with the observation at time t on child i. The distribution of the residuals can be written as

$$\varepsilon_{ti} \sim \mathcal{N}(0, \sigma^2)$$

We assume that the residuals are independent and identically distributed, conditional on the random effects, and that the residuals are independent of the random effects.

We do not include the specification of other models (e.g., Models 6.2 and 6.3) in this section. These models can be obtained by simplification of Model 6.1. For example, the $\boldsymbol{D}$ matrix in Model 6.2 has the following form, because the random intercepts are omitted from the model:

$$\boldsymbol{D} = \begin{pmatrix} \sigma^2_{age} & \sigma_{age,age\text{-}squared} \\ \sigma_{age,age\text{-}squared} & \sigma^2_{age\text{-}squared} \end{pmatrix}$$

6.3.2.2 Hierarchical Model Specification

We now present Model 6.1 in hierarchical form, using the same notation as in (6.1). The correspondence between this notation and the HLM software notation is defined in Table 6.3.

The hierarchical model has two components, reflecting contributions from the two levels of the Autism data: Level 1 (the time level), and Level 2 (the child level).

Level 1 Model (Time)

$$\text{VSAE}_{ti} = b_{0i} + b_{1i} \times \text{AGE\_2}_{ti} + b_{2i} \times \text{AGE\_2SQ}_{ti} + \varepsilon_{ti} \qquad (6.2)$$

where

$$\varepsilon_{ti} \sim \mathcal{N}(0, \sigma^2)$$

This model shows that at Level 1 of the data, we have a set of child-specific quadratic regressions of VSAE on AGE_2 and AGE_2SQ. The intercept (b_{0i}), the linear effect of AGE_2 (b_{1i}), and the quadratic effect of AGE_2SQ (b_{2i}) defined in the Level 2 model are allowed to vary between children, who are indexed by i.

The unobserved child-specific coefficients for the intercept, linear effect of age, and quadratic effect of age (b_{0i}, b_{1i}, and b_{2i}) in the Level 1 model depend on fixed effects associated with Level 2 covariates and random child effects, as shown in the Level 2 model below:

Level 2 Model (Child)

$$\begin{aligned} b_{0i} &= \beta_0 + \beta_3 \times \text{SICDEGP1}_i + \beta_4 \times \text{SICDEGP2}_i + u_{0i} \\ b_{1i} &= \beta_1 + \beta_5 \times \text{SICDEGP1}_i + \beta_6 \times \text{SICDEGP2}_i + u_{1i} \\ b_{2i} &= \beta_2 + \beta_7 \times \text{SICDEGP1}_i + \beta_8 \times \text{SICDEGP2}_i + u_{2i} \end{aligned} \qquad (6.3)$$

where

$$\boldsymbol{u_i} = \begin{pmatrix} u_{0i} \\ u_{1i} \\ u_{2i} \end{pmatrix} \sim \mathcal{N}(0, \boldsymbol{D})$$

The Level 2 model in (6.3) shows that the intercept (b_{0i}) for child i depends on the fixed overall intercept (β_0), the fixed effects (β_3 and β_4) of the child-level covariates SICDEGP1 and SICDEGP2, and a random effect (u_{0i}) associated with child i.

The child-specific linear effect of age (b_{1i}) depends on the overall fixed effect of age (β_1), the fixed effect of SICDEGP1 (β_5), the fixed effect of SICDEGP2 (β_6), and a random effect (u_{1i}) associated with child i. The equation for the child-specific quadratic effect of age (b_{2i}) for child i is defined similarly. The random effects in the Level 2 model allow the child-specific intercepts, linear effects of age, and quadratic effects of age to vary randomly between children. The variance-covariance matrix ($\boldsymbol{D}$) of the random effects is defined as in the general model specification.

The hierarchical specification of Model 6.1 is equivalent to the general specification for this model presented in Subsection 6.3.2.1. We can derive the model as specified in (6.1) by substituting the expressions for b_{0i}, b_{1i}, and b_{2i} from the Level 2 model (6.3) into the Level

1 model (6.2). The fixed effects associated with the child-specific covariates SICDEGP1 and SICDEGP2 in the Level 2 equations for b_{1i} and b_{2i} represent interactions between these covariates and AGE_2 and AGE_2SQ in the general model specification.

6.3.3 Hypothesis Tests

Hypothesis tests considered in the analysis of the Autism data are summarized in Table 6.4.

Hypothesis 6.1. The random effects associated with the quadratic effect of AGE can be omitted from Model 6.2.

We indirectly test whether these random effects can be omitted from Model 6.2. The null and alternative hypotheses are:

$$H_0: \boldsymbol{D} = \begin{pmatrix} \sigma_{age}^2 & 0 \\ 0 & 0 \end{pmatrix}$$

$$H_A: \boldsymbol{D} = \begin{pmatrix} \sigma_{age}^2 & \sigma_{age,age\text{-}squared} \\ \sigma_{age,age\text{-}squared} & \sigma_{age\text{-}squared}^2 \end{pmatrix}$$

To test Hypothesis 6.1, we use a REML-based likelihood ratio test. The test statistic is the value of the –2 REML log-likelihood value for Model 6.2A (the nested model excluding the random quadratic age effects) minus the value for Model 6.2 (the reference model). To obtain a p-value for this statistic, we refer it to a mixture of χ^2 distributions, with 1 and 2 degrees of freedom and equal weight 0.5.

Hypothesis 6.2. The fixed effects associated with the AGE-squared × SICDEGP interaction are equal to zero in Model 6.2.

The null and alternative hypotheses in this case are specified as follows:

$$H_0: \beta_7 = \beta_8 = 0$$

$$H_A: \beta_7 \neq 0 \text{ or } \beta_8 \neq 0$$

We test Hypothesis 6.2 using an ML-based likelihood ratio test. The test statistic is the value of the –2 ML log-likelihood for Model 6.3 (the nested model excluding the fixed effects associated with the interaction) minus the value for Model 6.2 (the reference model). To obtain a p-value for this statistic, we refer it to a χ^2 distribution with 2 degrees of freedom, corresponding to the 2 additional fixed-effect parameters in Model 6.2.

Hypothesis 6.3. The fixed effects associated with the AGE × SICDEGP interaction are equal to zero in Model 6.3.

The null and alternative hypotheses in this case are specified as follows:

$$H_0: \beta_5 = \beta_6 = 0$$

$$H_A: \beta_5 \neq 0 \text{ or } \beta_6 \neq 0$$

We test Hypothesis 6.3 using an ML-based likelihood ratio test. The test statistic is the value of the –2 ML log-likelihood for Model 6.4 (the nested model excluding the fixed effects associated with the interaction) minus the value for Model 6.3 (the reference model). To obtain a p-value for this statistic, we refer it to a χ^2 distribution with 2 degrees of freedom, corresponding to the 2 additional fixed-effect parameters in Model 6.3.

For the results of these hypothesis tests, see Section 6.5.

TABLE 6.4: Summary of Hypotheses Tested in the Autism Analysis

Label	Null (H_0)	Alternative (H_A)	Test	Models Compared		Est. Method	Test Stat. Dist. under H_0
				Nested Model (H_0)	Ref. Model (H_A)		
6.1	Drop u_{2i} (random effects associated with AGE-squared)	Retain u_{2i}	LRT	Model 6.2A	Model 6.2	REML	$0.5\chi_1^2 + 0.5\chi_2^2$
6.2	Drop fixed effects associated with AGE-squared by SICDEGP interaction ($\beta_7 = \beta_8 = 0$)	Either $\beta_7 \neq 0$, or $\beta_8 \neq 0$	LRT	Model 6.3	Model 6.2	ML	χ_2^2
6.3	Drop fixed effects associated with AGE by SICDEGP interaction ($\beta_5 = \beta_6 = 0$)	Either $\beta_5 \neq 0$, or $\beta_6 \neq 0$	LRT	Model 6.4	Model 6.3	ML	χ_2^2

6.4   Analysis Steps in the Software Procedures

In general, when fitting an LMM to longitudinal data using the procedures discussed in this book, all observations available for a subject at any time point are included in the analysis. For this approach to yield correct results, we assume that missing values are missing at random, or MAR (Little & Rubin, 2002). See Subsection 2.9.4 for a further discussion of the MAR concept, and how missing data are handled by software procedures that fit LMMs.

We compare results for selected models across the software procedures in Section 6.6.

6.4.1   SAS

We first import the comma-separated data file (`autism.csv`, assumed to be located in the `C:\temp` directory) into SAS, and create a temporary SAS data set named `autism`.

```
PROC IMPORT OUT = WORK.autism
DATAFILE="C:\temp\autism.csv"
DBMS=CSV REPLACE;
GETNAMES=YES;
DATAROW=2;
RUN;
```

Next, we generate a data set named `autism2` that contains the new variable AGE_2 (equal to AGE minus 2), and its square, AGE_2SQ.

```
data autism2;
set autism;
age_2 = age - 2;
age_2sq = age_2*age_2;
run;
```

Step 1: Fit a model with a "loaded" mean structure (Model 6.1).

The SAS syntax for Model 6.1 is as follows:

```
title "Model 6.1";
proc mixed data = autism2 covtest;
class childid sicdegp;
model vsae = age_2 age_2sq sicdegp age_2*sicdegp
age_2sq*sicdegp / solution ddfm = sat;
random int age_2 age_2sq / subject = childid type = un g v solution;
run;
```

The `model` statement specifies the dependent variable, VSAE, and lists the terms that have fixed effects in the model. The `ddfm = sat` option is used to request the Satterthwaite degrees of freedom approximation for the denominator in the F-tests of fixed effects (see Subsection 3.11.6 for a discussion of denominator degrees of freedom options in SAS).

The `random` statement lists the child-specific random effects associated with the intercept (`int`), the linear effect of age (`age_2`), and the quadratic effect of age (`age_2sq`). We specify the structure of the variance-covariance matrix of the random effects (called the *G matrix* by SAS; see Subsection 2.2.3) as unstructured (`type = un`). The g option requests that a single block of the estimated G matrix (the 3×3 D matrix in our notation) be displayed in the output, while the **v** option requests that the estimate of the implied marginal variance-covariance matrix for observations on the first child (the **subject** variable) be displayed as well. The **solution** option in the **random** statement instructs SAS to display the EBLUPs of the three random effects associated with each level of CHILDID. This option can be omitted to shorten the output.

Software Note: The following note is displayed in the SAS log after fitting Model 6.1:

NOTE: Estimated G matrix is not positive-definite.

This message is important, and should not be disregarded (even though SAS does not generate an error, which would cause the model fitting to terminate). When such a message is generated in the log, results of the model fit should be interpreted with extreme caution, and the model may need to be simplified or respecified.

The NOTE means that **proc mixed** has converged to an estimated solution for the covariance parameters in the G matrix that results in G being nonpositive-definite. One reason for G being nonpositive-definite is that a variance parameter estimate might be either very small (close to zero), or lie outside the parameter space (i.e., is estimated to be negative). There can be nonpositive-definite G matrices in which this is not the case (i.e., there are cases in which all variance estimates are positive, but the matrix is still not positive-definite). We note in the SAS output that the estimate for the variance of the random intercepts (σ_{int}^2) is set to zero.

One way to investigate the problem encountered with the estimation of the G matrix is to relax the requirement that it be positive-definite by using the **nobound** option in the **proc mixed** statement:

```
proc mixed data=autism2 nobound;
```

A single block of the G matrix obtained after specifying the **nobound** option (corresponding to the D matrix introduced in Subsection 6.3.2) is:

```
                    Estimated G Matrix

     Row      Effect childid      Col1     Col2     Col3

       1 Intercept        1   -10.5406   4.2760   0.1423
       2     age_2        1     4.2760  11.9673  -0.4038
       3   age_2sq        1     0.1423  -0.4038   0.1383
```

Note that this block of the estimated G matrix is symmetric as needed, but that the value of the entry that corresponds to the estimated variance of the random intercepts is negative (-10.54); G is therefore not a variance-covariance matrix. Consequently, we are not able to make a valid statement about the variance of the child-specific intercepts in the context of Model 6.1.

If we are not interested in making inferences about the between-child variability, this G matrix is valid in the context of the **marginal model** implied by Model 6.1. As long as the overall V matrix is positive-definite, we can use the marginal model to make valid inferences about the fixed-effect parameters. This alternative approach, which does not apply constraints to the D matrix (i.e., does not force it to be positive-definite) is not presently available in the other software procedures.

In spite of the problems encountered in fitting Model 6.1, we consider the results generated by `proc mixed` (using the original syntax, without the `nobound` option) in Section 6.6 so that we can make comparisons across the software procedures.

Step 2: Select a structure for the random effects (Model 6.2 vs. Model 6.2A).

We now fit Model 6.2, which has the same fixed effects as Model 6.1 but omits the random effects associated with the child-specific intercepts. We then decide whether to keep the remaining random effects in the model. The syntax for Model 6.2 is as follows:

```
title "Model 6.2";
proc mixed data = autism2 covtest;
class childid sicdegp;
model vsae = age_2 age_2sq sicdegp age_2*sicdegp
age_2sq*sicdegp / solution ddfm = sat;
random age_2 age_2sq / subject = childid type = un g;
run;
```

Note that the `int` (intercept) term has been removed from the `random` statement, which is the only difference between the syntax for Model 6.1 and Model 6.2. SAS does not indicate any problems with the estimation of the G matrix for Model 6.2.

We next carry out a likelihood ratio test of Hypothesis 6.1, to decide whether we need to keep the random effects associated with age-squared in Model 6.2. To test Hypothesis 6.1, we fit a nested model (Model 6.2A) using syntax much like the syntax for Model 6.1, but omitting the AGE_2SQ term from the `random` statement, as shown in the following text. We retain the `type = un` and `g` options below:

```
random age_2 / subject = childid type = un g;
```

We calculate a likelihood ratio test statistic for Hypothesis 6.1 by subtracting the –2 REML log-likelihood of Model 6.2 (the reference model, –2 REML LL = 4615.3) from that of Model 6.2A (the nested model, –2 REML LL = 4699.2). The p-value for the resulting test statistic is derived by referring it to a mixture of χ^2 distributions, with 1 and 2 degrees of freedom and weights equal to 0.5, as shown in the syntax that follows below. The p-value for this test will be displayed in the SAS log. Based on the significant result of this test ($p < 0.001$), we retain both the random linear and quadratic effects of age in Model 6.2.

```
title "p-value for Hypothesis 6.1";
data _null_;
lrtstat = 4699.2 - 4615.3;
pvalue = 0.5*(1 - probchi(lrtstat,1)) + 0.5*(1 - probchi(lrtstat,2));
format pvalue 10.8;
put pvalue =;
run;
```

Step 3: Reduce the model by removing nonsignificant fixed effects (Model 6.2 vs. Model 6.3).

In this step, we investigate whether we can reduce the number of fixed effects in Model 6.2 while maintaining the random linear and quadratic age effects. To test Hypothesis 6.2 (where the null hypothesis is that there is no AGE-squared × SICDEGP interaction), we fit Model 6.2 and Model 6.3 using maximum likelihood estimation, by including the `method = ML` option in the `proc mixed` statement:

```
title "Model 6.2 (ML)";
proc mixed data = autism2 covtest method = ML;
class childid sicdegp;
model vsae = age_2 age_2sq sicdegp age_2*sicdegp
age_2sq*sicdegp / solution ddfm = sat;
random age_2 age_2sq / subject = childid type = un;
run;
```

To fit a nested model, Model 6.3 (also using ML estimation), we remove the interaction term SICDEGP × AGE_2SQ from the `model` statement:

```
title "Model 6.3 (ML)";
proc mixed data = autism2 covtest method = ML;
class childid sicdegp;
model vsae = age_2 age_2sq sicdegp age_2*sicdegp
/ solution ddfm = sat;
random age_2 age_2sq / subject = childid type = un;
run;
```

We then compute a likelihood ratio test statistic for Hypothesis 6.2 by subtracting the -2 ML log-likelihood of Model 6.2 (the reference model, -2 LL = 4610.4) from that of Model 6.3 (the nested model, -2 LL = 4612.3). The SAS code for this likelihood ratio test is shown in the syntax that follows. Based on the nonsignificant test result ($p = 0.39$), we drop the fixed effects associated with the SICDEGP × AGE_2SQ interaction from Model 6.2 and obtain Model 6.3. Additional hypothesis tests for fixed effects (i.e., Hypothesis 6.3) do not suggest any further reduction of Model 6.3.

```
title "P-value for Hypothesis 6.2";
data _null_;
lrtstat = 4612.3 - 4610.4;
df = 2;
pvalue = 1 - probchi(lrtstat,df);
format pvalue 10.8;
put lrtstat= df= pvalue= ;
run;
```

We now refit Model 6.3 (our final model) using REML estimation. The `ods output` statement is included to capture the EBLUPs of the random effects in a data set, `eblup_dat`, and to get the conditional studentized residuals in another data set, `inf_dat`. The captured data sets can be used for checking model diagnostics.

```
title "Model 6.3 (REML)";
ods output influence = inf_dat solutionR = eblup_dat;
ods exclude influence solutionR;
proc mixed data = autism2 covtest;
class childid sicdegp;
model vsae = sicdegp age_2 age_2sq age_2*sicdegp
/ solution ddfm = sat influence;
random age_2 age_2sq / subject = childid solution g v vcorr type = un;
run;
```

The `ods exclude` statement requests that SAS not display the influence statistics for each observation or the EBLUPs for the random effects in the output, to save space. The `ods exclude` statement does not interfere with the `ods output` statement; influence statistics and EBLUPs are still captured in separate data sets, but they are omitted from the output. We must also include the `influence` option in the `model` statement and the `solution` option in the `random` statement for these data sets to be created. See Chapter 3 (Subsection 3.10.2) for information on obtaining influence statistics and graphics for the purposes of checking model diagnostics using SAS.

6.4.2 SPSS

We first import the raw comma-separated data file, `autism.csv`, from the `C:\temp` folder into SPSS:

```
GET DATA /TYPE = TXT
/FILE 'C:\temp\autism.csv'
/DELCASE = LINE
/DELIMITERS = ","
/ARRANGEMENT = DELIMITED
/FIRSTCASE = 2
/IMPORTCASE = ALL
/VARIABLES =
age F2.1
vsae F3.2
sicdegp F1.0
childid F2.1
.
CACHE.
EXECUTE.
```

Next, we compute the new AGE variable (AGE_2) and the squared version of this new variable, AGE_2SQ:

```
COMPUTE age_2 = age - 2 .
EXECUTE .
COMPUTE age_2sq = age_2*age_2 .
EXECUTE.
```

We now proceed with the analysis steps.

Step 1: Fit a model with a "loaded" mean structure (Model 6.1).

The SPSS syntax for Model 6.1 is as follows:

```
* Model 6.1 .
MIXED
vsae WITH age_2 age_2sq BY sicdegp
/CRITERIA = CIN(95) MXITER(100) MXSTEP(5) SCORING(1)
SINGULAR(0.000000000001) HCONVERGE(0, ABSOLUTE)
LCONVERGE(0, ABSOLUTE) PCONVERGE(0.000001, ABSOLUTE)
/FIXED = age_2 age_2sq sicdegp age_2*sicdegp
age_2sq*sicdegp | SSTYPE(3)
/METHOD = REML
/PRINT = G SOLUTION
/RANDOM INTERCEPT age_2 age_2sq | SUBJECT(CHILDID) COVTYPE(UN) .
```

The dependent variable, VSAE, is listed first after invocation of the MIXED command. The continuous covariates (AGE_2 and AGE_2SQ) appear after the WITH keyword. The categorical fixed factor, SICDEGP, appears after the BY keyword. The convergence criteria (listed after the /CRITERIA subcommand) are the defaults obtained when the model is set up using the SPSS menu system.

The FIXED subcommand identifies the terms with associated fixed effects in the model. The METHOD subcommand identifies the estimation method for the covariance parameters (the default REML method is used). The PRINT subcommand requests that the estimated G matrix be displayed (the displayed matrix corresponds to the D matrix that we defined for Model 6.1 in Subsection 6.3.2). We also request that estimates of the fixed effects (SOLUTION) be displayed in the output.

The RANDOM subcommand specifies that the model should include random effects associated with the intercept (INTERCEPT), the linear effect of age (AGE_2), and the quadratic effect of age (AGE_2SQ) for each level of CHILDID. The SUBJECT is specified as CHILDID in the RANDOM subcommand. The structure of the G matrix of variances and covariances of the random effects (COVTYPE) is specified as unstructured (UN) (see Subsection 6.3.2).

When we attempt to fit Model 6.1 in IBM SPSS Statistics (Version 21), the following warning message appears in the SPSS output:

```
                         Warnings

  Iteration was terminated but convergence has not been achieved.
  The MIXED procedure continues despite this warning. Subsequent results
  produced are based on the last iteration. Validity of the model fit
  is uncertain.
```

Although this is a warning message and does not appear to be a critical error (which would cause the model fitting to terminate), it should not be ignored and the model fit should be viewed with caution. It is always good practice to check the SPSS output for similar warnings when fitting a linear mixed model. Investigation of the "Estimates of Covariance Parameters" table in the SPSS output reveals problems.

```
        Estimates of Covariance Parameters (a)

  Parameter                            Estimate      Std. Error

  Residual                            36.945035        2.830969

  Intercept + age_2 +   UN (1,1)       0.000000  (b)  0.000000
  age_2sq [subject =    UN (2,1)     -15.014722        2.406356
  childid]              UN (2,2)      15.389867        3.258686
                        UN (3,1)       3.296464        0.237604
                        UN (3,2)      -0.676210        0.254689
                        UN (3,3)       0.135217        0.028072

  a. Dependent Variable: vsae.
  b. This covariance parameter is redundant.
```

The second footnote states that the variance of the random effects associated with the INTERCEPT for each child (labeled UN(1,1) in the table) is "redundant." This variance estimate is set to a value of 0.000000, with a standard error of zero. In spite of the estimation problems encountered, we display results from the fit of Model 6.1 in SPSS in Section 6.6, for comparison with the other software procedures.

Step 2: Select a structure for the random effects (Model 6.2 vs. Model 6.2A).

We now fit Model 6.2, which includes the same fixed effects as Model 6.1, but omits the random effects associated with the intercept for each CHILDID. The only change is in the RANDOM subcommand:

```
* Model 6.2 .
MIXED
vsae WITH age_2 age_2sq BY sicdegp
/CRITERIA = CIN(95) MXITER(100) MXSTEP(5) SCORING(1)
SINGULAR(0.000000000001) HCONVERGE(0, ABSOLUTE)
LCONVERGE(0, ABSOLUTE) PCONVERGE(0.000001, ABSOLUTE)
/FIXED = age_2 age_2sq sicdegp age_2*sicdegp
age_2sq*sicdegp | SSTYPE(3)
/METHOD = REML
/PRINT = G SOLUTION
/RANDOM age_2 age_2sq | SUBJECT(CHILDID) COVTYPE(UN) .
```

Note that the INTERCEPT term has been omitted from the RANDOM subcommand in the preceding code. The fit of Model 6.2 does not generate any warning messages.

To test Hypothesis 6.1, we fit a nested model (Model 6.2A) by modifying the RANDOM subcommand for Model 6.2 as shown:

```
* Model 6.2A modified RANDOM subcommand.
/RANDOM age_2 | SUBJECT(CHILDID).
```

Note that the AGE_2SQ term has been omitted. A likelihood ratio test can now be carried out by subtracting the –2 REML log-likelihood for Model 6.2 (the reference model) from that of Model 6.2A (the reduced model). The p-value for the test statistic is derived by referring it to a mixture of χ^2 distributions, with equal weight 0.5 and 1 and 2 degrees of freedom (see Section 6.5.1). Based on the significant result ($p < 0.001$) of this test, we retain the random effects associated with the quadratic (and therefore linear) effects of age in Model 6.2.

Step 3: Reduce the model by removing nonsignificant fixed effects (Model 6.2 vs. Model 6.3).

We now proceed to reduce the number of fixed effects in the model, while maintaining the random-effects structure specified in Model 6.2. To test Hypothesis 6.2, we first refit Model 6.2 using ML estimation (/METHOD = ML):

```
* Model 6.2 (ML) .
MIXED
vsae WITH age_2 age_2sq BY sicdegp
/CRITERIA = CIN(95) MXITER(100) MXSTEP(5) SCORING(1)
SINGULAR(0.000000000001) HCONVERGE(0, ABSOLUTE)
LCONVERGE(0, ABSOLUTE) PCONVERGE(0.000001, ABSOLUTE)
/FIXED = age_2 age_2sq sicdegp age_2*sicdegp
age_2sq*sicdegp | SSTYPE(3)
/METHOD = ML
/PRINT = G SOLUTION
/RANDOM age_2 age_2sq | SUBJECT(CHILDID) COVTYPE(UN) .
```

Next, we fit a nested model (Model 6.3) by removing the term representing the inter-action between SICDEGP and the quadratic effect of age, SICDEGP × AGE_2SQ, from the /FIXED subcommand. We again use /METHOD = ML (the other parts of the full MIXED command are not shown here):

```
/FIXED = age_2 age_2sq sicdegp age_2*sicdegp | SSTYPE(3)
/METHOD = ML
```

Based on the nonsignificant likelihood ratio test ($p = 0.39$; see Subsection 6.5.2), we con-clude that the fixed effects associated with this interaction can be dropped from Model 6.2, and we proceed with Model 6.3. Additional likelihood ratio tests (e.g., a test of Hypothesis 6.3) suggest no further reduction of Model 6.3.

We now fit Model 6.3 (the final model in this example) using REML estimation:

```
* Model 6.3 (REML).
MIXED
vsae WITH age_2 age_2sq BY sicdegp
/CRITERIA = CIN(95) MXITER(100) MXSTEP(5) SCORING(1)
SINGULAR(0.000000000001) HCONVERGE(0, ABSOLUTE)
LCONVERGE(0, ABSOLUTE) PCONVERGE(0.000001, ABSOLUTE)
/FIXED = age_2 age_2sq sicdegp age_2*sicdegp | SSTYPE(3)
/METHOD = REML
/PRINT = G SOLUTION
/SAVE = PRED RESID
/RANDOM age_2 age_2sq | SUBJECT(CHILDID) COVTYPE(UN) .
```

The SAVE subcommand is added to the syntax to save the conditional predicted values (PRED) in the data set. These predicted values are based on the fixed effects and the EBLUPs of the random AGE_2 and AGE_2SQ effects. We also save the conditional residuals (RESID) in the data set. These variables can be used for checking assumptions about the residuals for this model.

6.4.3 R

We start with the same data frame object (autism.updated) that was used for the initial data summary in R (Subsection 6.2.2), but we first create additional variables that will be used in subsequent analyses. Note that we create the new variable SICDEGP2, which has a value of zero for SICDEGP = 3, so that SICD group 3 will be considered the reference category (lowest value) for the SICDEGP2.F factor. We do this to be consistent with the output from the other software procedures.

```
> # Compute age.2 (AGE minus 2) and age.2sq (AGE2 squared).
> age.2 <- age - 2
> age.2sq <- age.2*age.2

> # Recode the SICDEGP factor for model fitting.
> sicdegp2 <- sicdegp
> sicdegp2[sicdegp == 3] <- 0
> sicdegp2[sicdegp == 2] <- 2
> sicdegp2[sicdegp == 1] <- 1
> sicdegp2.f <- factor(sicdegp2)
```

```
> # Omit two records with VSAE = NA, and add the recoded
> # variables to the new data frame object.
> autism.updated <- subset(data.frame(autism, sicdegp2.f, age.2),
    !is.na(vsae))
```

Alternatively, the new variable SICDEGP2 can be obtained using this syntax:

```
> sicdegp2 <- cut(sicdegp, breaks = 0:3, labels= FALSE)
```

6.4.3.1   Analysis Using the `lme()` Function

We first load the `nlme` library, so that we can utilize the `nlme` version of the `lme()` function for this example:

```
> library(nlme)
```

Next, we create a "grouped" data frame object named `autism.grouped`, using the `groupedData()` function, to define the hierarchical structure of the Autism data set. The arguments of this function indicate that (1) VSAE is the dependent variable, (2) AGE_2 is the primary covariate, and (3) CHILDID defines the "groups" of observations with which random effects are associated when fitting the models. Note that the `groupedData()` function is only available after loading the `nlme` library:

```
> autism.grouped <- groupedData(vsae ~ age.2 | childid,
data = autism.updated, order.groups = F)
```

The `order.groups = F` argument is specified to preserve the original order of the children in the input data set (children are sorted in descending order by SICDEGP, so SICDEGP = 3 is the first group in the data set). We now continue with the analysis steps.

Step 1: Fit a model with a loaded mean structure (Model 6.1).

We first fit Model 6.1 using the `lme()` function (note that the modified version of the AGE variable, AGE_2, is used):

```
> model6.1.fit <- lme(vsae ~ age.2 + I(age.2^2) + sicdegp2.f +
age.2:sicdegp2.f + I(age.2^2):sicdegp2.f,
random = ~ age.2 + I(age.2^2), method= "REML",
data = autism.grouped)
```

This specification of the `lme()` function is described as follows:

- The name of the created object that contains the results of the fitted model is `model6.1.fit`.

- The model formula, `vsae ~ age.2 + I(age.2^2) + sicdegp2.f +age.2:sicdegp2.f + I(age.2^2):sicdegp2.f`, defines the continuous response variable (VSAE) and the terms that have fixed effects in the model (including the interactions). The `lme()` function automatically creates the appropriate dummy variables for the categories of the SICDEGP2.F factor, treating the lowest-valued category (which corresponds to the original value of SICDEGP=3) as the reference. The `I()` function is used to inhibit R from interpreting the "^" character as an operator in the fixed-effects formula (as opposed to an arithmetic operator meaning "raised to the power of"). A fixed intercept is included by default.

- The second argument of the function, `random = ~ age.2 + I(age.2^2)`, indicates the variables that have random effects associated with them. A random effect associated with the intercept for each level of CHILDID is included by default. These random effects are associated with each level of CHILDID because of the definition of the grouped data frame object.

- The third argument of the function, `method= "REML"`, requests that the default REML estimation method is to be used.

- The fourth argument of the function, `data = autism.grouped`, indicates the "grouped" data frame object to be used.

By default, the `lme()` function uses an unstructured D matrix for the variance-covariance matrix of the random effects.

After fitting Model 6.1, the following message is displayed:

```
Error in lme.formula(vsae) ~ age.2 + I(age.2^2) +
sicdegp2.f + age.2:sicdegp2.f + :
nlminb problem, convergence error code = 1
iteration limit reached without convergence (10)
```

The estimation algorithm did not converge to a solution for the parameter estimates. As a result, the `model6.1.fit` object is not created, and estimates of the parameters in Model 6.1 cannot be obtained using the `summary()` function. We proceed to consider Model 6.2 as an alternative.

Step 2: Select a structure for the random effects (Model 6.2 vs. Model 6.2A).

We now fit Model 6.2, which includes the same fixed effects as in Model 6.1, but omits the random effects associated with the intercept for each child. We omit these random effects by including a `-1` in the random-effects specification:

```
> model6.2.fit <- lme(vsae ~ age.2 + I(age.2^2) + sicdegp2.f +
age.2:sicdegp2.f + I(age.2^2):sicdegp2.f,
random = ~ age.2 + I(age.2^2) - 1, method= "REML",
data = autism.grouped)
```

Results from the fit of Model 6.2 are accessible using `summary(model6.2.fit)`.

To decide whether to keep the random effects associated with the quadratic effects of age in Model 6.2, we test Hypothesis 6.1 using a likelihood ratio test. To do this, we fit a nested model (Model 6.2A) by removing AGE_2SQ (specifically, `I(age.2^2)`) from the random portion of the syntax:

```
> model6.2a.fit <- update(model6.2.fit, random = ~ age.2 - 1)
```

A likelihood ratio test is performed by subtracting the –2 REML log-likelihood for Model 6.2 (4615.3) from that of Model 6.2A (4699.2). This difference (83.9) follows a mixture of χ^2 distributions, with equal weight 0.5 and 1 and 2 degrees of freedom; more details for this test are included in Subsection 6.5.1. We calculate a p-value for this test statistic in R by making use of the `pchisq()` function:

```
> h6.1.pvalue <- 0.5*(1-pchisq(83.9,1)) + 0.5*(1-pchisq(83.9,2))
> h6.1.pvalue
```

The significant ($p < 0.001$) likelihood ratio test for Hypothesis 6.1 indicates that the random effects associated with the quadratic (and therefore linear) effects of age should be retained in Model 6.2 and in all subsequent models.

Step 3: Reduce the model by removing nonsignificant fixed effects (Model 6.2 vs. Model 6.3).

To test the fixed effects associated with the age-squared by SICD group interaction (Hypothesis 6.2), we first refit Model 6.2 (the reference model) using maximum likelihood estimation (method = "ML"):

```
> model6.2.ml.fit <- update(model6.2.fit, method = "ML")
```

Next, we consider a nested model (Model 6.3), with the interaction between age-squared and SICD group omitted. To fit Model 6.3 using ML estimation, we update the fixed part of Model 6.2 by omitting the interaction between the squared version of AGE_2 and SICDEGP2.F:

```
> model6.3.ml.fit <- update(model6.2.ml.fit,
fixed = ~ age.2 + I(age.2^2) + sicdegp2.f + age.2:sicdegp2.f)
```

We use the `anova()` function to perform a likelihood ratio test for Hypothesis 6.2:

```
> anova(model6.2.ml.fit, model6.3.ml.fit)
```

Based on the p-value for the test of Hypothesis 6.2 ($p = 0.39$; see Subsection 6.5.2), we drop the fixed effects associated with this interaction and obtain Model 6.3. An additional likelihood ratio test for the fixed effects associated with the age by SICD group interaction (i.e., Hypothesis 6.3) does not suggest that these fixed effects should be dropped from Model 6.3. We therefore refit our final model, Model 6.3, using REML estimation. To obtain Model 6.3 we `update` Model 6.2 with a previously used specification of the `fixed` argument:

```
> model6.3.fit <- update(model6.2.fit,
fixed = ~ age.2 + I(age.2^2) + sicdegp2.f + age.2:sicdegp2.f)
```

The results obtained by applying the `summary()` function to the `model6.3.fit` object are displayed in Section 6.6. Section 6.9 contains R syntax for checking the Model 6.3 diagnostics.

6.4.3.2 Analysis Using the `lmer()` Function

We first load the `lme4` library, so that we can utilize the `lmer()` function for this example:

```
> library(lme4)
```

Step 1: Fit a model with a loaded mean structure (Model 6.1).

Next, we fit Model 6.1 using the `lmer()` function (note that the modified version of the AGE variable, AGE_2, is used):

```
> model6.1.fit.lmer <- lmer(vsae ~ age.2 + I(age.2^2) + sicdegp2.f +
age.2*sicdegp2.f + I(age.2^2)*sicdegp2.f + (age.2 + I(age.2^2) | childid),
REML = T, data = autism.updated)
```

This specification of the `lmer()` function is described as follows:

- The name of the created object that contains the results of the fitted model is `model6.1.fit.lmer`.

- The first portion of the model formula, `vsae ~ age.2 + I(age.2^2) + sicdegp2.f + age.2*sicdegp2.f + I(age.2^2)*sicdegp2.f`, defines the continuous response variable (VSAE), and the terms that have fixed effects in the model (including the interactions). The `lmer()` function automatically creates the appropriate dummy variables for the categories of the SICDEGP2.F factor, treating the lowest-valued category (which corresponds to the original value of SICDEGP = 3) as the reference. The `I()` function is used to inhibit R from interpreting the "^" character as an operator in the fixed-effects formula (as opposed to an arithmetic operator meaning "raised to the power of"). A fixed intercept is included by default.

- The second portion of the model formula, `+ (age.2 + I(age.2^2) | childid)`, indicates the variables that have random effects associated with them (in parentheses). A random effect associated with the intercept for each level of CHILDID is included by default. These random effects are associated with each level of CHILDID by the use of `| childid` to "condition" the effects on CHILDID.

- The third argument of the function, `REML = T`, requests that the REML estimation method is to be used.

- The fourth argument of the function, `data = autism.updated`, indicates the "updated" data frame object to be used.

By default, the `lmer()` function uses an unstructured D matrix for the variance-covariance matrix of the random effects.

After fitting Model 6.1, we do not see an error message indicating lack of convergence, as was the case when performing the analysis using the `lme()` function. However, when applying the `summary()` function to the model fit object to examine the estimates of the parameters in this model, the estimate of the variance of the random CHILDID intercepts is displayed as $1.9381e - 10$, which indicates that this estimate is essentially equal to 0. Estimates of variance components that are set to zero generally indicate that the random-effects specification of a model should be reconsidered. We proceed to consider Model 6.2 as an alternative.

Step 2: Select a structure for the random effects (Model 6.2 vs. Model 6.2A).

We now fit Model 6.2, which includes the same fixed effects as in Model 6.1, but omits the random effects associated with the intercept for each child. We omit these random effects by including a `-1` in the random-effects specification:

```
> model6.2.fit.lmer <- lmer(vsae ~ age.2 + I(age.2^2) + sicdegp2.f +
age.2*sicdegp2.f + I(age.2^2)*sicdegp2.f +
(age.2 + I(age.2^2) - 1 | childid),
REML = T, data = autism.updated)
```

Results from the fit of Model 6.2 are accessible using `summary(model6.2.fit.lmer)`.

To decide whether to keep the random effects associated with the quadratic effects of age in Model 6.2, we test Hypothesis 6.1 using a likelihood ratio test. To do this, we fit a nested model (Model 6.2A) by removing AGE_2SQ (specifically, `I(age.2^2)`) from the random portion of the syntax:

```
> model6.2a.fit.lmer <- lmer(vsae ~ age.2 + I(age.2^2) + sicdegp2.f +
age.2*sicdegp2.f + I(age.2^2)*sicdegp2.f +
(age.2 - 1 | childid),
REML = T, data = autism.updated)
```

A likelihood ratio test is performed by subtracting the -2 REML log-likelihood for Model 6.2 (4615.3) from that of Model 6.2A (4699.2). This difference (83.9) follows a mixture of χ^2 distributions, with equal weight 0.5 and 1 and 2 degrees of freedom; more details for this test are included in Subsection 6.5.1. We calculate a p-value for this test statistic in R by making use of the `pchisq()` function:

```
> h6.1.pvalue <- 0.5*(1-pchisq(83.9,1)) + 0.5*(1-pchisq(83.9,2))
> h6.1.pvalue
```

The significant ($p < 0.001$) likelihood ratio test for Hypothesis 6.1 indicates that the random effects associated with the quadratic (and therefore linear) effects of age should be retained in Model 6.2 and in all subsequent models.

Step 3: Reduce the model by removing nonsignificant fixed effects (Model 6.2 vs. Model 6.3).

To test the fixed effects associated with the age-squared by SICD group interaction (Hypothesis 6.2), we first refit Model 6.2 (the reference model) using maximum likelihood estimation (note the use of `REML = F`):

```
> model6.2.ml.fit.lmer <- lmer(vsae ~ age.2 + I(age.2^2) + sicdegp2.f +
age.2*sicdegp2.f + I(age.2^2)*sicdegp2.f +
(age.2 + I(age.2^2) - 1 | childid),
REML = F, data = autism.updated)
```

Next, we consider a nested model (Model 6.3), with the interaction between age-squared and SICD group omitted. To fit Model 6.3 using ML estimation, we update the fixed part of Model 6.2 by omitting the interaction between these two terms:

```
> model6.3.ml.fit.lmer <- lmer(vsae ~ age.2 + I(age.2^2) + sicdegp2.f +
age.2*sicdegp2.f +
(age.2 + I(age.2^2) - 1 | childid),
REML = F, data = autism.updated)
```

We then use the `anova()` function to perform a likelihood ratio test for Hypothesis 6.2:

```
> anova(model6.2.ml.fit.lmer, model6.3.ml.fit.lmer)
```

Based on the p-value for the test of Hypothesis 6.2 ($p = 0.39$; see Subsection 6.5.2), we drop the fixed effects associated with this interaction and obtain Model 6.3. An additional likelihood ratio test for the fixed effects associated with the age by SICD group interaction (i.e., Hypothesis 6.3) does not suggest that these tested fixed effects should be dropped from Model 6.3. We therefore refit our final model, Model 6.3, using REML estimation (`REML = T`). To obtain Model 6.3, we update Model 6.2 with a previously used specification of the fixed effects:

```
> model6.3.fit.lmer <- lmer(vsae ~ age.2 + I(age.2^2) + sicdegp2.f +
age.2*sicdegp2.f + (age.2 + I(age.2^2) - 1 | childid),
REML = T, data = autism.updated)
```

The results obtained by applying the `summary()` function to the `model6.3.fit.lmer` object are displayed in Section 6.6. Section 6.9 contains R syntax for checking the Model 6.3 diagnostics.

6.4.4   Stata

We begin the analysis by importing the comma-separated values file (`autism.csv`) containing the Autism data into Stata:

```
. insheet using "C:\temp\autism.csv", comma
```

We generate the variable AGE_2 by subtracting 2 from AGE, and then square this new variable to generate AGE_2SQ:

```
. gen age_2 = age - 2
. gen age_2sq = age_2 * age_2
```

We now proceed with the analysis.

Step 1: Fit a model with a "loaded" mean structure (Model 6.1).

We first fit Model 6.1 using the `mixed` command, with the `reml` estimation option:

```
. * Model 6.1 (REML)
. mixed vsae ib3.sicdegp age_2 age_2sq
   ib3.sicdegp#c.age_2 ib3.sicdegp#c.age_2sq
   || childid: age_2 age_2sq, covariance(unstruct) variance reml
```

The `ib3.` notation causes Stata to generate the appropriate indicator variables for the SICDEGP factor, using the highest level of SICDEGP (3) as the reference category. The terms listed after the dependent variable (VSAE) represent the fixed effects that we wish to include in this model, and include the two-way interactions between SICDEGP (treated as a categorical variable with level 3 as the reference category, using `ib3.`) and both AGE_2 and AGE_2SQ (both specified as continuous variables in the interaction terms using the `c.` notation, which is necessary for correct specification of these interactions).

The random effects portion of the model is specified following the fixed effects portion, after two vertical lines (||), as follows:

```
   || childid: age_2 age_2sq, covariance(unstruct) variance
```

We specify CHILDID as the grouping variable that identifies the Level 2 units. The variables listed after the colon (:) are the Level 1 (time-varying) covariates with effects that vary randomly between children. We list AGE_2 and AGE_2SQ to allow both the linear and quadratic effects of age to vary from child to child. A random effect associated with the intercept for each child will be included by default, and does not need to be explicitly specified.

The structure of the variance-covariance matrix of the random effects (D) is specified as unstructured, by using the option `covariance(unstruct)`. The `variance` option is used so that Stata will display the estimated variances of the random effects and their standard errors in the output, rather than the default standard deviations of the random effects.

When attempting to fit this model in Stata, the following error message appears in red in the output:

```
Hessian is not negative semidefinite
conformability error
r(503)
```

While this error message is slightly cryptic, recall from Chapter 2 that the Hessian matrix is used to compute standard errors of the estimated parameters in a linear mixed model. This error message therefore indicates that there are problems with the Hessian matrix that was computed based on the specified model, and that these problems are preventing estimation of the standard errors (Stata does not provide any output for this model). This message should be considered seriously when it is encountered, and the model will need to be simplified or respecified. We now consider Model 6.2 as an alternative.

Step 2: Select a structure for the random effects (Model 6.2 vs. Model 6.2A).

We now fit Model 6.2, which includes the same fixed effects as Model 6.1, but excludes the random effects associated with the intercept for each child, by using the `noconst` option:

```
. * Model 6.2
. mixed vsae ib3.sicdegp age_2 age_2sq
   ib3.sicdegp#c.age_2 ib3.sicdegp#c.age_2sq
   || childid: age_2 age_2sq, noconst covariance(unstruct) variance reml
```

The resulting output does not indicate a problem with the computation of the standard errors. The information criteria associated with the fit of Model 6.2 can be obtained by submitting the `estat ic` command after estimation of the model has finished:

```
. estat ic
```

A likelihood ratio test of Hypothesis 6.1 is performed by fitting a nested model (Model 6.2A) that omits the random effects associated with AGE_2SQ:

```
. * Model 6.2A
. mixed vsae ib3.sicdegp age_2 age_2sq
   ib3.sicdegp#c.age_2 ib3.sicdegp#c.age_2sq
   || childid: age_2, noconst covariance(unstruct) variance reml
```

We calculate a test statistic by subtracting the –2 REML log-likelihood for Model 6.2 from that for Model 6.2A. We then refer this test statistic to a mixture of χ^2 distributions, with equal weight 0.5 and 1 and 2 degrees of freedom:

```
. di (-2 * -2349.6013) - (-2 * -2307.6378)
83.927
```

```
. di 0.5 * chi2tail(1,83.927) + 0.5 * chi2tail(2,83.927)
3.238e-19
```

The significant result ($p < 0.001$) of this likelihood ratio test indicates that the variance of the random effects associated with the quadratic effect of age (Hypothesis 6.1) is significant in Model 6.2, so both the quadratic and the linear random effects are retained in all subsequent models (see Subsection 6.5.1 for more details).

Step 3: Reduce the model by removing nonsignificant fixed effects (Model 6.2 vs. Model 6.3).

We now refit Model 6.2 using maximum likelihood estimation, so that we can carry out likelihood ratio tests for the fixed effects in the model (starting with Hypothesis 6.2). This is accomplished by specifying the `mle` option:

```
. * Model 6.2 (ML)
. mixed vsae ib3.sicdegp age_2 age_2sq
   ib3.sicdegp#c.age_2 ib3.sicdegp#c.age_2sq
   || childid: age_2 age_2sq, noconst covariance(unstruct) variance mle
```

After fitting Model 6.2 with ML estimation, we store the results in an object named `model6_2_ml`:

```
. est store model6_2_ml
```

We now omit the two fixed effects associated with the SICDEGP $\times$ AGE_2SQ interaction, and fit a nested model (Model 6.3):

```
. * Model 6.3 (ML)
. mixed vsae ib3.sicdegp age_2 age_2sq ib3.sicdegp#c.age_2
   || childid: age_2 age_2sq, noconst covariance(unstruct) variance mle
```

We store the results from the Model 6.3 fit in a second object named `model6_3_ml`:

```
. est store model6_3_ml
```

Finally, we perform a likelihood ratio test of Hypothesis 6.2 using the `lrtest` command:

```
. lrtest model6_2_ml model6_3_ml
```

The resulting test statistic (which follows a χ^2 distribution with 2 degrees of freedom, corresponding to the 2 fixed effects omitted from Model 6.2) is not significant ($p = 0.39$). We therefore simplify the model by excluding the fixed effects associated with the SICD group by age-squared interaction, and obtain Model 6.3.

Additional likelihood ratio tests can be performed for the remaining fixed effects in the model, beginning with the SICD group by age interaction (Hypothesis 6.3). No further model reduction is indicated; all remaining fixed effects are significant at the 0.05 level, and are retained. We fit the final model (Model 6.3) using REML estimation, and obtain the model fit criteria:

```
. * Model 6.3 (REML)
. mixed vsae ib3.sicdegp age_2 age_2sq ib3.sicdegp#c.age_2
   || childid: age_2 age_2sq, noconst covariance(unstruct) variance reml
. estat ic
```

6.4.5   HLM

6.4.5.1   Data Set Preparation

To perform the analysis of the Autism data using the HLM software package, two separate data sets need to be prepared:

1. The **Level 1 (time-level) data set** contains a single observation (row of data) for each observed age on each child, and is similar in structure to Table 6.2. The data set should include CHILDID, the time variable (i.e., AGE), a variable representing AGE-squared, and any other variables of interest measured over time for each child, including the response variable, VSAE. This data set must be sorted by CHILDID and by AGE within CHILDID.

2. The **Level 2 (child-level) data set** contains a single observation (row of data) for each level of CHILDID. The variables in this file represent measures that remain constant for a given child. This includes variables collected at baseline (e.g., SICDEGP) or demographic variables (e.g., GENDER) that do not change as the child gets older. This data set must also include the CHILDID variable, and be sorted by CHILDID.

Because the HLM program does not automatically create indicator variables for the levels of categorical predictors, we need to create two indicator variables representing the two nonreference levels of SICDEGP in the Level 2 data file prior to importing the data into HLM. For example, if the input data files were created in SPSS, the syntax used to compute the two appropriate indicator variables in the Level 2 file would look like this:

```
COMPUTE sicdegp1 = (sicdegp = 1) .
EXECUTE .
COMPUTE sicdegp2 = (sicdegp = 2) .
EXECUTE .
```

We also create a variable AGE_2, which is equal to the original AGE variable minus two (i.e., AGE_2 is equal to zero when the child is 2 years old), and create the squared version of this new variable (AGE_2SQ). Both of these variables are created in the Level 1 data set prior to importing it into HLM. SPSS syntax to compute these new variables is shown below:

```
COMPUTE age_2 = age - 2.
EXECUTE .
COMPUTE age_2sq = age_2 * age_2.
EXECUTE .
```

Once the Level 1 and Level 2 data sets have been created, we can proceed to prepare the multivariate data matrix (MDM) file in HLM.

6.4.5.2 Preparing the MDM File

We start by creating a new MDM file using the Level 1 and Level 2 data sets described above. In the main HLM menu, click **File**, **Make new MDM file**, and then **Stat package input**. In the window that opens, select **HLM2** to fit a two-level hierarchical linear model with random coefficients, and click **OK**. Select the **Input File Type** as **SPSS/Windows**.

To make the MDM file for Model 6.1, locate the **Level 1 Specification** area and **Browse** to the location of the Level 1 data set. Click the **Choose Variables** button and select the following variables in the Level 1 file to be used in the Model 6.1 analysis: CHILDID (click on "ID" for the CHILDID variable), AGE_2 (click "in MDM" for this time-varying variable), AGE_2SQ (click "in MDM"), and the time-varying dependent variable, VSAE (click "in MDM").

Next, locate the **Level 2 Specification** area, and **Browse** to the location of the Level 2 data set. Click the **Choose Variables** button to include CHILDID (click "ID") and the two indicator variables for the nonreference levels of SICDEGP (SICDEGP1 and SICDEGP2) in the MDM file. Click "in MDM" for these two indicator variables.

After making these choices, select the **longitudinal (occasions within persons)** radio option for the structure of this longitudinal data set (for notation purposes only). Also, select **Yes** for **Missing Data?** in the Level 1 data set (because some children have missing data), and make sure that the option to **Delete missing data when: running analyses** is selected. Enter a name for the MDM file with a .mdm extension in the upper-right corner of the MDM window, save the .mdmt template file under a new name (click **Save mdmt file**), and click **Make MDM**.

After HLM has processed the MDM file, click the **Check Stats** button to view descriptive statistics for the variables in the Level 1 and Level 2 files (this is not optional). Be sure that the desired number of records has been read into the MDM file and that there are no unusual values for the variables. Click **Done** to begin building Model 6.1.

Step 1: Fit a model with a loaded mean structure (Model 6.1).

In the model-building window, select VSAE from the list of variables, and then click on **Outcome variable**. This will cause the initial "unconditional" model without covariates (or with intercepts only) to be displayed, broken down into the Level 1 and the Level 2 models. To add more informative subscripts to the model specification, click **File** and **Preferences**, and then choose **Use level subscripts**.

We now set up the Level 1 portion of Model 6.1 by adding the effects of the time-varying covariates AGE_2 and AGE_2SQ. Click the **Level 1** button in the model-building window, and then click the AGE_2 variable. Choose **add variable uncentered**. The Level 1 model shows that the AGE_2 covariate has been added, along with its child-specific coefficient, π_{1i}. Repeat this process, adding the uncentered version of the AGE_2SQ variable to the Level 1 model:

Model 6.1: Level 1 Model

$$\text{VSAE}_{ti} = \pi_{0i} + \pi_{1i}(\text{AGE\_2}_{ti}) + \pi_{2i}(\text{AGE\_2SQ}_{ti}) + e_{ti}$$

At this point, the Level 2 equation for the child-specific intercept (π_{0i}) contains a random effect for each child (r_{0i}). However, the coefficients for AGE_2 and AGE_2SQ (π_{1i} and π_{2i}, respectively) are simply defined as constants, and do not include any random effects. We can change this by clicking the r_{1i} and r_{2i} terms in the Level 2 model to add random child-specific effects to the coefficients of AGE_2 and AGE_2SQ, as shown in the following preliminary Level 2 model:

Model 6.1: Level 2 Model (Preliminary)

$$\pi_{0i} = \beta_{00} + r_{0i}$$
$$\pi_{1i} = \beta_{10} + r_{1i}$$
$$\pi_{2i} = \beta_{20} + r_{2i}$$

This specification implies that the child-specific intercept (π_{0i}) depends on the overall intercept (β_{00}) and the child-specific random effect associated with the intercept (r_{0i}). The effects of AGE_2 and AGE_2SQ (π_{1i} and π_{2i}, respectively) depend on the overall fixed effects of AGE_2 and AGE_2SQ (β_{10} and β_{20}) and the random effects associated with each child (r_{1i} and r_{2i}, respectively). As a result, the initial VSAE value at age 2 and the trajectory of VSAE are both allowed to vary randomly between children.

To complete the specification of Model 6.1, we need to add the uncentered versions of the indicator variables for the first two levels of SICDEGP to the Level 2 equations for the intercept (π_{0i}), the linear effect of age (π_{1i}), and the quadratic effect of age (π_{2i}). Click the **Level 2** button in the model-building window. Then, click on each Level 2 equation, and click on the two indicator variables to add them to the equations (uncentered). The completed Level 2 model now appears as follows:

Model 6.1: Level 2 Model (Final)

$$\pi_{0i} = \beta_{00} + \beta_{01}(\text{SICDEGP1}_i) + \beta_{02}(\text{SICDEGP2}_i) + r_{0i}$$
$$\pi_{1i} = \beta_{10} + \beta_{11}(\text{SICDEGP1}_i) + \beta_{12}(\text{SICDEGP2}_i) + r_{1i}$$
$$\pi_{2i} = \beta_{20} + \beta_{21}(\text{SICDEGP1}_i) + \beta_{22}(\text{SICDEGP2}_i) + r_{2i}$$

Adding SICDEGP1_i and SICDEGP2_i to the Level 2 equation for the child-specific intercept (π_{0i}) shows that the main effects (β_{01} and β_{02}) of these indicator variables represent changes in the intercept (i.e., the expected VSAE response at age 2) for SICDEGP groups 1 and 2 relative to the reference category (SICDEGP = 3).

By including the SICDEGP1_i and SICDEGP2_i indicator variables in the Level 2 equations for the effects of AGE_2 and AGE_2SQ (π_{1i} and π_{2i}), we imply that the interactions between these indicator variables and AGE_2 and AGE_2SQ will be included in the overall linear mixed model, as shown below. We can view the overall linear mixed model by clicking the **Mixed** button in the model-building window:

Model 6.1: Overall Mixed Model

$$\begin{aligned}
\text{VSAE}_{ti} = \quad & \beta_{00} + \beta_{01} * \text{SICDEGP1}_i + \beta_{02} * \text{SICDEGP2}_i + \beta_{10} * \text{AGE\_2}_{ti} \\
& + \beta_{11} * \text{SICDEGP1}_i * \text{AGE\_2}_{ti} + \beta_{12} * \text{SICDEGP2}_i * \text{AGE\_2}_{ti} \\
& + \beta_{20} * \text{AGE\_2SQ}_{ti} + \beta_{21} * \text{SICDEGP1}_i * \text{AGE\_2SQ}_{ti} \\
& + \beta_{22} * \text{SICDEGP2}_i * \text{AGE\_2SQ}_{ti} \\
& + r_{0i} + r_{1i} * \text{AGE\_2}_{ti} + r_{2i} * \text{AGE\_2SQ}_{ti} + e_{ti}
\end{aligned}$$

This model is the same as the general specification of Model 6.1 (see (6.1)) introduced in Subsection 6.3.2.1, although the notation is somewhat different. The correspondence between the HLM notation and the general notation that we used in Subsection 6.3.2 is displayed in Table 6.3. Note that we can also derive this form of the overall linear mixed model by substituting the expressions for the child-specific effects in the Level 2 model into the Level 1 model.

After specifying Model 6.1, click **Basic Settings** to enter a title for this analysis (such as "Autism Data: Model 6.1") and a name for the output (.html) file that HLM generates when it fits this model. We do not need to alter the outcome variable distribution setting, because the default is **Normal (Continuous)**. Click **OK** to return to the model-building window, and then click **File** and **Save As** to save this model specification in a new .hlm file. Finally, click **Run Analysis** to fit the model. HLM by default uses REML estimation to estimate the covariance parameters in Model 6.1.

We see the following message generated by HLM when it attempts to fit Model 6.1:

> The maximum number of iterations has been reached, but the analysis has not converged.
>
> Do you want to continue until convergence?

At this point, one can request that the iterative estimation procedure continue by typing a "y" and hitting enter. After roughly 1600 iterations, the analysis finishes running. The large number of iterations required for the REML estimation algorithm to converge to a solution indicates a potential problem in fitting the model. We can click on **File** and **View Output** to see the results from this model fit.

Despite the large number of iterations required to fit Model 6.1, the HLM results for this model are displayed in Section 6.6 for comparison with the other software procedures.

Step 2: Select a structure for the random effects (Model 6.2 vs. Model 6.2A).

We now fit Model 6.2, which includes the same fixed effects as Model 6.1 but does not have child-specific random effects associated with the intercept. To remove the random effects associated with the intercept from the model, simply click the r_{0i} term in the Level 2 equation for the child-specific intercept. The new Level 2 equation for the intercept is:

Model 6.2: Level 2 Equation for Child-Specific Intercept

$$\pi_{0i} = \beta_{00} + \beta_{01}(\text{SICDEGP1}_i) + \beta_{02}(\text{SICDEGP2}_i)$$

Model 6.2 implies that the intercept for a given child is a function of their SICD group, but does not vary randomly from child to child. Click **Basic Settings** to enter a different title for this analysis (such as "Autism Data: Model 6.2") and to change the name of the output file. Then click **File** and **Save As** to save the new model specification in a different .hlm file, so that the previous .hlm file is not overwritten.

Click **Run Analysis** to fit Model 6.2. The REML algorithm converges to a solution in only 19 iterations. The default χ^2 tests for the covariance parameters reported by HLM in the output for Model 6.2 suggest that there is significant variability in the child-specific linear and quadratic effects of age (rejecting the null hypothesis for Hypothesis 6.1), so we retain the random effects associated with AGE_2 and AGE_2SQ in all subsequent models. Note that the tests reported by HLM are not likelihood ratio tests, with p-values based on a mixture of χ^2 distributions, as we reported in the other software procedures; see pages 63-64 of Raudenbush & Bryk (2002) for more details on these tests.

We do not illustrate how to carry out a REML-based likelihood ratio test of Hypothesis 6.1 using HLM2.[2] We can perform this likelihood ratio test (with a p-value based on a mixture of χ^2 distributions) in HLM2 by calculating the difference in the deviance statistics reported for a reference (Model 6.2) and a nested (Model 6.2A) model, as long as at least one random effect is retained in the Level 2 models that are being compared. Subsection 6.5.1 provides more detail on the likelihood ratio test for the random quadratic age effects considered in this example.

Step 3: Reduce the model by removing nonsignificant fixed effects (Model 6.2 vs. Model 6.3).

In this step, we test the fixed effects in the model, given that the random child effects associated with the linear and quadratic effects of age are included. We begin by testing the SICD group by age-squared interaction (Hypothesis 6.2) in Model 6.2.

We first refit Model 6.2 using maximum likelihood (ML) estimation. To do this, click **Other Settings** and then **Estimation Settings** in the model-building window. Select the **Full maximum likelihood** option (as opposed to REML estimation, which is the default), and click **OK**. Then click **Basic Settings**, save the output file under a different name, and enter a different title for the analysis (such as "Model 6.2: Maximum Likelihood"). Save the .hlm file under a new name, and click **Run Analysis** to refit the Model 6.2 using ML estimation.

---

[2]HLM uses chi-square tests for covariance parameters by default (see Chapter 3). Likelihood ratio tests may also be calculated, as long as at least one random effect is retained in the Level 2 model for both the reference and nested models.

We now fit a nested model (Model 6.3) that omits the fixed effects associated with the SICD group by age-squared interaction. To do this, we remove the SICDEGP1 and SICDEGP2 terms from the Level 2 model for the child-specific effect of AGE_2SQ (π_{2i}). Click on this Level 2 equation in the model-building window, click the SICDEGP1 variable in the variable list, and then select **Delete variable from model**. Do the same for the SICDEGP2 variable. The equation for the child-specific quadratic effect of age (i.e., the effect of AGE_2SQ) now appears as follows:

Model 6.3: Level 2 Equation for Child-Specific Quadratic Effect of Age

$$\pi_{2i} = \beta_{20} + r_{2i}$$

This reduced model implies that the child-specific effect of AGE_2SQ depends on an overall fixed effect (β_{20}) and a random effect associated with the child (r_{2i}). The child-specific effect of AGE_2SQ no longer depends on the SICD group of the child in this reduced model.

After removing these fixed effects, click on **Basic Settings**, and save the output in a different file, so that the original ML fit of Model 6.2 is not overwritten. To perform a likelihood ratio test comparing the fit of the nested model (Model 6.3) to the fit of Model 6.2, we locate the deviance (i.e., the –2 ML log-likelihood) and number of parameters associated with the ML fit of Model 6.2 in the previous output file (4610.44 and 13, respectively). Click **Other Settings**, and then click **Hypothesis Testing**. Enter the deviance and number of parameters from the ML fit of Model 6.2 (deviance = 4610.44 and number of parameters = 13) in the window that opens, and click **OK**. HLM will now compare the deviance associated with the ML fit of Model 6.3 to the deviance of the ML fit of Model 6.2, and perform the appropriate likelihood ratio test. Save the .hlm file under a new name, and fit the model by clicking **Run Analysis**.

HLM provides the result of the likelihood ratio test for Hypothesis 6.2 at the bottom of the resulting output file, which can be viewed by clicking **File** and **View Output**. The test is not significant in this case (HLM reports $p > .50$ for the resulting χ^2 statistic), suggesting that the fixed effects associated with the SICDEGP × AGE_2SQ interaction can be dropped from the model. We refer to the model obtained after removing these fixed effects as Model 6.3. Additional likelihood ratio tests can be performed for other fixed effects (e.g., Hypothesis 6.3) in a similar manner. Based on these tests, we conclude that the Model 6.3 is our final model.

We now refit Model 6.3 using REML estimation. This model has the same setup as Model 6.2, but without the fixed effects associated with the SICD group by age-squared interaction. To do this, the **Estimation Settings** need to be reset to REML, and the title of the output, as well as the output file name, should also be reset.

In the **Basic Settings** window, files containing the Level 1 and Level 2 residuals can be generated for the purpose of checking assumptions about the residuals and random child effects in the model, by clicking the **Level 1 Residual File** and **Level 2 Residual File** buttons. We choose to generate SPSS versions of these residual files. The Level 1 residual file will contain the conditional residuals associated with the longitudinal measures (labeled L1RESID) and the conditional predicted values (labeled FITVAL). The Level 2 residual file will contain the EBLUPs for the child-specific random effects associated with AGE_2 and AGE_2SQ (labeled EBAGE_2 and V9, because EBAGE_2SQ is more than eight characters long).

TABLE 6.5: Summary of Hypothesis Test Results for the Autism Analysis

Hypo-thesis Label	Test	Estima-tion Method	Models Compared (Nested vs. Reference)	Test Statistic Values (Calculation)	p-Value
6.1	LRT	REML	6.2A vs. 6.2	$\chi^2(1:2) = 83.9$ $(4699.2 - 4615.3)$	$< .001$
6.2	LRT	ML	6.3 vs. 6.2	$\chi^2(2) = 1.9$ $(4612.3 - 4610.4)$	0.39
6.3	LRT	ML	6.4 vs. 6.3	$\chi^2(2) = 23.4$ $(4635.7 - 4612.3)$	$< .001$

Note: See Table 6.4 for null and alternative hypotheses and distributions of test statistics under H_0.

6.5   Results of Hypothesis Tests

6.5.1   Likelihood Ratio Test for Random Effects

In Step 2 of the analysis we used a likelihood ratio test to test Hypothesis 6.1, and decide whether to retain the random quadratic (and therefore linear) effects of age in Model 6.2. These likelihood ratio tests were carried out based on REML estimation in all software packages except HLM.

Hypothesis 6.1. The child-specific quadratic random effects of age can be omitted from Model 6.2.

We tested the need for the quadratic random effects of age indirectly, by carrying out tests for the corresponding elements in the D matrix. The null and alternative hypotheses for Hypothesis 6.1 are defined in terms of the D matrix, and shown in Subsection 6.3.3.

We calculated the likelihood ratio test statistic for Hypothesis 6.1 by subtracting the –2 REML log-likelihood value for Model 6.2 (the reference model) from the value for Model 6.2A (the nested model). The resulting test statistic is equal to 83.9 (see Table 6.5). The asymptotic distribution of the likelihood ratio test statistic under the null hypothesis is a mixture of χ_1^2 and χ_2^2 distributions with equal weights of 0.5 rather than the usual χ_2^2 distribution, because the null hypothesis value of one of the parameters ($\sigma_{age\text{-}squared}^2 = 0$) is on the boundary of the parameter space (Verbeke & Molenberghs, 2000).

The p-value for this test statistic is computed as follows:

$$p\text{-value} = 0.5 \times P(\chi_2^2 > 83.9) + 0.5 \times P(\chi_1^2 > 83.9) < 0.001$$

We therefore decided to retain the random quadratic age effects in this model and in all subsequent models. We also retain the random linear age effects as well, so that the model is well formulated in a hierarchical sense (Morrell et al., 1997). In other words, because we keep the higher-order quadratic effects, the lower-order linear effects are also kept in the model.

The child-specific linear and quadratic effects of age are in keeping with what we observed in Figure 6.1 in our initial data summary, in which we noted marked differences in the individual VSAE trajectories of children as they grew older. The random effects in the model capture the variability between these trajectories.

6.5.2 Likelihood Ratio Tests for Fixed Effects

In Step 3 of the analysis we carried out likelihood ratio tests for selected fixed effects using ML estimation in all software packages. Specifically, we tested Hypotheses 6.2 and 6.3.

Hypothesis 6.2. The age-squared by SICD group interaction effects can be dropped from Model 6.2 ($\beta_7 = \beta_8 = 0$).

To perform a test of Hypothesis 6.2, we used maximum likelihood (ML) estimation to fit Model 6.2 (the reference model) and Model 6.3 (the nested model with the AGE_2SQ × SICDEGP interaction term omitted). The likelihood ratio test statistic was calculated by subtracting the –2 ML log-likelihood for Model 6.2 from the value for Model 6.3. The asymptotic null distribution of the test statistic is a χ^2 with 2 degrees of freedom. The 2 degrees of freedom arise from the two fixed effects omitted in Model 6.3. The result of the test was not significant ($p = 0.39$), so we dropped the AGE_2SQ × SICDEGP interaction term from Model 6.2.

Hypothesis 6.3. The age by SICD group interaction effects can be dropped from Model 6.3 ($\beta_5 = \beta_6 = 0$).

To test Hypothesis 6.3 we used ML estimation to fit Model 6.3 (the reference model) and Model 6.4 (a nested model without the AGE_2 × SICDEGP interaction). The test statistic was calculated by subtracting the –2 ML log-likelihood for Model 6.3 from that of Model 6.4. The asymptotic null distribution of the test statistic again was a χ^2 with 2 degrees of freedom. The p-value for this test was significant ($p < 0.001$). We concluded that the linear effect of age on VSAE does differ for different levels of SICD group, and we kept the AGE_2 × SICDEGP interaction term in Model 6.3.

6.6 Comparing Results across the Software Procedures

6.6.1 Comparing Model 6.1 Results

Table 6.6 shows a comparison of selected results obtained by fitting Model 6.1 to the Autism data, using four of the six software procedures (results were not available when using the lme() function in R or the mixed command in Stata, because of problems encountered when fitting this model). We present results for SAS, SPSS, the lmer() function in R, and HLM, despite the problems encountered when fitting Model 6.1 using each of the procedures, to highlight the differences and similarities across the procedures. Both warning and error messages were produced by the procedures in SAS, SPSS, R, and Stata, and a large number of iterations were required to fit this model when using the HLM2 procedure. See the data analysis steps for each software procedure in Section 6.4 for details on the problems encountered when fitting Model 6.1. Because of the estimation problems, the results in Table 6.6 should be regarded with a great deal of caution.

TABLE 6.6: Comparison of Results for Model 6.1

	SAS: proc mixed	SPSS: MIXED	R: lmer() function	HLM2
Estimation Method	REML	REML	REML	REML
Warning Message	G Matrix Not Positive-Definite	Lack of Convergence	None	1603 Iterations
Fixed-Effect Parameter	*Estimate (SE)*	*Estimate (SE)*	*Estimate (SE)*	*Estimate (SE)*
β_0 (Intercept)	13.78(0.81)	13.76(0.79)	13.77(0.81)	13.79(0.82)
β_1 (AGE_2)	5.61(0.79)	5.60(0.80)	5.60(0.79)	5.60(0.79)
β_2 (AGE_2SQ)	0.20(0.09)	0.21(0.08)	0.20(0.08)	0.20(0.09)
β_3 (SICDEGP1)	−5.43(1.10)	−5.41(1.07)	−5.42(1.09)	−5.44(1.11)
β_4 (SICDEGP2)	−4.03(1.03)	−4.01(1.01)	−4.04(1.03)	−4.04(1.05)
β_5 (AGE_2 × SICDEGP1)	−3.29(1.08)	−3.25(1.10)	−3.30(1.09)	−3.28(1.08)
β_6 (AGE_2 × SICDEGP2)	−2.77(1.02)	−2.76(1.03)	−2.75(1.03)	−2.75(1.02)
β_7 (AGE_2SQ × SICDEGP1)	−0.14(0.12)	−0.14(0.11)	−0.13(0.11)	−0.14(0.12)
β_8 (AGE_2SQ × SICDEGP2)	−0.13(0.11)	−0.13(0.11)	−0.13(0.11)	−0.13(0.11)
Covariance Parameter	*Estimate (SE)*	*Estimate (SE)*	*Estimate (n.c.)*	*Estimate (n.c.)*
σ^2_{int}	0.00(n.c.)	0.00(0.00)[a]	0.00	1.48
$\sigma_{int,age}$	0.62(2.29)	−15.01(2.41)	0.00	0.25
$\sigma_{int,age-sq}$	0.57(0.22)	3.30(0.24)	0.00	0.42
σ^2_{age}	14.03(3.09)	15.39(3.26)	14.67	14.27
$\sigma_{age,age-sq}$	−0.64(0.26)	−0.68(0.25)	−0.32(corr.)	−0.59
σ^2_{age-sq}	0.17(0.03)	0.14(0.03)	0.13	0.16
σ^2	38.71	36.95	38.50	37.63

TABLE 6.6: (Continued)

Estimation Method	SAS: proc mixed	R: lmer() function	SPSS: MIXED	HLM2
	REML	REML	REML	REML
Model Information Criteria				
−2 REML log-likelihood	4604.7	4610.0	4618.8	4606.2
AIC	4616.7	4647.0	4632.8	n.c.
BIC	4635.1	4718.0	4663.6	n.c.

Note: (n.c.) = not computed

Note: 610 Longitudinal Measures at Level 1; 158 Children at Level 2

[a]This covariance parameter is reported to be "redundant" by the MIXED command in SPSS.

TABLE 6.7: Comparison of Results for Model 6.2

Estimation Method	SAS: proc mixed REML	SPSS: MIXED REML	R: lme() function REML	R: lmer() function REML	Stata: mixed REML	HLM: HLM2 REML
Fixed-Effect Parameter	*Estimate (SE)*	*Estimate (SE)*	*Estimate (SE)*	*Estimate (SE)*	*Estimate (SE)*	*Estimate (n.c.)*
β_0 (Intercept)	13.77(0.81)	13.77(0.81)	13.77(0.81)	13.77(0.81)	13.77(0.81)	13.77(0.81)
β_1 (AGE_2)	5.60(0.79)	5.60(0.79)	5.60(0.79)	5.60(0.79)	5.60(0.79)	5.60(0.79)
β_2 (AGE_2SQ)	0.20(0.08)	0.20(0.08)	0.20(0.08)	0.20(0.08)	0.20(0.08)	0.20(0.08)
β_3 (SICDEGP1)	−5.42(1.09)	−5.42(1.09)	−5.42(1.09)	−5.42(1.09)	−5.42(1.09)	−5.42(1.09)
β_4 (SICDEGP2)	−4.04(1.03)	−4.04(1.03)	−4.04(1.03)	−4.04(1.03)	−4.04(1.03)	−4.04(1.03)
β_5 (AGE_2 × SICDEGP1)	−3.30(1.09)	−3.30(1.09)	−3.30(1.09)	−3.30(1.09)	−3.30(1.09)	−3.30(1.09)
β_6 (AGE_2 × SICDEGP2)	−2.75(1.03)	−2.75(1.03)	−2.75(1.03)	−2.75(1.03)	−2.75(1.03)	−2.75(1.03)
β_7 (AGE_2SQ × SICDEGP1)	−0.13(0.11)	−0.13(0.11)	−0.13(0.11)	−0.13(0.11)	−0.13(0.11)	−0.13(0.11)
β_8 (AGE_2SQ × SICDEGP2)	−0.13(0.11)	−0.13(0.11)	−0.13(0.11)	−0.13(0.11)	−0.13(0.11)	−0.13(0.11)
Covariance Parameter	*Estimate (SE)*	*Estimate (SE)*	*Estimate (n.c.)*	*Estimate (n.c.)*	*Estimate (SE)*	*Estimate (n.c.)*
σ^2_{age}	14.67(2.63)	14.67(2.63)	14.67	14.67	14.67(2.63)	14.67
$\sigma_{age,age\text{-}sq}$	−0.44(0.21)	−0.44(0.21)	−0.32(corr.)	−0.32	−0.44(0.21)	−0.44
$\sigma^2_{age\text{-}sq}$	0.13(0.03)	0.13(0.03)	0.13	0.13	0.13(0.03)	0.13
σ^2	38.50	38.50	38.50	38.50	38.50	38.50
Model Information Criteria						
−2 RE/ML log-likelihood	4615.3	4615.3	4615.3	4615.3	4615.3	4613.4
AIC	4623.3	4623.3	4641.3	4641.0	4641.3	n.c.
BIC	4635.5	4640.9	4698.5	4699.0	4698.7	n.c.

Note: (n.c.) = not computed
Note: 610 Longitudinal Measures at Level 1; 158 Children at Level 2

The major differences in the results for Model 6.1 across the software procedures are in the covariance parameter estimates and their standard errors. These differences are due to the violation of positive-definite constraints for the D matrix. Despite these differences, the fixed-effect parameter estimates and their standard errors are similar. The –2 REML log-likelihoods, which are a function of the fixed-effect and covariance parameter estimates, also differ across the software procedures.

In general, warning messages that can result in these types of discrepancies depending on the software used should *never* be ignored when fitting linear mixed models with multiple random effects. Given the likely lack of variability in the intercepts in this particular model, we consider results from Model 6.2 (with random intercepts excluded) next.

6.6.2   Comparing Model 6.2 Results

Selected results obtained by fitting Model 6.2 to the Autism data using each of the six software procedures are displayed in Table 6.7. The only difference between Models 6.1 and 6.2 is that the latter does not contain the random child-specific effects associated with the intercept. The difficulties in estimating the covariance parameters that were encountered when fitting Model 6.1 were not experienced when fitting this model.

The five procedures agree very closely in terms of the estimated fixed effects, the covariance parameter estimates, and their standard errors. The –2 REML log-likelihoods reported by the procedures in SAS, SPSS, R, and Stata all agree. The –2 REML log-likelihood reported by HLM differs, perhaps because of differences in default convergence criteria (see Subsection 3.6.1). The other model information criteria (AIC and BIC), not reported by HLM, differ because of differences in the calculation formulas used across the software procedures (see Section 3.6 for a discussion of these differences).

6.6.3   Comparing Model 6.3 Results

Table 6.8 compares the results obtained by fitting the final model, Model 6.3, across the six software procedures. As we noted in the comparison of the Model 6.2 results, there is agreement between the six procedures in terms of both the fixed-effect and covariance parameter estimates and their standard errors (when reported). The –2 REML log-likelihoods agree across the procedures in SAS, SPSS, R and Stata. The HLM value of the –2 REML log-likelihood is again different from that reported by the other procedures. Other differences in the model information criteria (e.g., AIC and BIC) are due to differences in the calculation formulas, as noted in Subsection 6.6.2.

6.7   Interpreting Parameter Estimates in the Final Model

We now use the results obtained by using the `lme()` function in R to interpret the parameter estimates for Model 6.3.

6.7.1   Fixed-Effect Parameter Estimates

We show a portion of the output for Model 6.3 below. This output includes the fixed-effect parameter estimates, their corresponding standard errors, the degrees of freedom, the t-test values, and the corresponding p-values. The output is obtained by applying the `summary()` function to the object `model6.3.fit`, which contains the results of the model fit.

TABLE 6.8: Comparison of Results for Model 6.3

Estimation Method	SAS: proc mixed REML	SPSS: MIXED REML	R: lme() function REML	R: lmer() function REML	Stata: mixed REML	HLM: HLM2 REML
Fixed-Effect Parameter	*Estimate (SE)*	*Estimate (SE)*	*Estimate (SE)*	*Estimate (SE)*	*Estimate (SE)*	*Estimate (n.c.)*
β_0(Intercept)	13.46(0.78)	13.46(0.78)	13.46(0.78)	13.46(0.78)	13.46(0.78)	13.46(0.78)
β_1(AGE_2)	6.15(0.69)	6.15(0.69)	6.15(0.69)	6.15(0.69)	6.15(0.69)	6.15(0.69)
β_2(AGE_2SQ)	0.11(0.04)	0.11(0.04)	0.11(0.04)	0.11(0.04)	0.11(0.04)	0.11(0.04)
β_3(SICDEGP1)	−4.99(1.04)	−4.99(1.04)	−4.99(1.04)	−4.99(1.04)	−4.99(1.04)	−4.99(1.04)
β_4(SICDEGP2)	−3.62(0.98)	−3.62(0.98)	−3.62(0.98)	−3.62(0.98)	−3.62(0.98)	−3.62(0.98)
β_5(AGE_2 × SICDEGP1)	−4.07(0.88)	−4.07(0.88)	−4.07(0.88)	−4.07(0.88)	−4.07(0.88)	−4.07(0.88)
β_6(AGE_2 × SICDEGP2)	−3.50(0.83)	−3.50(0.83)	−3.50(0.83)	−3.50(0.83)	−3.50(0.83)	−3.50(0.83)
Covariance Parameter	*Estimate (SE)*	*Estimate (SE)*	*Estimate (n.c.)*	*Estimate (n.c.)*	*Estimate (SE)*	*Estimate (n.c.)*
σ^2_{age}	14.52(2.61)	14.52(2.61)	14.52	14.52	14.52(2.61)	14.52
$\sigma_{age,age\text{-}sq}$	−0.42(0.20)	−0.42(0.20)	−0.31[a]	−0.31	−0.42(0.20)	−0.42
$\sigma^2_{age\text{-}squared}$	0.13(0.03)	0.13(0.03)	0.13	0.13	0.13(0.03)	0.13
σ^2	38.79	38.79	38.79	38.79	38.79	38.79
Model Information Criteria						
−2 RE/ML log-likelihood	4611.6	4611.6	4611.6	4612.0	4611.6	4609.7
AIC	4619.6	4619.6	4633.6	4634.0	4633.6	n.c.
BIC	4631.8	4637.2	4682.0	4682.0	4682.2	n.c.

Note: (n.c.) = not computed
Note: 610 Longitudinal Measures at Level 1; 158 Children at Level 2
[a] (correlation).

```
Fixed effects: vsae ~ age.2 + I(age.2 ^ 2) + sicdegp2.f + age.2:sicdegp2.f

                      Value      Std.Error   DF    t-value    p-value
(Intercept)          13.463533   0.7815177   448   17.227419  0.0000
age.2                 6.148750   0.6882638   448    8.933711  0.0000
I(age.2 ^ 2)          0.109008   0.0427795   448    2.548125  0.0112
sicdegp2.f1          -4.987639   1.0379064   155   -4.805480  0.0000
sicdegp2.f2          -3.622820   0.9774516   155   -3.706394  0.0003
age.2:sicdegp2.f1    -4.068041   0.8797676   448   -4.623995  0.0000
age.2:sicdegp2.f2    -3.495530   0.8289509   448   -4.216812  0.0000
```

The (Intercept) (= 13.46) represents the estimated mean VSAE score for children at 2 years of age in the reference category of SICDEGP2.F (i.e., Level 3 of SICDEGP: the children who had the highest initial expressive language scores). The value reported for sicdegp2.f1 represents the estimated difference between the mean VSAE score for 2-year-old children in Level 1 of SICDEGP vs. the reference category. In this case, the estimate is negative (-4.99), which means that the mean initial VSAE score for children in Level 1 of SICDEGP is 4.99 units lower than that of children in the reference category. Similarly, the effect of sicdegp2.f2 represents the estimated difference in the mean VSAE score for children at age 2 in Level 2 of SICDEGP vs. Level 3. Again, the value is negative (-3.62), which means that the children in Level 2 of SICDEGP are estimated to have an initial mean VSAE score at age 2 years that is 3.62 units lower than children in the reference category.

The parameter estimates for age.2 and I(age.2^2) (6.15 and 0.11, respectively) indicate that both coefficients defining the quadratic regression model for children in the reference category of SICD group (SICDEGP = 3) are positive and significant, which suggests a trend in VSAE scores that is consistently accelerating as a function of age. The value associated with the interaction term age2:sicdegp2.f1 represents the difference in the linear effect of age for children in Level 1 of SICDEGP vs. Level 3. The linear coefficient of age for children in SICDEGP = 1 is estimated to be 4.07 units less than that of children in SICDEGP = 3. However, the estimated linear effect of age for children in SICDEGP = 1 is still positive: $6.15 - 4.07 = 2.08$. The value for the interaction term age2:sicdegp2.f2 represents the difference in the linear coefficient of age for children in Level 2 of SICDEGP vs. Level 3, which is again negative. Despite this, the linear trend for age for children in Level 2 of SICDEGP is also estimated to be positive and very similar to that for children in SICDEGP Level 1: $6.15 - 3.50 = 2.65$.

6.7.2   Covariance Parameter Estimates

In this subsection, we discuss the covariance parameter estimates for the child-specific linear and quadratic age effects for Model 6.3. Notice in the output from R below that the estimated standard deviations (StdDev) and correlation (Corr) of the two random effects are reported in the R output, rather than their variances and covariances, as shown in Table 6.8. We remind readers that this output is based on the lme() function in R; when using the lmer() function, estimates of the variances and corresponding standard deviations will be displayed by the summary() function.

```
Random effects:
Formula: ~age.2 + I(age.2 ^ 2) - 1 | childid
Structure: General positive-definite, Log-Cholesky
parametrization
                    StdDev       Corr
age.2               3.8110274    age.2
I(age.2 ^ 2)        0.3556805    -0.306
Residual            6.2281389
```

To calculate the estimated variance of the random linear age effects, we square the reported `StdDev` value for AGE.2 ($3.81 \times 3.81 = 14.52$). We also square the reported `StdDev` of the random quadratic effects of age to obtain their estimated variance ($0.36 \times 0.36 = 0.13$). The correlation of the random linear age effects and the random quadratic age effects is estimated to be -0.31. The residual variance is estimated to be $6.23 \times 6.23 = 38.81$. There is no entry in the `Corr` column for the `Residual`, because we assume that the residuals are independent of the random effects in the model.

We use the `intervals()` function to obtain the estimated standard errors for the covariance parameter estimates, and approximate 95% confidence intervals for the parameters. R calculates these estimates for the standard deviations and correlation of the random effects, rather than for their variances and covariance.

```
> intervals(model6.3.fit)
```

The approximate 95% confidence intervals for the standard deviations of the random linear and quadratic age effects do not contain zero. However, these confidence intervals are based on the asymptotic normality of the covariance parameter estimates, as are the Wald tests for covariance parameters produced by SAS and SPSS. Because these confidence intervals are only approximate, they should be interpreted with caution.

For formal tests of the need for the random effects in the model (Hypothesis 6.1), we recommend likelihood ratio tests, with p-values calculated using a mixture of χ^2 distributions, as discussed in Subsection 6.5.1. Based on the likelihood ratio test results, we concluded that there is significant between-child variability in the quadratic effects of age on VSAE score.

We noted in the initial data summary (Figure 6.2) that the variability of the individual VSAE scores increased markedly with age. The marginal $\boldsymbol{V}_i$ matrix ($= \boldsymbol{Z}_i \boldsymbol{D} \boldsymbol{Z}_i' + \boldsymbol{R}_i$) for the i-th child implied by Model 6.3 can be obtained for the first child by using the R syntax shown below. We note in this matrix that the estimated marginal variances of the VSAE scores (shown on the diagonal of the matrix) increase dramatically with age.

```
> getVarCov(model6.3.fit, individual= "1", type= "marginal")
```

```
CHILDID1
Marginal variance covariance matrix
        1       2        3        4        5
1   38.79   0.000    0.000    0.000     0.00
2    0.00  52.610   39.728   84.617   120.27
3    0.00  39.728  157.330  273.610   425.24
4    0.00  84.617  273.610  769.400  1293.00
5    0.00 120.270  425.240 1293.000  2543.20
```

The fact that the implied marginal covariances associated with age 2 years (in the first row and first column of the $\boldsymbol{V}_i$ matrix) are zero is a direct result of our choice to delete the random effects associated with the intercepts from Model 6.3, and to use AGE − 2 as a covariate. The values in the first row of the $\mathbf{Z}_i$ matrix correspond to the values of AGE − 2 and AGE − 2 squared for the first measurement (age 2 years). Because we used AGE − 2 as a covariate, the $\mathbf{Z}_i$ matrix has values of 0 in the first row.

$$\mathbf{Z}_i = \begin{pmatrix} 0 & 0 \\ 1 & 1 \\ 3 & 9 \\ 7 & 49 \\ 11 & 121 \end{pmatrix}$$

In addition, because we did not include a random intercept in Model 6.3, the only nonzero component corresponding to the first time point (age = 2) in the upper-left corner of the $\mathbf{V}_i$ matrix is contributed by the $\mathbf{R}_i$ matrix, which is simply $\sigma^2 \mathbf{I}_{ni}$ (see Subsection 6.3.2 in this example). This means that the implied marginal variance at age 2 is equal to the estimated residual variance (38.79), and the corresponding marginal covariances are zero. This reflects our decision to attribute all the variance in VSAE scores at age 2 to residual variance.

6.8 Calculating Predicted Values

6.8.1 Marginal Predicted Values

Using the estimates of the fixed-effect parameters obtained by fitting Model 6.3 in R (Table 6.8), we can write a formula for the marginal predicted VSAE score at visit t for child i, as shown in (6.4):

$$
\begin{aligned}
\widehat{\mathrm{VSAE}}_{ti} = \; & 13.46 + 6.15 \times \mathrm{AGE\_2}_{ti} + 0.11 \times \mathrm{AGE\_2SQ}_{ti} \\
& - 4.99 \times \mathrm{SICDEGP1}_i - 3.62 \times \mathrm{SICDEGP2}_i - 4.07 \times \mathrm{AGE\_2}_{ti} \times \mathrm{SICDEGP1}_i \\
& - 3.50 \times \mathrm{AGE\_2}_{ti} \times \mathrm{SICDEGP2}_i
\end{aligned} \tag{6.4}
$$

We can use the values in (6.4) to write three separate formulas for predicting the marginal VSAE scores for children in the three levels of SICDEGP. Recall that SICDEGP1 and SICDEGP2 are dummy variables that indicate whether a child is in the first or second level of SICDEGP. The marginal predicted values are the same for all children at a given age who share the same level of SICDEGP.

FOR SICDEGP = 1:

$$
\begin{aligned}
\widehat{\mathrm{VSAE}}_{ti} &= (13.46 - 4.99) + (6.15 - 4.07) \times \mathrm{AGE\_2}_{ti} + 0.11 \times \mathrm{AGE\_2SQ}_{ti} \\
&= 8.47 + 2.08 \times \mathrm{AGE\_2}_{ti} + 0.11 \times \mathrm{AGE\_2SQ}_{ti}
\end{aligned}
$$

FOR SICDEGP = 2:

$$
\begin{aligned}
\widehat{\mathrm{VSAE}}_{ti} &= (13.46 - 3.62) + (6.15 - 3.50) \times \mathrm{AGE\_2}_{ti} + 0.11 \times \mathrm{AGE\_2SQ}_{ti} \\
&= 9.84 + 2.65 \times \mathrm{AGE\_2}_{ti} + 0.11 \times \mathrm{AGE\_2SQ}_{ti}
\end{aligned}
$$

FOR SICDEGP = 3:

$$
\widehat{\mathrm{VSAE}}_{ti} = 13.46 + 6.15 \times \mathrm{AGE\_2}_{ti} + 0.11 \times \mathrm{AGE\_2SQ}_{ti}
$$

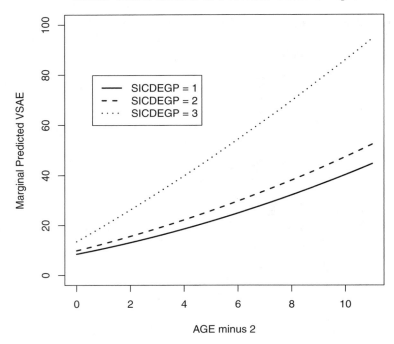

FIGURE 6.4: Marginal predicted VSAE trajectories in the three SICDEGP groups for Model 6.3.

The marginal intercept for children in the highest expressive language group at 2 years of age (SICDEGP = 3) is higher than that of children in group 1 or group 2. The marginal linear effect of age is also less for children in SICDEGP Level 1 and Level 2 than for children in Level 3 of SICDEGP, but the quadratic effect of age is assumed to be the same for the three levels of SICDEGP.

Figure 6.4 graphically shows the marginal predicted values for children in each of the three levels of SICDEGP at each age, obtained using the following R syntax:

```
> curve(0.11*x^2 + 6.15*x + 13.46, 0, 11,
    xlab = "AGE minus 2", ylab = "Marginal Predicted VSAE",
    lty = 3, ylim=c(0,100), lwd = 2)

> curve(0.11*x^2+ 2.65*x + 9.84, 0, 11, add=T, lty = 2, lwd = 2)

> curve(0.11*x^2 + 2.08*x + 8.47, 0, 11, add=T, lty = 1, lwd = 2)

> # Add a legend to the plot; R will wait for the user to click
> # on the point in the plot where the legend is desired.

> legend(locator(1),
    c("SICDEGP = 1", "SICDEGP = 2", "SICDEGP = 3"),
    lty = c(1, 2, 3), lwd = c(2, 2, 2))
```

The different intercepts for each level of SICDEGP are apparent in Figure 6.4, and the differences in the predicted trajectories for each level of SICDEGP can be easily visualized.

Children in SICDEGP = 3 are predicted to start at a higher initial level of VSAE at age 2 years, and also have predicted mean VSAE scores that increase more quickly as a function of age than children in the first or second SICD group.

6.8.2 Conditional Predicted Values

We can also write a formula for the predicted VSAE score at visit t for child i, conditional on the random linear and quadratic age effects in Model 6.3, as follows:

$$
\begin{aligned}
\widehat{\text{VSAE}}_{ti} =\ & 13.46 + 6.15 \times \text{AGE\_2}_{ti} + 0.11 \times \text{AGE\_2SQ}_{ti} \\
& - 4.99 \times \text{SICDEGP1}_i - 3.62 \times \text{SICDEGP2}_i \\
& - 4.07 \times \text{AGE\_2}_{ti} \times \text{SICDEGP1}_i - 3.50 \times \text{AGE\_2}_{ti} \times \text{SICDEGP2}_i \\
& + \hat{u}_{1i} \times \text{AGE\_2}_{ti} + \hat{u}_{2i} \times \text{AGE\_2SQ}_{ti}
\end{aligned}
\tag{6.5}
$$

In general, the intercept will be the same for all children in a given level of SICDEGP, but their individual trajectories will differ, because of the random linear and quadratic effects of age that were included in the model.

For the i-th child, the predicted values of u_{1i} and u_{2i} are the realizations of the EBLUPs of the random linear and quadratic age effects, respectively. The formula below can be used to calculate the conditional predicted values for a given child i in SICDEGP = 3:

$$
\begin{aligned}
\widehat{\text{VSAE}}_{ti} =\ & 13.46 + 6.15 \times \text{AGE\_2}_{ti} + 0.11 \times \text{AGE\_2SQ}_{ti} \\
& + \hat{u}_{1i} \times \text{AGE\_2}_{ti} + \hat{u}_{2i} \times \text{AGE\_2SQ}_{ti}
\end{aligned}
\tag{6.6}
$$

For example, we can write a formula for the predicted value of VSAE at visit t for CHILDID = 4 (who is in SICDEGP = 3) by substituting the predicted values of the EBLUPs generated by R using the `random.effects()` function for the fourth child into the formula above. The EBLUP for u_{14} is 2.31, and the EBLUP for u_{24} is 0.61:

$$
\begin{aligned}
\widehat{\text{VSAE}}_{t4} =\ & 13.46 + 6.15 \times \text{AGE\_2}_{t4} + 0.11 \times \text{AGE\_2SQ}_{t4} \\
& + 2.31 \times \text{AGE\_2}_{t4} + 0.61 \times \text{AGE\_2SQ}_{t4} \\
=\ & 13.46 + 8.46 \times \text{AGE\_2}_{t4} + 0.72 \times \text{AGE\_2SQ}_{t4}
\end{aligned}
\tag{6.7}
$$

The conditional predicted VSAE value for child 4 at age 2 is 13.46, which is the same for all children in SICDEGP = 3. The predicted linear effect of age specific to child 4 is positive (8.46), and is larger than the predicted marginal effect of age for all children in SICDEGP = 3 (6.15). The quadratic effect of age for this child (0.72) is also much larger than the marginal quadratic effect of age (0.11) for all children in SICDEGP = 3. See the third panel in the bottom row of Figure 6.5 for a graphical depiction of the individual trajectory of CHILDID = 4.

We graph the child-specific predicted values of VSAE for the first 12 children in SICDEGP = 3, along with the marginal predicted values for children in SICDEGP = 3, using the following R syntax:

```
> # Load the lattice package.
> # Set the trellis graphics device to have no color.

> library(lattice)
```

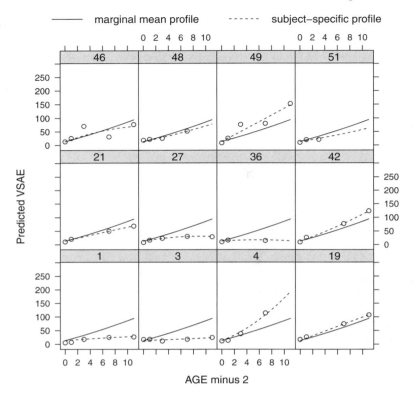

FIGURE 6.5: Conditional (dashed lines) and marginal (solid lines) trajectories, for the first 12 children with SICDEGP = 3.

```
> trellis.device(color=F)

> # Use the augPred function in the nlme package to plot
> # conditional predicted values for the first twelve children
> # with SICDEGP = 3, based on the fit of Model 6.3 (note that
> # this requires the autism.csv data set to be sorted
> # in descending order by SICDEGP, prior to
> # being imported into R).

> plot(augPred(model6.3.fit, level = 0:1),
   layout = c(4, 3, 1), xlab = "AGE minus 2", ylab = "Predicted VSAE",
   key = list(
     lines = list(lty = c(1, 2), col = c(1, 1), lwd = c(1, 1) ),
     text = list(c("marginal mean profile", "subject-specific profile")),
     columns = 2))
```

We can clearly see the variability in the fitted trajectories for different children in the third level of SICDEGP in Figure 6.5.

In general, the `fitted()` function can be applied to a model fit object (e.g., `model6.3.fit`) to obtain conditional predicted values in the R software, and the `random.effects()` function (in the `nlme` package) can be applied to a model fit object to obtain EBLUPs of random effects. Refer to Section 6.9 for additional R syntax that can be used to obtain and plot conditional predicted values.

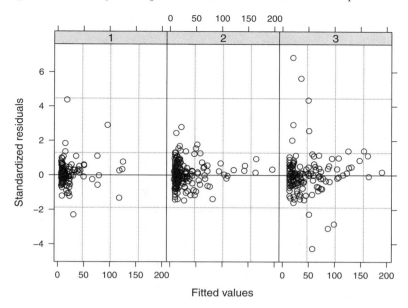

FIGURE 6.6: Residual vs. fitted plot for each level of SICDEGP, based on the fit of Model 6.3.

6.9 Diagnostics for the Final Model

We now check the assumptions for Model 6.3, fitted using REML estimation, using informal graphical procedures in the R software. Similar plots can be generated in the other four software packages after saving the conditional residuals, the conditional predicted values, and the EBLUPs of the random effects based on the fit of Model 6.3 (see the book's web page in Appendix A).

6.9.1 Residual Diagnostics

We first assess the assumption of constant variance for the residuals in Model 6.3. Figure 6.6 presents a plot of the standardized conditional residuals vs. the conditional predicted values for each level of SICDEGP.

```
> library(lattice)
> trellis.device(color= F)
> plot(model6.3.fit,
    resid(., type = "p") ~ fitted(.) | factor(sicdegp),
    layout=c(3,1), aspect=2, abline=0)
```

The variance of the residuals appears to decrease for larger fitted values, and there are some possible outliers that may warrant further investigation. The preceding syntax may be modified by adding the `id = 0.05` argument to produce a plot (not shown) that identifies outliers at the 0.05 significance level:

```
> plot(model6.3.fit,
    resid(., type= "p") ~ fitted(.) | factor(sicdegp),
    id = 0.05, layout = c(3,1), aspect = 2, abline = 0)
```

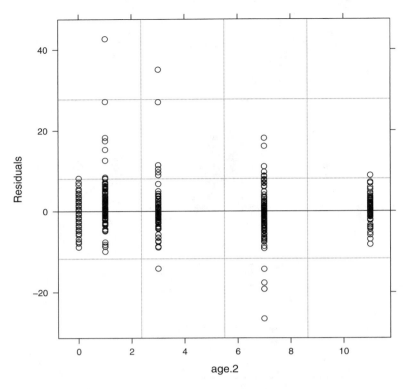

FIGURE 6.7: Plot of conditional raw residuals versus AGE.2.

Next, we investigate whether the residual variance is constant as a function of AGE.2.

```
> plot(model6.3.fit, resid(.) ~ age.2, abline = 0)
```

Figure 6.7 suggests that the variance of the residuals is fairly constant across the values of AGE − 2. We again note the presence of outliers.

Next, we assess the assumption of normality of the residuals using Q–Q plots within each level of SICDEGP, and request that unusual points be identified by CHILDID using the `id = 0.05` argument:

```
> qqnorm(model6.3.fit,
    ~resid(.) | factor(sicdegp) ,
    layout = c(3,1), aspect = 2, id = 0.05)
```

Figure 6.8 suggests that the assumption of normality for the residuals seems acceptable. However, the presence of outliers in each level of SICDEGP (e.g., CHILDID = 46 in SICDEGP = 3) may warrant further investigation.

6.9.2   Diagnostics for the Random Effects

We now check the distribution of the random effects (EBLUPs) generated by fitting Model 6.3 to the Autism data. Figure 6.9 presents Q–Q plots for the two sets of random effects. Significant outliers at the 0.10 level of significance are identified by CHILDID in this graph (`id = 0.10`):

```
> qqnorm(model6.3.fit, ~ranef(.) , id = 0.10)
```

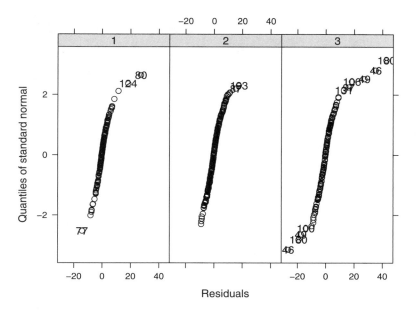

FIGURE 6.8: Normal Q–Q Plots of conditional residuals within each level of SICDEGP.

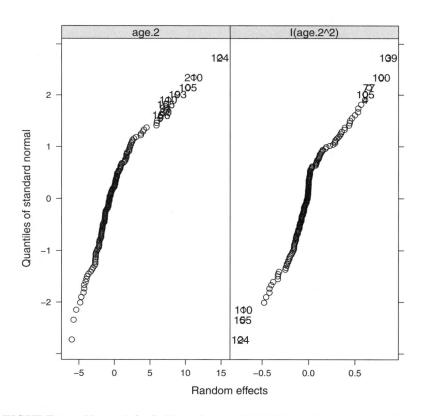

FIGURE 6.9: Normal Q–Q Plots for the EBLUPs of the random effects.

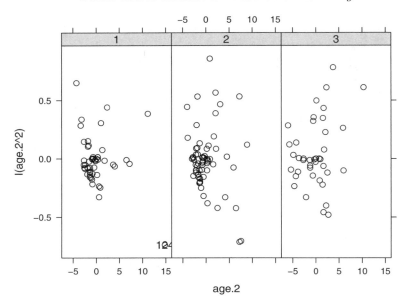

FIGURE 6.10: Scatter plots of EBLUPs for age-squared vs. age by SICDEGP.

We note that CHILDID = 124 is an outlier in terms of both random effects. The children indicated as outliers in these plots should be investigated in more detail to make sure that there is nothing unusual about their observations.

Next, we check the joint distribution of the random linear and quadratic age effects across levels of SICDEGP using the `pairs()` function:

```
> pairs(model6.3.fit,
    ~ranef(.) | factor(sicdegp),
    id = ~childid == 124, layout = c(3, 1), aspect = 2)
```

The form of these plots is not suggestive of a very strong relationship between the random effects for age and age-squared, although R reported a modest negative correlation ($r = -0.31$) between them in Model 6.3 (see Table 6.8).

The distinguishing features of these plots are the outliers, which give the overall shape of the plots a rather unusual appearance. The EBLUPs for CHILDID = 124 are again unusual in Figure 6.10. Investigation of the values for children with unusual EBLUPs would be useful at this point, and might provide insight into the reasons for the outliers; we do not pursue such an investigation here.

We also remind readers (from Chapter 2) that selected influence diagnostics can also be computed when using the `nlmeU` or `HLMdiag` packages in R; see Section 20.3 of Galecki & Burzykowski (2013) or Loy & Hofmann (2014) for additional details on the computational steps involved.

6.9.3 Observed and Predicted Values

Finally, we check for agreement between the conditional predicted values based on the fit of Model 6.3 and the actual observed VSAE scores. Figure 6.11 displays scatter plots of the observed VSAE scores vs. the conditional predicted VSAE scores for each level of SICDEGP, with possible outliers once again identified:

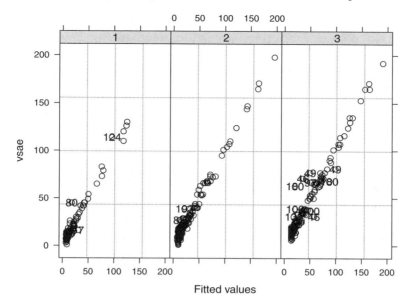

FIGURE 6.11: Agreement of observed VSAE scores with conditional predicted VSAE scores for each level of SICDEGP, based on Model 6.3.

```
> plot(model6.3.fit, vsae ~ fitted(.) | factor(sicdegp),
    id = 0.05, layout = c(3,1) , aspect = 2)
```

We see relatively good agreement between the observed and predicted values within each SICDEGP group, with the exception of some outliers.

We refit Model 6.3 after excluding the observations for CHILDID = 124 (in SICDEGP = 1) and CHILDID = 46 (in SICDEGP = 3):

```
> autism.grouped2 <- autism.grouped[(autism.grouped$childid != 124 &
                      autism.grouped$childid != 46),]
```

```
> model6.3.fit.out <- update(model6.3.fit, data = autism.grouped2)
```

Applying the `summary()` and `intervals()` functions to the `model6.3.fit.out` object indicates that the primary results in Model 6.3 did not change substantially after excluding the outliers.

6.10   Software Note: Computational Problems with the D Matrix

The major differences between the software procedures in this analysis were encountered when attempting to estimate the variance-covariance matrix of the random effects (the D matrix) in Model 6.1. This model included three random effects for each child, associated with the intercept, the linear effect of age, and the quadratic effect of age. Each software procedure reacted differently to the estimation problems that occurred when fitting Model 6.1.

We summarize these differences in Table 6.6. SAS `proc mixed` produced a note in the log stating that the estimated G matrix (i.e., the block-diagonal matrix with blocks defined by the 3×3 **D** matrix for a single child) was not positive-definite. The estimated value of σ^2_{int} was reported to be zero in the output, and its standard error was not reported. SPSS **MIXED** produced a warning in the output stating that model convergence was not achieved, and that the validity of the model fit was uncertain. SPSS also reported the estimated value of σ^2_{int} and its standard error to be zero. The lme() function in R reported problems with convergence of the estimation algorithm, and estimates of the parameters in the model were not available. Similar to the procedures in SAS and SPSS, the lmer() function in R reported an estimate of 0 for σ^2_{int}. The mixed command in Stata reported an error message and did not produce any output. HLM did not report any problems in fitting the model, but required more than 1600 iterations to converge.

To investigate the estimation problems encountered in fitting Model 6.1, we used the **nobound** option in SAS **proc mixed**. This option allowed us to fit the implied marginal model without requiring that the **G** matrix be positive-definite. We found that the estimated value for what would have been σ^2_{int} in the unconstrained matrix was actually negative (-10.54). Subsequent models were simplified by omitting the random effect associated with the intercept. SAS **proc mixed** is currently the only software procedure that provides an option to relax the requirement that the **D** matrix be positive-definite.

6.10.1   Recommendations

We recommend carefully checking the covariance parameter estimates and their standard errors. Models with multiple random effects, like those fitted in this chapter, may need to be simplified or respecified if problems are encountered. For this analysis, we decided to remove the child-specific random effects associated with the intercept from Model 6.1 because of the estimation problems, and because there was little variability in the initial VSAE scores for these autistic children at 2 years of age, as illustrated in our initial data summary (see Figures 6.1 and 6.2). This had implications for the marginal covariance matrix, as illustrated in Subsection 6.7.2. But the remedies for this problem will depend on the subject matter under study and the model specification. We find that this issue arises quite often when analysts attempt to include too many random effects in a given linear mixed model (and accordingly request that the software provide estimates of several variances and covariances). Careful examination of the covariance parameter estimates and possible reduction of the number of random effects included in the model (for example, were we really interested in the variance of the intercepts in the model for the Autism data?) will generally prevent these estimation issues.

6.11   An Alternative Approach: Fitting the Marginal Model with an Unstructured Covariance Matrix

Fitting a marginal model with an "unstructured" covariance matrix is a plausible alternative for the Autism data, because a relatively small number of observations are made on each child, and the observations are made at the same ages. We use the gls() function in R to fit a marginal model having the same fixed effects as in Model 6.3, but with no random effects, and an "unstructured" covariance structure for the marginal residuals.

To fit a marginal model with an **unstructured** covariance matrix (i.e., an unstructured **R_i** matrix) for the residuals using the R software, we need to specify a "General" correlation structure within each level of CHILDID. This correlation structure (specified with the `correlation = corSymm()` argument) is characterized by completely general (uncon-strained) correlations. In order to have general (unconstrained) variances as well (resulting in a specification consistent with the "unstructured" covariance structure in SAS and SPSS), we also need to make use of the `weights =` argument.

The `correlation = corSymm()` argument requires the specification of an index variable with consecutive integer values, to identify the ordering of the repeated measures in a longitudinal data set. We first create this variable as follows:

```
> index <- age.2
> index[age.2 ==  0] <- 1
> index[age.2 ==  1] <- 2
> index[age.2 ==  3] <- 3
> index[age.2 ==  7] <- 4
> index[age.2 == 11] <- 5
```

We then add the index variable to the original `autism` data frame object:

```
> autism.updated <- subset(data.frame(
    autism, sicdegp2.f, age.2, index), !is.na(vsae))
```

We now specify this correlation structure in the `gls()` function using the `correlation=corSymm(form = ~ index | childid)` argument in the following syntax, and allow for unequal variances at each level of AGE_2 by using the `weights = varIdent(form = ~ 1 | age.2)` argument. Refer to Pinheiro & Bates (1996) for more information about this structure; the SAS and SPSS procedures allow users to select this "unstructured" covariance structure for the R_i matrix directly as opposed to specifying `correlation` and `weights` arguments separately.

```
> marg.model.fit <- gls(
    vsae ~ age. 2 + I(age.2^2) + sicdegp2.f + age.2:sicdegp2.f,
    correlation=corSymm(form = ~ index | childid),
    weights = varIdent(form = ~ 1 | age.2) , autism.updated)
```

The estimated fixed effects in the marginal model can be obtained by applying the `summary()` function to the model fit object:

```
> summary(marg.model.fit)
```

```
Coefficients
                         Value    Std.Error    t-value   p-value
 (Intercept)         12.471584   0.5024636   24.820869   0.000
 age.2                7.373850   0.6042509   12.203293   0.000
 I(age.2^2)          -0.004881   0.0373257   -0.130761   0.896
 sicdegp2.f1         -5.361092   0.6691472   -8.011828   0.000
 sicdegp2.f2         -3.725960   0.6307950   -5.906769   0.000
 age.2:sicdegp2.f1   -4.332097   0.7787670   -5.562764   0.000
 age.2:sicdegp2.f2   -3.867433   0.7334745   -5.272757   0.000
```

We note that the effect of AGE_2 squared (`I(age.2^2)`) for the reference SICD group (SICDEGP = 3) is not significant in the marginal model ($p = 0.896$), whereas it is positive and significant ($p = 0.01$) in Model 6.3. In general, the estimates of the fixed effects and their standard errors in the marginal model are different from those estimated for Model 6.3, because the estimated V_i matrix for the marginal model differs from the implied V_i matrix for Model 6.3.

The covariance matrix from the marginal model is also part of the output generated by the `summary()` function:

```
Correlation Structure: General
Formula: ~index | childid
Parameter estimate(s):
Correlation:
         1        2        3        4
2   0.365
3   0.214    0.630
4   0.263    0.448    0.574
5   0.280    0.543    0.664    0.870
```

The correlations of the marginal residuals at time 1 (corresponding to 2 years of age) with residuals at later times are in general smaller (ranging from about 0.21 to 0.37) than the correlations of the residuals at times 2 through 5 with the residuals for later time points. Recall that the marginal correlations of observations at 2 years of age with other observations implied by Model 6.3 were zero (see Subsection 6.7.2), because we had omitted the random intercept and used AGE – 2 as a predictor (resulting in zeroes for the Z_i matrix values corresponding to 2 years of age).

We report the estimated marginal variance-covariance matrix V_i for a single child i as follows, using additional R code that can be found on the book's web page (see Appendix A):

```
            0          1          3          7         11
[1,]   10.811330   8.858858   8.310693   24.59511   43.05836
[2,]    8.858858  54.509537  55.123024   94.33787  187.58256
[3,]    8.310693  55.123024 140.582610  194.00964  368.23192
[4,]   24.595114  94.337868 194.009635  811.74023 1160.52408
[5,]   43.058363 187.582559 368.231923 1160.52408 2191.80398
```

Software Note: The `getVarCov()` function, when applied to the `marg.model.fit` object obtained by using `gls()`, returns an incorrect result for the estimated V_i matrix (not shown). This function was primarily designed to work with model fit objects obtained using the `lme()` function. According to the R documentation, this function should also work for a `gls()` object. However, in our example, which involves the weights argument, the `getVarCov()` function does not return a correct marginal variance-covariance matrix.

The estimated marginal variances, indicated on the diagonal of the matrix above, increase with age, similar to the marginal variances implied by Model 6.3 (see Subsection 6.7.2).

We can also fit the preceding marginal model using the following syntax in SAS:

```
title "Marginal Model w/ Unstructured Covariance Matrix";
proc mixed data = autism2 noclprint covtest;
class childid sicdegp age;
model vsae = sicdegp age_2 age_2sq age_2*sicdegp / solution;
repeated age / subject = childid type = un r rcorr;
run;
```

The `repeated` statement indicates that observations on the same CHILDID are indexed by the original AGE variable (identified in the `class` statement), and that an unstructured covariance matrix (`type = un`) is to be used. The `r` option requests that SAS display the estimated marginal variance-covariance matrix in the output, and the `rcorr` option requests the corresponding correlation matrix. SPSS users can also fit a similar model using only a REPEATED statement in the MIXED syntax.

6.11.1 Recommendations

Fitting the marginal model directly with an "unstructured" residual covariance matrix allows us to get a better sense of the marginal variances and covariances in the observed data than from the marginal model implied by the random coefficient model (Model 6.3). When analyzing data sets from longitudinal studies with balanced designs (i.e., where all subjects are measured at the same time points), we recommend fitting the marginal model in this manner, because it allows us to get an idea of whether the random coefficient model being fitted implies a reasonable marginal covariance structure for the data. However, fitting the marginal model directly does not allow us to answer research questions about the between-child variability in the VSAE trajectories, because the marginal model does not explicitly include any random child effects.

Because these models do not include explicit random effects, they should *not* be referred to as linear mixed models. Linear mixed models, by definition, include a mix of fixed and random effects. When describing these models in technical reports or academic publications, they should be referred to as general linear models with correlated errors, and the type of covariance structure used for the errors should be clearly defined.

7

Models for Clustered Longitudinal Data: The Dental Veneer Example

7.1 Introduction

In this chapter we illustrate fitting linear mixed models (LMMs) to **clustered longitudinal data**, in which **units of analysis** are nested within **clusters**, and **repeated measures** are collected on the units of analysis over time. Each cluster may have a different number of units of analysis, and the time points at which the dependent variable is measured can differ for each unit of analysis. Such data sets can be considered to have three levels. Level 3 represents the clusters of units, Level 2 the units of analysis, and Level 1 represents the longitudinal (repeated) measures made over time.

In Table 7.1 we illustrate examples of clustered longitudinal data in different research settings. Such data structures might arise in an educational setting when student achievement scores are measured over time, with students clustered within a sample of classrooms. A clustered longitudinal data structure might also be encountered in an environmental setting in a study of the weekly oxygen yield for a sample of trees measured over the course of a growing season. In this case, trees are clustered within a sample of plots, and the trees are measured repeatedly over time. In the Dental Veneer data set analyzed in this chapter, the dependent variable, gingival crevicular fluid (GCF), was measured at two post-treatment time points for each tooth, with teeth clustered within patients.

Figure 7.1 illustrates an example of the structure of a clustered longitudinal data set, using the first patient in the Dental Veneer data set. Note that patient 1 (who represents a cluster of units) had four teeth (the units of analysis) included in the study, but other patients could have a different number of treated teeth. Measurements were made on each tooth at two follow-up time points (3 months and 6 months).

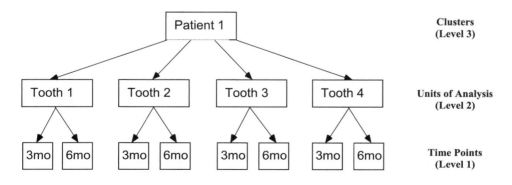

FIGURE 7.1: Structure of the clustered longitudinal data for the first patient in the Dental Veneer data set.

TABLE 7.1: Examples of Clustered Longitudinal Data in Different Research Settings

| Level of Data | | Research Setting | | |
		Environment	Education	Dentistry
Cluster of Units (Level 3)	**Cluster ID variable (random factor)**	Plot	Classroom	Patient
	Covariates	Soil minerals, tree crown density in the plot	Teacher's years of experience, classroom size	Gender, age
Unit of Analysis (Level 2)	**Unit of Analysis ID variable (random factor)**	Tree	Student	Tooth
	Covariates	Tree size	Gender, age, baseline score	Treatment, tooth type
Time (Level 1)	**Time variable**	Week	Marking period	Month
	Dependent variable	Oxygen yield	Test score	Gingival crevicular fluid (GCF)
	Time-varying covariates	Sunlight exposure, precipitation	Attendance	Frequency of tooth brushing

LMMs for clustered longitudinal data are "hybrids" of the models that we have used for the examples in previous chapters. Specifically, these models include random effects associated with both the clusters (e.g., plots, classrooms, and patients) and the units of analysis nested within these clusters (e.g., trees, students, and teeth) to take the clustered structure of the data into account, and also allow the residuals associated with the longitudinal measures on the same unit of analysis to be correlated.

The variables indicating the units of analysis (Level 2) and the clusters of units (Level 3) are assumed to be random factors, with levels (e.g., students and classrooms) sampled from a larger population. However, convenience samples, which are easy to obtain but do not arise from a probability sample, are commonly used in practice. The time variable itself can be considered to be either a categorical fixed factor or a continuous predictor. If each unit is measured at the same time points, time is *crossed* with the random factors defined at Level 2 and Level 3. However, we do not necessarily require each unit of analysis to be measured at the same time points.

In this chapter, we highlight features in the Stata software package.

7.2 The Dental Veneer Study

7.2.1 Study Description

The Dental Veneer data were collected by researchers at the University of Michigan Dental School, in a study investigating the impact of veneer placement on subsequent gingival (gum) health among adult patients (Ocampo, 2005). Ceramic veneers were applied to selected teeth to hide discoloration. The treatment process involved removing some of the surface of each treated tooth, and then attaching the veneer to the tooth with an adhesive. The veneer was placed to match the original contour of the tooth as closely as possible. The investigators were interested in studying whether varying amounts of contour difference (CDA) due to placement of the veneer might affect gingival health in the treated teeth over time. One measure of gingival health was the amount of GCF in pockets of the gum adjacent to the treated teeth. GCF was measured for each tooth at visits 3 months and 6 months post-treatment.

A total of 88 teeth in 17 patients were prepared for veneer placement, and a baseline measure of GCF was collected for each tooth. We consider only the 55 treated teeth located in the maxillary arches of 12 patients in this example to avoid duplication of results from the original authors. Each patient could have different numbers of treated teeth, and the particular teeth that were treated could differ by patient. Table 7.2 presents a portion of the Dental Veneer data set in the "long" format appropriate for analysis using the `mixed` command in Stata, and the LMM procedures in SAS, SPSS, and R.

The following variables are included in the Dental Veneer data set:

Patient (Level 3) Variables

- **PATIENT** = Patient ID variable (Level 3 ID)

- **AGE** = Age of patient when veneer was placed; constant for all observations on the same patient

Tooth (Level 2) Variables

- **TOOTH** = Tooth number (Level 2 ID)

- **BASE_GCF** = Baseline measure of GCF for the tooth; constant for all observations on the same tooth

- **CDA** = Average contour difference in the tooth after veneer placement; constant for all observations on the same tooth

Time-Varying (Level 1) Variables

- **TIME** = Time points of longitudinal measures (3 = 3 Months, 6 = 6 Months)

- **GCF** = Gingival crevicular fluid adjacent to the tooth, collected at each time point (dependent variable)

TABLE 7.2: Sample of the Dental Veneer Data Set

Patient (Level 3)		Tooth (Level 2)			Longitudinal Measures (Level 1)	
Cluster ID	Co-variate	Unit ID	Covariate		Time Variable	Dependent Variable
PATIENT	AGE	TOOTH	BASE_GCF	CDA	TIME	GCF
1	46	6	17	4.67	3	11
1	46	6	17	4.67	6	68
1	46	7	22	4.67	3	13
1	46	7	22	4.67	6	47
...						
1	46	11	17	5.67	3	11
1	46	11	17	5.67	6	53
3	32	6	3	7.67	3	28
3	32	6	3	7.67	6	23
3	32	7	4	11.00	3	17
3	32	7	4	11.00	6	15
...						

Note: "..." indicates portion of the data not displayed.

7.2.2 Data Summary

The data summary for this example was generated using Stata (Release 13). A link to the syntax and commands that can be used to perform similar analyses in the other software packages is included on the book's web page (see Appendix A).

We begin by importing the tab-delimited raw data file (veneer.dat) into Stata from the C:\temp directory:

```
. insheet using "C:\temp\veneer.dat", tab
```

We create line graphs of the GCF values across time for all teeth within each patient. To create these graphs, we need to restructure the data set to a "wide" format using the reshape wide command:

```
. keep patient tooth age time gcf
. reshape wide gcf, i(patient time) j(tooth)
```

Software Note: The Dental Veneer data set is restructured here only for the purpose of generating Figure 7.2. The original data set in the "long" form displayed in Table 7.2 will be used for all mixed model analyses in Stata.

The following output is generated as a result of applying the reshape command to the original data set. As indicated in this output, the restructured data set contains just 24 observations (2 per patient, corresponding to the 2 time points) and 9 variables: PATIENT, TIME, AGE, and GCF6 through GCF11. The new GCF variables are indexed by levels of the original TOOTH variable (GCF6 corresponds to tooth 6, etc.).

```
(note: j = 6 7 8 9 10 11)


Data                                 long    ->   wide
------------------------------------------------------------------------
Number of obs.                        110    ->      24
Number of variables                     5    ->       9
j variable (6 values)                tooth   ->   (dropped)
xij variables:
                                      gcf    ->   gcf6 gcf7 ... gcf11
------------------------------------------------------------------------
```

The first four observations in the restructured data set are as follows:

`. list in 1/4`

```
+--------------------------------------------------------------------+
| patient   time   gcf6   gcf7   gcf8   gcf9   gcf10   gcf11   age |
|--------------------------------------------------------------------|
1. |    1       3     11     13     14     10      14      11     46 |
2. |    1       6     68     47     58     57      44      53     46 |
3. |    3       3     28     17     19     34      54      38     32 |
4. |    3       6     23     15     32     46      39      19     32 |
+--------------------------------------------------------------------+
```

We now label the six new GCF variables to indicate the tooth numbers to which they correspond and label the AGE variable:

```
. label var gcf6 "6"
. label var gcf7 "7"
. label var gcf8 "8"
. label var gcf9 "9"
. label var gcf10 "10"
. label var gcf11 "11"
. label var age "Patient Age"
```

Next, we generate Figure 7.2 using the `twoway line` plotting command to create a single figure containing multiple line graphs:

```
. twoway line gcf6 gcf7 gcf8 gcf9 gcf10 gcf11 time, ///
lcolor(black black black black black black) ///
lpattern(solid dash longdash vshortdash shortdash dash_dot) ///
ytitle(GCF) by(age)
```

We use the `///` symbols in this command because we are splitting a long command into multiple lines.

The `lcolor` option is used to set the color of the six possible lines (corresponding to teeth within a given patient) within a single graph to be black. The pattern of the lines is set with the `lpattern` option. The plots in this graph are ordered by the age of each patient, displayed at the top of each panel.

In Figure 7.2, we observe that the GCF values for all teeth within a given patient tend to follow the same trend over time (the lines for the treated teeth appear to be roughly parallel within each patient). In some patients, the GCF levels tend to increase, whereas in others the GCF levels tend to decrease or remain relatively constant over time. This pattern suggests that an appropriate model for the data might include random patient-specific time slopes. The GCF levels of the teeth also tend to differ by patient, suggesting

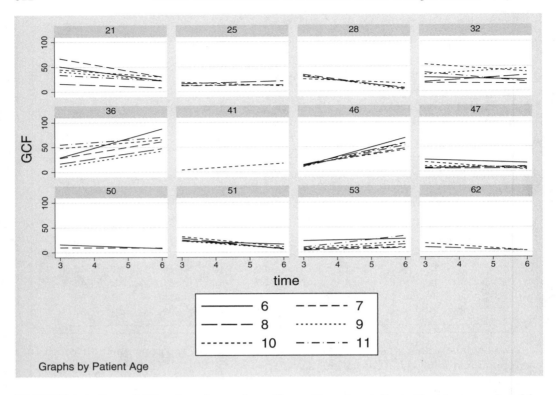

FIGURE 7.2: Raw GCF values for each tooth vs. time, by patient. Panels are ordered by patient age.

that a model should also include random patient-specific intercepts. There is also evidence in most patients that the level of GCF tends to differ by tooth, suggesting that we may want to include random tooth-specific intercepts in the model.

7.3   Overview of the Dental Veneer Data Analysis

For the analysis of the Dental Veneer data, we use the "top-down" modeling strategy discussed in Subsection 2.7.1 of Chapter 2. In Subsection 7.3.1, we outline the analysis steps and informally introduce related models and hypotheses to be tested. In Subsection 7.3.2 we present more detailed specifications of selected models. The hypotheses tested are detailed in Subsection 7.3.3. To follow the analysis steps outlined in this section, refer to the schematic diagram presented in Figure 7.3.

7.3.1   Analysis Steps

Step 1: Fit a model with a "loaded" mean structure (Model 7.1).

Fit a three-level model with a "loaded" mean structure, and random effects associated with both the intercept and slope for patients and with individual teeth within patients.

In Model 7.1, we include fixed effects associated with all covariates under consideration (TIME, BASE_CGF, CDA, and AGE) and the two-way interactions between TIME and each of the other covariates. We consider TIME to be a continuous predictor in all models, even though it only has two levels; this allows us to interpret the fixed effect of TIME as the expected change in GCF over a one-month period. Based on inspection of the initial graphs in Figure 7.2, we add the following random components to this model: random patient-specific effects associated with both the intercept and slope (i.e., the effect of time) for each patient, and random effects associated with the intercept for each tooth nested within a patient. We assume that the residuals in Model 7.1 are independent and identically distributed, with constant variance across the time points.

Step 2: Select a structure for the random effects (Model 7.1 vs. Model 7.1A).

Decide whether to keep the random tooth-specific intercepts in the model.

In this step we fit Model 7.1A, which excludes the random intercepts associated with teeth nested within patients, and test whether we need to keep these random effects in the model (Hypothesis 7.1). Based on the results of this hypothesis test, we decide to retain the nested random tooth effects in the model (and therefore the random patient effects as well, to preserve the hierarchical structure of the data in the model specification) and keep Model 7.1 as our preferred model at this stage.

Step 3: Select a covariance structure for the residuals (Models 7.1, 7.2A, 7.2B, or 7.2C).

Fit models with different covariance structures for the residuals associated with observations on the same tooth.

In this step, we investigate different covariance structures for the residuals, while maintaining the same fixed and random effects as in Model 7.1. Because there are only two repeated measures on each tooth, we can only consider a limited number of residual covariance structures, and we will note that parameters in some of these covariance structures are aliased (or not identifiable; see Subsection 2.9.3). In short, the tooth-specific random intercepts are inducing marginal correlations between the repeated measures on each tooth, so an additional correlation parameter for the residuals is not needed.

Model 7.2A: Unstructured covariance matrix for the residuals.

The most flexible residual covariance structure involves three parameters: different residual variances at 3 and 6 months, and the covariance between the residuals at the two visits. This is an appropriate structure to use if GCF at 3 months has a residual variance different from that at 6 months. We include a residual covariance because we expect the two residuals for a given tooth to be positively related.

Unfortunately, specifying an unstructured covariance matrix for the residuals together with the variance for the random tooth intercepts leads to an aliasing problem with the covariance parameters. Preferably, aliasing should be detected during model specification. If aliasing is unnoticed during model specification it leads to an estimation problem and usually can be indirectly detected during estimation of model parameters, as will be shown in the software-specific sections.

Although in practice we would not consider this model, we present it (and Model 7.2B following) to illustrate what happens in each software procedure when aliasing of covariance parameters occurs.

Model 7.2B: Compound symmetry covariance structure for the residuals.

We next fitted a simpler model with equal (homogeneous) residual variances at 3 months and at 6 months, and correlated residuals. This residual covariance structure has only two parameters, but the residual covariance parameter is again aliased with the variance of the random tooth-specific intercepts in this model.

Model 7.2C: Heterogeneous (unequal) residual variances at each time point, with zero covariance for the residuals.

This residual covariance structure also has two parameters, representing the different residual variances at 3 months and at 6 months. However, the covariance of the residuals at the two time points is constrained to be zero. In this model, the residual covariance parameters are no longer aliased with the variance of the random tooth-specific intercepts.

In Model 7.1, we assume that the residuals have constant variance at each time point, but this restriction is relaxed in Model 7.2C. We test Hypothesis 7.2 in this step to decide whether to use a heterogeneous residual covariance structure, Model 7.2C, or retain the simpler residual covariance structure of Model 7.1. Because of the nonsignificant result of this hypothesis test, we keep Model 7.1 as our preferred model at this stage of the analysis.

Step 4: Reduce the model by removing fixed effects associated with the two-way interactions (Model 7.1 vs. Model 7.3) and check model diagnostics.

We now fit Model 7.3, which omits the fixed effects associated with the two-way interactions between TIME and the other fixed covariates, using maximum likelihood (ML) estimation. We test the significance of these fixed effects (Hypothesis 7.3) and decide that Model 7.3 is our preferred model. We decide to retain all the main fixed effects associated with the covariates in our final model, so that the primary research questions can be addressed.

We refit the final model using REML estimation to obtain unbiased estimates of the covariance parameters. We informally examine diagnostics for Model 7.3 in Section 7.10, using Stata.

Figure 7.3 provides a schematic guide to the model selection process in the analysis of the Dental Veneer data (see Section 3.3.1 for details on how to interpret this figure).

7.3.2 Model Specification

The general specification of the models in Subsection 7.3.2.1 corresponds closely to the syntax used to fit the models in SAS, SPSS, Stata, and R. In Subsection 7.3.2.2, we discuss the hierarchical specification of the models, which corresponds closely to the model setup in HLM. Selected models considered in the analysis of the Dental Veneer data are summarized in Figure 7.3.

7.3.2.1 General Model Specification

The general form of Models 7.1 through 7.2C for an individual GCF response at visit t (t = 1, 2, corresponding to months 3 and 6) on tooth i nested within patient j (denoted by GCF_{tij}) is as follows:

$$
\begin{aligned}
\text{GCF}_{tij} = \ & \left.
\begin{aligned}
& \beta_0 + \beta_1 \times \text{TIME}_t + \beta_2 \times \text{BASE\_GCF}_{ij} \\
& + \beta_3 \times \text{CDA}_{ij} + \beta_4 \times \text{AGE}_j + \beta_5 \times \text{TIME}_t \times \text{BASE\_GCF}_{ij} \\
& + \beta_6 \times \text{TIME}_t \times \text{CDA}_{ij} + \beta_7 \times \text{TIME}_t \times \text{AGE}_j
\end{aligned}
\right\} \text{fixed} \\
& \hspace{2.5cm} + u_{0j} + u_{1j} \times \text{TIME}_t + u_{0i|j} + \varepsilon_{tij} \quad \left.\vphantom{\int}\right\} \text{random} \quad (7.1)
\end{aligned}
$$

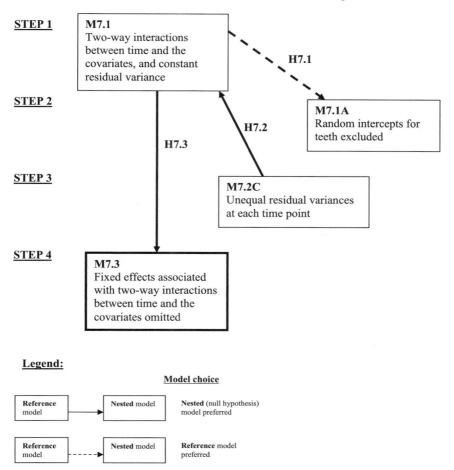

FIGURE 7.3: Guide to model selection and related hypotheses for the analysis of the Dental Veneer data.

The parameters β_0 through β_7 represent the fixed effects associated with the intercept, TIME, the patient-level and tooth-level covariates, and their two-way interactions; u_{0j} and u_{1j} are random patient effects associated with the intercept and time slope, respectively; $u_{0i|j}$ is the random effect associated with a tooth nested within a patient; and ε_{tij} represents a residual. We write the joint distribution of the two random effects associated with patient j as:

$$u_j = \begin{pmatrix} u_{0j} \\ u_{1j} \end{pmatrix} \sim \mathcal{N}(\mathbf{0}, \mathbf{D}^{(2)})$$

where the variance-covariance matrix $\mathbf{D}^{(2)}$ is defined as:

$$\mathbf{D}^{(2)} = \begin{pmatrix} Var(u_{0j}) & cov(u_{0j}, u_{1j}) \\ cov(u_{0j}, u_{1j}) & Var(u_{1j}) \end{pmatrix}$$

For all models considered in this analysis, we specify an unstructured $\mathbf{D}^{(2)}$ matrix defined by three covariance parameters, $\sigma^2_{int:patient}$, $\sigma_{int,time:patient}$, and $\sigma^2_{time:\,patient}$, as shown in the following matrix:

$$\boldsymbol{D}^{(2)} = \left(\begin{array}{cc} \sigma^2_{int:patient} & \sigma_{int,time:\,patient} \\ \sigma_{int,time:\,patient} & \sigma^2_{time:\,patient} \end{array} \right)$$

The distribution of the random effects associated with tooth i nested within patient j is

$$u_{0i|j} \sim \mathcal{N}(0, \boldsymbol{D}^{(1)}),$$

where the 1×1 matrix $\boldsymbol{D}^{(1)}$ contains the variance, $\sigma^2_{int:tooth(patient)}$, of the nested random tooth effects $u_{0i|j}$.

$$\boldsymbol{D}^{(1)} = \mathrm{Var}(u_{0i|j}) = \sigma^2_{int:\,tooth(patient)}$$

The distribution of the residuals, $\boldsymbol{\varepsilon}_{ij}$, associated with observations on the same tooth is

$$\boldsymbol{\varepsilon}_{ij} = \left(\begin{array}{c} \varepsilon_{1ij} \\ \varepsilon_{2ij} \end{array} \right) \sim \mathcal{N}(0, \boldsymbol{R}_{ij})$$

where the variance-covariance matrix $\boldsymbol{R}_{ij}$ for the residuals is defined as

$$\boldsymbol{R}_{ij} = \left(\begin{array}{cc} \mathrm{Var}(\varepsilon_{1ij}) & cov(\varepsilon_{1ij}, \varepsilon_{2ij}) \\ cov(\varepsilon_{1ij}, \varepsilon_{2ij}) & \mathrm{Var}(\varepsilon_{2ij}) \end{array} \right)$$

Different structures for $\boldsymbol{R}_{ij}$ will be specified for different models in the analysis.

We assume that the random effects associated with patients, u_{0j} and u_{1j}, are independent of the random effects associated with teeth nested within patients, $u_{0i|j}$, and that all random effects are independent of the residuals.

In Model 7.1, the $\boldsymbol{R}_{ij}$ matrix is equal to $\sigma^2 \boldsymbol{I}_2$ and involves only one parameter, σ^2, as shown below:

$$\boldsymbol{R}_{ij} = \left(\begin{array}{cc} \sigma^2 & 0 \\ 0 & \sigma^2 \end{array} \right) = \sigma^2 \boldsymbol{I_2}$$

In this structure, the residual variance at 3 months is equal to that at 6 months, and there is no covariance between the residuals at 3 and at 6 months (they are independent).

In Model 7.2A, we have an unstructured (UN) $\boldsymbol{R}_{ij}$ matrix

$$\boldsymbol{R}_{ij} = \left(\begin{array}{cc} \sigma^2_{t1} & \sigma_{t1,t2} \\ \sigma_{t1,t2} & \sigma^2_{t2} \end{array} \right)$$

defined by three covariance parameters: $\sigma^2_{t1}, \sigma^2_{t2}$, and $\sigma_{t1,t2}$. In this structure, the residual variance is allowed to be different at 3 and at 6 months, and we allow the covariance between the residuals at 3 and at 6 months to be different from zero (i.e., the residuals are not independent).

In Model 7.2B, we have a compound symmetric (CS) $\boldsymbol{R}_{ij}$ matrix involving two covariance parameters, σ^2 and $\sigma_{tl,t2}$:

$$\boldsymbol{R}_{ij} = \left(\begin{array}{cc} \sigma^2 + \sigma_{t1,t2} & \sigma_{t1,t2} \\ \sigma_{t1,t2} & \sigma^2 + \sigma_{t1,t2} \end{array} \right)$$

In this structure, the residual variance is equal at 3 and at 6 months, and there is a covariance between the residuals at the two time points.

In Model 7.2C, we have a heterogeneous (HET) $\boldsymbol{R}_{ij}$ matrix with two parameters, σ^2_{t1} and σ^2_{t2}:

$$\boldsymbol{R}_{ij} = \left(\begin{array}{cc} \sigma^2_{t1} & 0 \\ 0 & \sigma^2_{t2} \end{array} \right)$$

In this structure, the residual variances are different at 3 and at 6 months, and the covariance is assumed to be zero.

Finally, Model 7.3 has the same random effects and residual covariance structure as in Model 7.1, but omits the fixed effects associated with the two-way interactions.

7.3.2.2 Hierarchical Model Specification

We now consider an equivalent specification of the model defined in (7.1), corresponding to the hierarchical specification in the HLM software. The correspondence between the notation used in HLM and our notation is defined in Table 7.3.

The hierarchical model has three components, reflecting contributions from the three levels of the Dental Veneer data (time, tooth, and patient). First, we write the **Level 1** component as:

Level 1 Model (Time)

$$\text{GCF}_{tij} = b_{0i|j} + b_{1i|j} \times \text{TIME}_t + \varepsilon_{tij} \tag{7.2}$$

where

$$\begin{pmatrix} \varepsilon_{1ij} \\ \varepsilon_{2ij} \end{pmatrix} \sim \mathcal{N}(0, \boldsymbol{R}_{ij})$$

The model in (7.2) implies that at the most basic level of the data set (the GCF measures at each time point), we have a set of simple linear regressions of GCF on TIME. The unobserved regression coefficients, i.e., the tooth-specific intercepts ($b_{0i|j}$) and TIME slopes ($b_{1i|j}$), depend on other fixed and random effects, as shown in the **Level 2** model below:

Level 2 Model (Tooth)

$$\begin{aligned} b_{0i|j} &= b_{0j} + \beta_2 \times \text{BASE\_GCF}_{ij} + \beta_3 \times \text{CDA}_{ij} + u_{0i|j} \\ b_{1i|j} &= b_{1j} + \beta_5 \times \text{BASE\_GCF}_{ij} + \beta_6 \times \text{CDA}_{ij} \end{aligned} \tag{7.3}$$

where

$$u_{0i|j} \sim \mathcal{N}(0, \sigma^2_{int:tooth(patient)})$$

The **Level 2** model in (7.3) implies that the intercept, $b_{0i|j}$, for tooth i nested within patient j depends on the intercept specific to the j-th patient, b_{0j}, the tooth-specific covariates (BASE_GCF and CDA), and a random effect associated with the tooth, $u_{0i|j}$. The tooth-specific slope for time, $b_{1i|j}$, depends on the patient-specific time effect, b_{1j}, and the tooth-specific covariates. Note that we do not include a random effect for the tooth-specific slope for TIME.

The **Level 3** model for the patient-specific contributions to the intercept and slope is shown below:

Level 3 Model (Patient)

$$\begin{aligned} b_{0j} &= \beta_0 + \beta_4 \times \text{AGE}_j + u_{0j} \\ b_{1j} &= \beta_1 + \beta_7 \times \text{AGE}_j + u_{1j} \end{aligned} \tag{7.4}$$

where

$$u_j = \begin{pmatrix} u_{0j} \\ u_{1j} \end{pmatrix} \sim \mathcal{N}(\boldsymbol{0}, \boldsymbol{D}^{(2)})$$

TABLE 7.3: Summary of Models Considered for the Dental Veneer Data

	Term/Variable	General Notation	HLM Notation	Model 7.1	Model 7.2A[a]	Model 7.2B[a]	Model 7.2C	Model 7.3
Fixed effects	Intercept	β_0	γ_{000}	✓	✓	✓	✓	✓
	TIME	β_1	γ_{100}	✓	✓	✓	✓	✓
	BASE_GCF	β_2	γ_{010}	✓	✓	✓	✓	✓
	CDA	β_3	γ_{020}	✓	✓	✓	✓	✓
	AGE	β_4	γ_{001}	✓	✓	✓	✓	✓
	TIME × BASE_GCF	β_5	γ_{110}	✓	✓	✓	✓	
	TIME × CDA	β_6	γ_{120}	✓	✓	✓	✓	
	TIME × AGE	β_7	γ_{101}	✓	✓	✓	✓	
Random effects	Patient (j): Intercept	u_{0j}	u_{00k}	✓	✓	✓	✓	✓
	Tooth (i) within Patient (j): TIME	u_{1j}	u_{10k}	✓	✓		✓	✓
	Tooth (i) within Patient (j): Intercept	u_{0ij}	r_{0jk}	✓	✓	✓	✓	✓
Residuals	Visit (t)	ε_{tij}	ε_{ijk}	✓	✓	✓	✓	✓
Covariance Parameters (θ_D) for D Matrix	Patient level: Variance of intercepts	$\sigma^2_{int:pat}$	$\tau_\beta[1,1]$	✓	✓	✓	✓	✓
	Variance of slopes	$\sigma^2_{time:pat}$	$\tau_\beta[2,2]$	✓	✓		✓	✓
	Covariance of intercepts, slopes	$\sigma_{int,time:pat}$	$\tau_\beta[2,1]$	✓	✓		✓	✓

TABLE 7.3: (Continued)

	Term/Variable	General Notation	HLM Notation	Model				
				7.1	7.2A[a]	7.2B[a]	7.2C	7.3
Tooth Level	**Structure**[b]	$D^{(2)}$	τ_β	UN	UN	UN	UN	UN
	Variance of intercepts	$\sigma^2_{int:tooth(pat)}$	τ_π	✓	✓	✓	✓	✓
Covariance Parameters $(\boldsymbol{\theta}_R)$ for R_{ij} Matrix — **Time Level**	**Variances at Time 1, Time 2**	σ^2_{t1} σ^2_{t2}	σ^2_1 σ^2_2	Equal	Unequal	Equal	Unequal	Equal
	Covariance of Time 1, Time 2	$\sigma_{t1,t2}$	Varies[c]	0	✓	✓	0	0
	Structure[b]	R_{ij}	S	$\sigma^2 I_2$	UN	CS	HET	$\sigma^2 I_2$

[a] In Models 7.2A and 7.2B, the residual covariance parameters are aliased with the variance of the random tooth-level intercepts.

[b] UN = unstructured, CS = compound symmetry, HET = diagonal with heterogeneous variances.

[c] The notation for this covariance parameter varies in HLM, depending on the structure specified.

The **Level 3** model in (7.4) implies that the unobserved patient-specific intercept, b_{0j}, depends on the overall fixed intercept, β_0, the single covariate measured at the patient level (AGE), and a random effect, u_{0j}, associated with patient j. The Level 3 model also implies that the patient-specific TIME effect, b_{1j}, depends on the overall fixed TIME effect, β_1, the patient's age, and the random effect of TIME, u_{1j}, associated with patient j.

By substituting the values for b_{0j} and b_{1j} from the Level 3 model back into the Level 2 model, and then substituting the values for $b_{0i|j}$ and $b_{1i|j}$ from the Level 2 model into the Level 1 model, we obtain the general LMM specified in (7.1).

7.3.3 Hypothesis Tests

Hypothesis tests considered in the analysis of the Dental Veneer data are summarized in Table 7.4.

We consider three-level models for the Dental Veneer data, with random effects at Level 3 (the patient level) and at Level 2 (the tooth level). To preserve the hierarchy of the data, we first test the significance of the random effects beginning at Level 2 (the tooth level), while retaining those at Level 3 (the patient level).

Hypothesis 7.1. The nested random effects, $u_{0i|j}$, associated with teeth within the same patient can be omitted from Model 7.1.

We decide whether we can remove the random effects associated with teeth indirectly by testing null and alternative hypotheses about the variance of these random effects, as follows:

$$H_0\colon \sigma^2_{int:tooth(patient)} = 0$$
$$H_A\colon \sigma^2_{int:tooth(patient)} > 0$$

We use a REML-based likelihood ratio test for Hypothesis 7.1. The test statistic is calculated by subtracting the –2 REML log-likelihood for Model 7.1 (the reference model) from that of Model 7.1A (the nested model, excluding the nested random tooth effects). To obtain a p-value for this test statistic, we refer it to a mixture of χ^2 distributions, with 0 and 1 degrees of freedom, and equal weights of 0.5.

Based on the significant result of this test, we retain the nested random tooth effects in the model. Having made this choice, we do not test for the need of the random patient effects, u_{0j} and u_{1j}, to preserve the hierarchical structure of the data.

Hypothesis 7.2. The variance of the residuals is constant (homogeneous) across the time points in Model 7.2C.

We test Hypothesis 7.2 to decide whether we should include heterogeneous residual variances at the two time points (Model 7.2C) vs. having homogeneous residual variances. We write the null and alternative hypotheses as follows:

$$H_0\colon \sigma^2_{t1} = \sigma^2_{t2}$$
$$H_A\colon \sigma^2_{t1} \neq \sigma^2_{t2}$$

The test statistic is calculated by subtracting the –2 REML log-likelihood for Model 7.2C (the reference model, with heterogeneous residual variances) from that of Model 7.1 (the nested model, with homogeneous residual variance). The distribution of the test statistic under the null hypothesis is a χ^2 with 1 degree of freedom.

Hypothesis 7.3. The fixed effects associated with the two-way interactions between TIME and the patient- and tooth-level covariates can be omitted from Model 7.1.

TABLE 7.4: Summary of Hypotheses Tested for the Dental Veneer Data

| | Hypothesis Specification | | | Hypothesis Test | | | |
| | | | | Models Compared | | | |
Label	Null (H_0)	Alternative (H_A)	Test	Nested Model (H_0)	Ref. Model (H_A)	Est. Method	Test Stat. Dist. under H_0
7.1	Drop $u_{0i|j}$, random tooth-specific intercepts ($\sigma^2_{int:tooth(pat)} = 0$)	Retain $u_{0i|j}$ ($\sigma^2_{int:tooth(pat)} > 0$)	LRT	Model 7.1A	Model 7.1	REML	$0.5\chi^2_0 + 0.5\chi^2_1$
7.2	Constant residual variance: $\sigma^2_{t1} = \sigma^2_{t2}$	$\sigma^2_{t1} \neq \sigma^2_{t2}$	LRT	Model 7.1	Model 7.2C	REML	χ^2_1
7.3	Drop fixed effects associated with all two-way interactions ($\beta_5 = \beta_6 = \beta_7 = 0$)	$\beta_5 \neq 0$, or $\beta_6 \neq 0$, or $\beta_7 \neq 0$	LRT	Model 7.3	Model 7.1	ML	χ^2_3

In Hypothesis 7.3, we test whether the fixed effects associated with the two-way interactions between TIME and the covariates BASE_CGF, CDA, and AGE are all equal to zero. The null and alternative hypotheses are specified as follows:

$$H_0\colon \beta_5 = \beta_6 = \beta_7 = 0$$
$$H_A\colon \beta_5 \neq 0, \text{ or } \beta_6 \neq 0, \text{ or } \beta_7 \neq 0$$

We test Hypothesis 7.3 using a likelihood ratio test, based on ML estimation. The test statistic is calculated by subtracting the –2 ML log-likelihood for Model 7.1 (the reference model) from that of Model 7.3 (the nested model, excluding the fixed effects associated with the two-way interactions). The distribution of the test statistic under the null hypothesis is a χ^2 with 3 degrees of freedom, corresponding to the 3 fixed effects set to zero under the null hypothesis.

For the results of these hypothesis tests, see Section 7.5.

7.4   Analysis Steps in the Software Procedures

In this section, we demonstrate the analysis steps described earlier in the chapter using the procedures in SAS, SPSS, R, Stata, and HLM. We compare results for selected models across the procedures in Section 7.6.

7.4.1   SAS

Step 1: Fit a model with a "loaded" mean structure (Model 7.1).

To fit Model 7.1 using SAS `proc mixed`, we use a temporary SAS data set named `veneer`, assumed to have the data structure shown in Table 7.2. Note that we sort the data by levels of PATIENT, TOOTH, and TIME before fitting the model, to make portions of the output easier to read (this sorting is optional and does not affect the analysis):

```
proc sort data = veneer;
by patient tooth time;
run;


title "Model 7.1";
proc mixed data = veneer noclprint covtest;
class patient tooth;
model gcf = time base_gcf cda age time*base_gcf
time*cda time*age / solution outpred = resids;
random intercept time / subject = patient type = un solution
v = 1 vcorr = 1;
random intercept / subject = tooth(patient) solution;
run;
```

We include the `noclprint` option in the `proc mixed` statement to save space by preventing SAS from displaying all levels of the class variables in the output. The `covtest` option is only used to display the estimated standard errors of the covariance parameters in the output for comparison with the estimates from the other software procedures. Because no estimation method is specified, the default REML method will be used.

The `class` statement identifies the categorical variables used in the `random` statements. PATIENT denotes clusters of units, and TOOTH denotes the units of analysis.

The `model` statement specifies the terms that have associated fixed effects in the model. We include TIME, BASE_GCF, CDA, AGE, and the two-way interactions between TIME and the other covariates. The `solution` option is specified so that SAS displays the estimated values of the fixed-effect parameters in the output. The `outpred =` option is used to obtain conditional residuals (conditional on the values of all of the appropriate random effects) for each observation in a new data set called `resids`.

The first `random` statement (`random intercept time / subject = patient`) identifies the intercept and the slope for TIME as two random effects associated with each patient. The `solution` option requests that the EBLUPs for the random effects be displayed in the output. The `type = un` option specifies that the individual 2×2 blocks of the G covariance matrix of random effects for each subject (which we denote as the $D^{(2)}$ matrix) are unstructured. We also request that the variance-covariance matrix and the corresponding correlation matrix for the marginal model implied by Model 7.1 be displayed for the first subject by using the `v = 1` and `vcorr = 1` options.

The second `random` statement (`random intercept / subject = tooth(patient)`) specifies that we wish to include a random effect associated with the intercept for each tooth nested within a patient. We also use the `solution` option to request that the EBLUPs for these random effects be displayed in the output.

Because we do not include a `repeated` statement in the syntax for Model 7.1, SAS assumes that the residuals associated with all observations, ε_{tij}, are independent and identically distributed, with constant variance σ^2.

Step 2: Select a structure for the random effects (Model 7.1 vs. Model 7.1A).

To test Hypothesis 7.1, we fit Model 7.1A, which omits the random effects associated with the intercept for each tooth nested within a patient. To do this, we remove the second `random` statement from the syntax for Model 7.1. We then subtract the –2 REML log-likelihood for Model 7.1 (the reference model) from the –2 REML log-likelihood for Model 7.1A (the nested model, excluding the nested random tooth effects). The p-value for the resulting test statistic ($858.3 - 847.1 = 11.2$) can be obtained in the SAS log by using the following syntax:

```
data _null_;
p_value = 0.5*(1-probchi(11.2,1));
put p_value;
run;
```

We use the `probchi()` function to obtain the appropriate p-value for the χ_1^2 distribution and weight it by 0.5. Note that the χ_0^2 distribution is not included in the syntax, because it contributes zero to the resulting p-value.

We keep Model 7.1 as our preferred model at this stage of the analysis based on the result of this test.

Step 3: Select a covariance structure for the residuals (Model 7.1 and Models 7.2A through 7.2C).

Next, we investigate different covariance structures for the residuals associated with observations on the same tooth, by adding a `repeated` statement to the SAS syntax. We first create a new variable in the `veneer` data set named CATTIME, which has the same values as the original TIME variable and will be used to index the repeated measures:

```
data veneer;
set veneer;
cattime = time;
run;
```

We fit Model 7.2A, which has an **unstructured** covariance matrix for the residuals associated with observations on the same tooth:

```
title "Model 7.2A";
proc mixed data = veneer noclprint covtest;
class patient tooth cattime;
model gcf = time base_gcf cda age time*base_gcf
time*cda time*age / solution outpred = resids;
random intercept time / subject = patient type = un solution
v = 1 vcorr = 1;
random intercept / subject = tooth(patient) solution;
repeated cattime / subject = tooth(patient) type = un r rcorr;
run;
```

Note that the `proc mixed` code used to fit this model is similar to the code used for Model 7.1; the only differences are the presence of the `repeated` statement and the inclusion of CATTIME in the `class` statement.

The `repeated` statement specifies CATTIME as an index for the time points, allowing SAS to identify the repeated measures on a given tooth correctly and to identify appropriately the row and column elements in the R_{ij} matrix. The `subject =` option identifies the units of analysis on which repeated observations were measured and, in this case, the subject is each tooth nested within a patient: `subject = tooth(patient)`. We specify the covariance structure (or type) of the R_{ij} matrix as unstructured by using the `type = un` option. We also request that the estimated 2×2 R_{ij} matrix for the first tooth in the data set be displayed in the output along with the corresponding correlation matrix by specifying the `r` and `rcorr` options.

Software Note: `proc mixed` requires that the variable used to define the index for the time points in the `repeated` statement be included in the `class` statement, so that `proc mixed` considers the variable as categorical. As long as there are no missing values on any of the repeated measures for any subjects and the data have been sorted (e.g., by PATIENT, TOOTH, and TIME), the use of such an index variable is not necessary. Index variables are also not necessary if a covariance structure that does not require ordering of the repeated measures, such as compound symmetry, is specified for the R_{ij} matrix.

The following note is produced in the SAS log after fitting Model 7.2A using SAS Release 9.3:

```
NOTE: Convergence criteria met but final Hessian is not
positive-definite.
```

The Hessian matrix is used to compute the standard errors of the estimated covariance parameters. Because we specified the `covtest` option in the syntax, we can inspect the estimated standard errors of the covariance parameters. As shown in the following SAS output, the estimated standard error of covariance parameter UN(2,2) for TOOTH(PATIENT) is zero. This is the estimated standard error of the estimated residual variance at time 2.

```
            Covariance Parameter Estimates

                                    Standard      Z
Cov Parm     Subject      Estimate     Error    Value    Pr Z

UN(1,1)      PATIENT        546.61    279.34     1.96    0.0252
UN(2,1)      PATIENT       -148.64    74.3843   -2.00    0.0457
UN(2,2)      PATIENT        44.6420   21.0988    2.12    0.0172
Intercept    TOOTH(PATIENT)  7.7452   18.4261    0.42    0.3371
UN(1,1)      TOOTH(PATIENT) 101.55    26.4551    3.84    <.0001
UN(2,1)      TOOTH(PATIENT)  39.1711  15.2980    2.56    0.0105
UN(2,2)      TOOTH(PATIENT)  76.1128        0      .        .
```

The nonpositive-definite Hessian matrix encountered in fitting Model 7.2A is a consequence of the aliasing of the residual covariance parameters and the variance of the random tooth effects.

We fit Model 7.2B by modifying the `repeated` statement to specify that residuals associated with observations on the same tooth have a simpler **compound symmetry** covariance structure:

```
repeated cattime / subject = tooth(patient) type = cs rcorr r;
```

The only difference in the SAS syntax here is the specification of `type = cs` in the `repeated` statement, which requests a compound symmetry structure for the R_{ij} matrix. When attempting to fit Model 7.2B in SAS Version 9.3, we see the following warning in the SAS log:

```
NOTE: Stopped because of too many likelihood evaluations.
```

The REML estimation procedure does not converge to a valid solution, and this is another consequence of the aliasing of the residual covariance and the variance of the random tooth effects. We do not see any estimates printed in the SAS output as a result.

We now fit Model 7.2C, which has a diagonal R_{ij} matrix that allows the variance of the residuals at 3 and at 6 months to differ, and has zero covariance between the two time points. We use the following `repeated` statement for this model:

```
repeated cattime / subject = tooth(patient) type = un(1) rcorr r;
```

The option `type = un(1)` specifies that the residuals in the R_{ij} matrix are uncorrelated and that the residual variances differ for observations at different time points. This model now has no accompanying error messages, and the covariance parameter estimates and their standard errors seem appropriate. There is no apparent problem with aliasing of covariance parameters in this model.

```
             Covariance Parameter Estimates

                                    Standard        Z
   Cov Parm    Subject     Estimate    Error    Value    Pr Z

   UN(1,1)     PATIENT       546.60   279.33     1.96   0.0252
   UN(2,1)     PATIENT      -148.64   74.3839   -2.00   0.0457
   UN(2,2)     PATIENT      44.6417   21.0987    2.12   0.0172
   Intercept   TOOTH(PATIENT) 46.9155 16.5274    2.84   0.0023
   UN(1,1)     TOOTH(PATIENT) 62.3774 18.8107    3.32   0.0005
   UN(2,1)     TOOTH(PATIENT)       0       .       .      .
   UN(2,2)     TOOTH(PATIENT) 36.9482 15.3006    2.41   0.0079
```

Note in the output above that the UN(2,1) parameter equals zero, because it is constrained to be zero by the specification of the `type = un(1)` structure for the R_{ij} matrix. Model 7.2C is the only possible model other than Model 7.1 to consider at this stage of the analysis, because the other models have problems with aliasing of the covariance parameters.

We use a likelihood ratio test for Hypothesis 7.2 to decide if we wish to keep the heterogeneous residual variances, as in Model 7.2C, or if we should have constant residual variance as in Model 7.1. We calculate the test statistic by subtracting the –2 REML log-likelihood for Model 7.2C (the reference model with heterogeneous residual variances) from that of Model 7.1 (the nested model). The test statistic is equal to 0.9. The p-value for this test is calculated by referring the test statistic to a χ^2 distribution with 1 degree of freedom. We do not use a mixture of χ^2 distributions for this test, because the null hypothesis value of the test statistic is not at the boundary of the parameter space. SAS code for computing the p-value for this test and displaying it in the log is as follows:

```
data _null_;
p_value = 1 - probchi(0.9, 1);
put p_value;
run;
```

Because the test is not significant ($p = 0.34$), we keep Model 7.1, with homogeneous residual variance, as our preferred model at this stage of the analysis.

Step 4: Reduce the model by removing nonsignificant fixed effects (Model 7.1 vs. Model 7.3).

We test Hypothesis 7.3 to decide whether we want to keep the fixed effects of the two-way interactions between TIME and the other covariates in Model 7.1. We first refit Model 7.1, using ML estimation, by including the `method = ml` option in the `prox mixed` statement, as shown in the following code:

```
title "Model 7.1: ML Estimation";
proc mixed data = veneer noclprint covtest method = ml;
class patient tooth;
model gcf = time base_gcf cda age time*base_gcf
time*cda time*age / solution outpred = resids;
random intercept time / subject = patient type = un solution
v = 1 vcorr = 1;
random intercept / subject = tooth(patient) solution;
run;
```

Next, we fit Model 7.3, also using ML estimation, by removing all two-way interactions from the `model` statement and refitting the model:

```
model gcf = time base_gcf cda age / solution outpred = resids;
```

We calculate a likelihood ratio test statistic by subtracting the –2 ML log-likelihood for Model 7.1 (the reference model) from that for Model 7.3 (the nested model, without any two-way interactions). The test statistic has a value of 1.84, and the null distribution of this test statistic is a χ^2 with 3 degrees of freedom, corresponding to the 3 fixed-effect parameters that we omitted from Model 7.1. The *p*-value for this likelihood ratio test is calculated and displayed in the log using the following SAS syntax:

```
data _null_;
p_value = 1 - probchi(1.84, 3);
put p_value;
run;
```

Because this test statistic is not significant, we conclude that we can omit the fixed effects associated with the two-way interactions from the model, and select Model 7.3 as our final model.

We now refit Model 7.3 using the default REML estimation method, to obtain unbiased estimates of the covariance parameters:

```
title "Model 7.3: REML Estimation";
proc mixed data = veneer noclprint covtest;
class patient tooth;
model gcf = time base_gcf cda age / solution outpred = resids;
random intercept time / solution subject = patient type = un
v = 1 vcorr = 1;
random intercept / solution subject = tooth(patient);
run;
```

7.4.2 SPSS

Step 1: Fit a model with a "loaded" mean structure (Model 7.1).

We assume that an SPSS data set having the format displayed in Table 7.2 is currently open, and begin by specifying the SPSS syntax to set up Model 7.1:

```
* Model 7.1 .
MIXED
gcf WITH time base_gcf cda age
/CRITERIA = CIN(95) MXITER(100) MXSTEP(5) SCORING(1)
SINGULAR(0.000000000001) HCONVERGE(0, ABSOLUTE) LCONVERGE(0,
ABSOLUTE) PCONVERGE(0.000001, ABSOLUTE)
/FIXED = time base_gcf cda age time*base_gcf time*cda time*age | SSTYPE(3)
/METHOD = REML
/PRINT = G SOLUTION TESTCOV
/SAVE = PRED RESID
/RANDOM INTERCEPT time | SUBJECT(Patient) COVTYPE(UN)
/RANDOM INTERCEPT | SUBJECT(tooth*Patient) .
```

This syntax is very similar to that used to fit the three-level models discussed in Chapter 4. The first RANDOM subcommand sets up the two random effects for each patient: INTERCEPT and TIME. The subject is specified as PATIENT, and the covariance structure for the patient-specific random effects is unstructured, as indicated by the COVTYPE(UN) option.

The second RANDOM subcommand sets up the random effects associated with the INTERCEPT for each tooth. The subject identified in this subcommand is (TOOTH*PATIENT), which actually indicates tooth nested within patient, although the syntax appears to be specifying TOOTH crossed with PATIENT. We would run this syntax to fit Model 7.1.

Step 2: Select a structure for the random effects (Model 7.1 vs. Model 7.1A).

To test Hypothesis 7.1, we fit a nested model, Model 7.1A, which excludes the random effects associated with the teeth nested within patients, by omitting the second RANDOM subcommand entirely from the syntax used for Model 7.1:

```
* Model 7.1A .
MIXED
gcf WITH time base_gcf cda age
/CRITERIA = CIN(95) MXITER(100) MXSTEP(5) SCORING(1)
SINGULAR(0.000000000001) HCONVERGE(0, ABSOLUTE)
LCONVERGE(0,ABSOLUTE) PCONVERGE(0.000001, ABSOLUTE)
/FIXED = time base_gcf cda age time*base_gcf time*cda time*age | SSTYPE(3)
/METHOD = REML
/PRINT = G SOLUTION TESTCOV
/SAVE = PRED RESID
/RANDOM INTERCEPT time | SUBJECT(Patient) COVTYPE(UN) .
```

The likelihood ratio test statistic is calculated by subtracting the –2 REML log-likelihood for Model 7.1 (reported in the SPSS output) from that of Model 7.1A. The p-value is calculated by referring the test statistic to a mixture of χ^2 distributions with degrees of freedom of 0 and 1, and equal weights of 0.5 (see Section 7.5).

Step 3: Select a covariance structure for the residuals (Model 7.1, and Models 7.2A through 7.2C).

Once the structure for the random effects has been set up, we consider different covariance structures for the residuals. But before fitting these models, we sort the data set in ascending order by PATIENT, TOOTH, and TIME to ensure that all desired covariance structures will be displayed in a way that is easy to read in the output (this sorting is recommended but not essential for the analysis):

```
SORT CASES BY
patient (A) tooth (A) time (A) .
```

We first fit Model 7.2A with an **unstructured** covariance structure for the residuals associated with observations over time on the same tooth:

```
* Model 7.2A .
MIXED
gcf WITH time base_gcf cda age
/CRITERIA = CIN(95) MXITER(100) MXSTEP(5) SCORING(1)
```

```
SINGULAR(0.000000000001) HCONVERGE(0, ABSOLUTE)
LCONVERGE(0, ABSOLUTE) PCONVERGE(0.000001, ABSOLUTE)
/FIXED = time base_gcf cda age time*base_gcf time*cda time*age | SSTYPE(3)
/METHOD = REML
/PRINT = G R SOLUTION TESTCOV
/SAVE = PRED RESID
/RANDOM INTERCEPT time | SUBJECT(Patient) COVTYPE(UN)
/RANDOM INTERCEPT | SUBJECT(tooth*Patient)
/REPEATED time | SUBJECT(tooth*Patient) COVTYPE(UN) .
```

This syntax is identical to that used to fit Model 7.1, with the exception of the `PRINT` and `REPEATED` subcommands. In the `PRINT` subcommand, we have requested that the estimated 2×2 $\boldsymbol{R}_{ij}$ variance-covariance matrix be displayed in the SPSS output by using the `R` keyword (the corresponding correlation matrix is not available).

The `REPEATED` subcommand is used to identify the units of analysis (or subjects) that have measurements made on them over time, the time index (TIME), and the structure of the covariance matrix for the residuals. In this case, we again specify the subject as (TOOTH*PATIENT), which denotes teeth nested within patients, and identify the covariance structure as `UN` (i.e., unstructured).

After fitting Model 7.2A, the following warning message is displayed in the SPSS output:

```
                              Warnings

Iteration was terminated but convergence has not been achieved. The MIXED
procedure continues despite this warning. Subsequent results produced are
based on the last iteration. Validity of the model fit is uncertain.
```

The `MIXED` command fails to converge to a solution because the residual covariance parameters are aliased with the variance of the random tooth effects in Model 7.2A. Results from the model fit should not be interpreted if this warning message appears.

Next, we use SPSS syntax to fit Model 7.2B, which has a compound symmetry residual covariance structure. The `REPEATED` subcommand in the syntax used to fit Model 7.2A is modified to include the `COVTYPE(CS)` option:

```
/REPEATED time | SUBJECT(tooth*patient) COVTYPE(CS) .
```

Model 7.2B also has a problem with aliasing of the covariance parameters, and the same warning message about the model failing to converge is displayed in the output. The results of this model fit should also not be interpreted.

We now fit Model 7.2C using the `COVTYPE(DIAG)` option, which specifies that the two residuals associated with observations on the same tooth are independent and have different variances:

```
/REPEATED time | SUBJECT(tooth*patient) COVTYPE(DIAG) .
```

In the diagonal covariance structure defined for the $\boldsymbol{R}_{ij}$ matrix in this model, the covariance between observations on the same tooth is set to zero, and the residual variances along the diagonal of the matrix are heterogeneous.

We test Hypothesis 7.2 by carrying out a likelihood ratio test to decide if we wish to have heterogeneous residual variances in the model. The test statistic is calculated by subtracting the -2 REML log-likelihood for Model 7.2C (the reference model with heterogeneous residual variances) from that of Model 7.1 (the nested model). The appropriate p-value for the test statistic is based on a χ^2 distribution with 1 degree of freedom. Because this test is not significant, we keep Model 7.1 (with homogeneous residual variance) as our preferred model at this stage of the analysis.

Step 4: Reduce the model by removing nonsignificant fixed effects (Model 7.1 vs. Model 7.3).

Finally, we test Hypothesis 7.3 to see if we can remove the fixed effects associated with all the two-way interactions from the model. To do this, we first refit Model 7.1 using ML estimation by including the /METHOD = ML subcommand:

```
* Model 7.1 (ML) .
MIXED
gcf WITH time base_gcf cda age
/CRITERIA = CIN(95) MXITER(100) MXSTEP(5) SCORING(1)
SINGULAR(0.000000000001) HCONVERGE(0, ABSOLUTE)
LCONVERGE(0, ABSOLUTE) PCONVERGE(0.000001, ABSOLUTE)
/FIXED = time base_gcf cda age time*base_gcf time*cda time*age | SSTYPE(3)
/METHOD = ML
/PRINT = G SOLUTION TESTCOV
/SAVE = PRED RESID
/RANDOM INTERCEPT time | SUBJECT(Patient) COVTYPE(UN)
/RANDOM INTERCEPT | SUBJECT(tooth*Patient) .
```

Next, we fit Model 7.3 by removing all two-way interactions listed in the FIXED subcommand:

```
/FIXED = time base_gcf cda age | SSTYPE(3)
```

We can now carry out a likelihood ratio test for Hypothesis 7.3. To do this, we subtract the –2 ML log-likelihood for Model 7.1 (the reference model) from that of Model 7.3 (the nested model). Because this test is not significant (see Subsection 7.5.3), we choose Model 7.3 as our preferred model at this stage of the analysis.

Finally, we refit Model 7.3 using the default REML estimation method, to obtain unbiased estimates of the covariance parameters:

```
* Model 7.3 .
MIXED
gcf WITH time base_gcf cda age
/CRITERIA = CIN(95) MXITER(100) MXSTEP(5) SCORING(1)
SINGULAR(0.000000000001) HCONVERGE(0, ABSOLUTE)
LCONVERGE(0, ABSOLUTE) PCONVERGE(0.000001, ABSOLUTE)
/FIXED = time base_gcf cda age | SSTYPE(3)
/METHOD = REML
/PRINT = G SOLUTION TESTCOV
/SAVE = PRED RESID
/RANDOM INTERCEPT time | SUBJECT(Patient) COVTYPE(UN)
/RANDOM INTERCEPT | SUBJECT(tooth*Patient) .
```

7.4.3   R

We begin the analysis of the Dental Veneer data using R by reading the tab-delimited raw data file, which has the structure described in Table 7.2 with variable names in the first row, into a data frame object:

```
> veneer <- read.delim("c:\\temp\\veneer.dat", h = T)
```

7.4.3.1 Analysis Using the `lme()` Function

We first load the `nlme` package, so that the `lme()` function can be used in the analysis:

```
> library(nlme)
```

Step 1: Fit a model with a "loaded" mean structure (Model 7.1).

We now fit Model 7.1 using the `lme()` function:

```
> model7.1.fit <- lme(gcf ~ time + base_gcf + cda + age +
time:base_gcf + time:cda + time:age,
random = list(patient = ~time, tooth = ~1),
data = veneer, method = "REML")
```

The syntax used for the `lme()` function is discussed in detail below:

- `model7.1.fit` is the name of the object that contains the results from the fit of Model 7.1.

- The first argument of the function is a formula, which defines the continuous response variable (GCF), and the covariates and interaction terms that have fixed effects in the model.

- The second argument of the function, `random = list(patient = ~time, tooth = ~1)`, indicates the random effects to be included in the model. Note that a "list" has been declared to identify the specific structure of the random effects, and each random factor (i.e., PATIENT and TOOTH) in the list needs to have at least one associated random effect. This syntax implies that levels of the TOOTH variable are nested within levels of the PATIENT variable, because the PATIENT variable is the first argument of the list function. We include a patient-specific TIME slope (listed after the ~ symbol), and a random intercept associated with patients is included by default. Next, we explicitly specify that a random effect associated with the intercept (1) should be included for each level of TOOTH. By default, the `lme()` function chooses an unstructured 2×2 $\boldsymbol{D}^{(2)}$ matrix for the random patient-specific effects.

- The third argument of the function, `data = veneer`, indicates the name of the data frame object to be used.

- The final argument, `method = "REML"`, specifies that REML estimation, which is the default method for the `lme()` function, should be used.

The estimates from the model fit can be obtained by using the `summary()` function:

```
> summary(model7.1.fit)
```

Confidence intervals for the covariance parameters in Model 7.1 can be obtained using the `intervals()` function:

```
> intervals(model7.1.fit)
```

The EBLUPs for each of the random effects in the model associated with the patients and the teeth within patients are obtained by using the `random.effects()` function:

```
> random.effects(model7.1.fit)
```

Unfortunately, the `getVarCov()` function cannot be used to obtain the estimated marginal variance-covariance matrices for given individuals in the data set, because this function has not yet been implemented in the `nlme` package for analyses with multiple levels of nested random effects.

Step 2: Select a structure for the random effects (Model 7.1 vs. Model 7.1A).

We carry out a likelihood ratio test of Hypothesis 7.1 by fitting Model 7.1 and Model 7.1A using REML estimation. Model 7.1 was fitted in Step 1 of the analysis. Model 7.1A is specified by omitting the random tooth effects from Model 7.1:

```
> model7.1A.fit <- lme(gcf ~ time + base_gcf + cda + age +
time:base_gcf + time:cda + time:age,
random = list(patient = ~ time),
data = veneer, method = "REML")
```

We then subtract the –2 REML log-likelihood for Model 7.1 from that of the nested model, Model 7.1A. The p-value for the resulting test statistic can be obtained by referring it to a mixture of χ^2 distributions with degrees of freedom of 0 and 1, and equal weights of 0.5. Users of R can use the `anova()` function to calculate the likelihood ratio test statistic for Hypothesis 7.1, but the reported p-value should be divided by 2, because of the distribution that the statistic follows under the null hypothesis.

```
> anova(model7.1.fit, model7.1A.fit)
```

We have strong evidence in favor of retaining the nested random tooth effects in the model.

Step 3: Select a covariance structure for the residuals (Model 7.1, and Models 7.2A through 7.2C).

Next, we fit models with less restrictive covariance structures for the residuals associated with observations on the same tooth. We first attempt to fit a model with an unstructured residual covariance matrix (Model 7.2A). We have included additional arguments in the `lme()` function, as shown in the following syntax:

```
> model7.2A.fit <- lme(gcf ~ time + base_gcf + cda + age +
time:base_gcf + time:cda + time:age,
random = list(patient = ~ time, tooth = ~1),
corr = corCompSymm(0.5, form = ~1 | patient/tooth),
weights = varIdent(form = ~1 | time),
data = veneer, method = "REML")
```

We specify the **unstructured** residual covariance structure in two parts: the correlation of the residuals and their variances. First, we specify that the correlation structure is compound symmetric, by using the `corr=corCompSymm()` argument. An arbitrary starting value of 0.5 is used to estimate the correlation; the `patient/tooth` argument is used to identify the units of analysis to which the correlation structure applies (i.e., teeth nested within patients). Next, we use the `weights = varIdent(form = ~1 | time)` argument, to identify a covariance structure for the residuals that allows the residuals at the two time points to have different variances.

We obtain results from this model fit by using the `summary()` function:

```
> summary(model7.2A.fit)
```

R does not produce a warning message when fitting this model. We attempt to use the `intervals()` function to obtain 95% confidence intervals for the covariance parameters:

```
> intervals(model7.2A.fit)
```

Unlike the other software procedures, the `lme()` function in R computes approximate confidence intervals for each of the parameters being estimated (including the covariance parameters). However, we see that the estimated standard deviation of the residuals at Time 3 is $8.39 \times 1.000 = 8.39$, with a 95% confidence interval of $(0.0008, 90591.87)$, while the estimated standard deviation of the residuals at Time 6 is $8.39 \times 0.799 = 6.70$, with a 95% confidence interval of $(0.004, 143.73)$ (see Section 3.4.3 for more details about how these estimates are reported by the `lme()` function). Furthermore, the 95% confidence interval for the correlation of the two random effects at the patient level is $(-1, 1)$. These results suggest that the estimates are extremely unstable, despite the apparently valid solution. We therefore consider simpler versions of this model.

Next, we specify Model 7.2B with a more parsimonious **compound symmetry** covariance structure for the residuals associated with observations on the same tooth:

```
> model7.2B.fit <- lme(gcf ~ time + base_gcf + cda + age +
time:base_gcf + time:cda + time:age,
random = list(patient = ~ time, tooth = ~1),
corr = corCompSymm(0.5, form = ~1 | patient/tooth),
data = veneer, method = "REML")
```

This syntax is changed from that used to fit Model 7.2A by omitting the `weights()` argument, so that the residual variances at the two time points are constrained to be equal. We obtain results for this model fit by using the `summary()` function and attempt to obtain confidence intervals for the covariance parameters by using the `intervals()` function:

```
> summary(model7.2B.fit)
> intervals(model7.2B.fit)
```

We once again see extremely wide intervals suggesting that the estimates of the covariance parameters are very unstable, despite the fact that the solution appears to be valid. This is due to the fact that the covariance of the residuals associated with the same tooth is aliased with the variance of the random tooth effects.

We now fit Model 7.2C, which allows the residuals to have different variances at the two time points but assumes that the residuals at the two time points are *uncorrelated* (the **heterogeneous**, or diagonal, structure):

```
> model7.2C.fit <- lme(gcf ~ time + base_gcf + cda + age +
time:base_gcf + time:cda + time:age,
random = list(patient = ~ time, tooth = ~1),
weights = varIdent(form = ~1 | time),
data = veneer, method = "REML")
```

In this syntax, we omit the `corr = corCompSymm()` option and include the `weights` option, which specifies that the residual variances at each time point differ (`form = ~1 | time`). We obtain results of the model fit by using the `summary()` function, and we no longer see extremely wide confidence intervals when using the `intervals()` function to obtain approximate 95% confidence intervals for the covariance parameters:

```
> summary(model7.2C.fit)
> intervals(model7.2C.fit)
```

We test Hypothesis 7.2 by subtracting the –2 REML log-likelihood for Model 7.2C (the reference model, with heterogeneous residual variances) from that of Model 7.1 (the nested model). This likelihood ratio test can be easily implemented using the `anova()` function within R:

```
> anova(model7.2C.fit, model7.1.fit)
```

Because the test is not significant ($p = 0.33$), we keep Model 7.1 as our preferred model at this stage of the analysis.

Step 4: Reduce the model by removing nonsignificant fixed effects (Model 7.1 vs. Model 7.3).

We test whether we can omit the fixed effects associated with the two-way interactions between TIME and the other covariates in the model (Hypothesis 7.3) using a likelihood ratio test. First, we refit Model 7.1 using maximum likelihood estimation by including the `method = "ML"` option in the following syntax:

```
> model7.1.ml.fit <- lme(gcf ~ time + base_gcf + cda + age +
time:base_gcf + time:cda + time:age,
random = list(patient = ~ time, tooth = ~1),
data = veneer, method = "ML")
```

Next, we fit Model 7.3, without the fixed effects of any of the two-way interactions, also using ML estimation:

```
> model7.3.ml.fit <- lme(gcf ~ time + base_gcf + cda + age,
random = list(patient = ~ time, tooth = ~1),
data = veneer, method = "ML")
```

After fitting these two models using ML estimation, we use the `anova()` function to perform a likelihood ratio test of Hypothesis 7.3:

```
> anova(model7.1.ml.fit, model7.3.ml.fit)
```

The likelihood ratio test is nonsignificant ($p = 0.61$), so we keep Model 7.3 as our final model.

We now refit Model 7.3 using REML estimation to obtain unbiased estimates of the covariance parameters.

```
> model7.3.fit <- lme(gcf ~ time + base_gcf + cda + age,
random = list(patient = ~ time, tooth = ~1),
data = veneer, method = "REML")
> summary(model7.3.fit)
```

7.4.3.2   Analysis Using the `lmer()` Function

We first load the `lme4` package, so that the `lmer()` function can be used in the analysis:

```
> library(lme4)
```

Next, because the values of the TOOTH variable are not unique to a given patient and we need to specify that teeth are nested within patients, we create a new version of the TOOTH variable, TOOTH2, that does have unique values for each patient:

```
> veneer$tooth2 <- as.numeric(paste(factor(veneer$patient),
factor(veneer$tooth),sep=""))
```

Because of the way that models are specified when using the `lmer()` function, a failure to do this recoding would result in the random effects of TOOTH being *crossed* with the random patient effects (i.e., TOOTH = 6 for the first patient and TOOTH = 6 for the second patient would be interpreted as the *same tooth* for model-fitting purposes). This would result in unnecessary estimation complications, and incorrectly capture the nested structure of these data. We now proceed with model fitting using this recoded version of the TOOTH variable.

Step 1: Fit a model with a "loaded" mean structure (Model 7.1).

We first fit Model 7.1 using the `lmer()` function:

```
> model7.1.fit.lmer <- lmer(gcf ~ time + base_gcf + cda + age +
time*base_gcf + time*cda + time*age +
(time | patient) + (1 | tooth2),
data = veneer, REML = T)
```

The syntax used for the `lmer()` function is discussed in detail below:

- `model7.1.fit.lmer` is the name of the object that contains the results from the fit of Model 7.1.

- The first argument of the function is a formula, which first defines the continuous response variable (GCF), and the covariates and interaction terms that have fixed effects in the model. Note that asterisks are used to indicate the interaction terms.

- The formula also defines the random effects that are included in this model. The first set of random effects is specified with (`time | patient`), which indicates that the effect of time on GCF is allowed to randomly vary across levels of patient, in addition to the intercept (where random intercepts are once again included by default when the effect of a covariate is allowed to randomly vary). The covariance structure for these two random patient effects will be unstructured. The next random effect is specified with (`1 | tooth2`), which indicates that the intercept (represented by the constant value of 1) is also allowed to randomly vary across teeth *within* a patient (given that the recoded TOOTH2 identifier is being used).

- The second argument of the function, `data = veneer`, indicates the name of the data frame object to be used.

- The final argument, `REML = T`, specifies that REML estimation should be used.

The estimates from the model fit can be obtained by using the `summary()` function:

```
> summary(model7.1.fit.lmer)
```

The `intervals()` function that was available for objects created using the `lme()` function is not available for objects created using the `lmer()` function, so we do not generate approximate 95% confidence intervals for the parameters in Model 7.1 here.

The EBLUPs for each of the random effects in the model associated with the patients and the teeth within patients are obtained by using the `ranef()` function:

```
> ranef(model7.1.fit.lmer)$patient
> ranef(model7.1.fit.lmer)$tooth2
```

Step 2: Select a structure for the random effects (Model 7.1 vs. Model 7.1A).

We carry out a likelihood ratio test of Hypothesis 7.1 by fitting Model 7.1 and Model 7.1A using REML estimation. Model 7.1 was fitted in Step 1 of the analysis. Model 7.1A is specified by omitting the random tooth effects from Model 7.1:

```
> model7.1A.fit.lmer <- lmer(gcf ~ time + base_gcf + cda + age +
time*base_gcf + time*cda + time*age +
(time | patient),
data = veneer, REML = T)
```

We then subtract the –2 REML log-likelihood for Model 7.1 from that of the nested model, Model 7.1A. The p-value for the resulting test statistic can be obtained by referring it to a mixture of χ^2 distributions with degrees of freedom of 0 and 1, and equal weights of 0.5. Users of R can use the `anova()` function to calculate the likelihood ratio test statistic for Hypothesis 7.1, but the reported p-value should be divided by 2, because of the distribution that the statistic follows under the null hypothesis.

```
> anova(model7.1.fit.lmer, model7.1A.fit.lmer)
```

We have strong evidence in favor of retaining the nested random tooth effects in the model.

Step 3: Select a covariance structure for the residuals (Model 7.1, and Models 7.2A through 7.2C).

The current implementation of the `lmer()` function does not allow users to fit models with specified covariance structures for the conditional residuals, and assumes that residuals have constant variance and zero covariance (conditional on the random effects included in a given model). Given the problems with aliasing that we encountered when attempting to fit Models 7.2A through 7.2C in the other software procedures, this is not a serious limitation for this case study; however, some models may be improved by allowing for heterogeneous residual variances across groups, and this could be a limitation in other studies (e.g., Chapter 3). We therefore proceed with testing the fixed effects of the interactions in the model that retains all of the random effects associated with the patients and the teeth within patients.

Step 4: Reduce the model by removing nonsignificant fixed effects (Model 7.1 vs. Model 7.3).

We test whether we can omit the fixed effects associated with the two-way interactions between TIME and the other covariates in the model (Hypothesis 7.3) using a likelihood ratio test. First, we refit Model 7.1 using maximum likelihood estimation by including the `REML = F` option in the following syntax:

```
> model7.1.ml.fit.lmer <- lmer(gcf ~ time + base_gcf + cda + age +
time*base_gcf + time*cda + time*age +
(time | patient) + (1 | tooth2),
data = veneer, REML = F)
```

Next, we fit Model 7.3, without the fixed effects of any of the two-way interactions, also using ML estimation:

```
> model7.3.ml.fit.lmer <- lmer(gcf ~ time + base_gcf + cda + age +
(time | patient) + (1 | tooth2),
data = veneer, REML = F)
```

After fitting these two models using ML estimation, we use the `anova()` function to perform a likelihood ratio test of Hypothesis 7.3:

```
> anova(model7.1.ml.fit.lmer, model7.3.ml.fit.lmer)
```

The likelihood ratio test is nonsignificant ($p = 0.61$), so we keep Model 7.3 as our final model.

We now refit Model 7.3 using REML estimation to obtain unbiased estimates of the covariance parameters.

```
> model7.3.fit.lmer <- lmer(gcf ~ time + base_gcf + cda + age +
(time | patient) + (1 | tooth2),
data = veneer, REML = T)
> summary(model7.3.fit.lmer)
```

As mentioned in earlier chapters, we recommend use of the `lmerTest` package in R for users interested in testing hypotheses about individual fixed-effect parameters estimated using the `lmer()` function. Likelihood ratio tests can be used, as demonstrated here, but these tests rely on asymptotic assumptions that may not apply for smaller data sets like the Dental Veneer data.

7.4.4   Stata

Before we begin the analysis using Stata, we illustrate how to import the raw tab-delimited Dental Veneer data from the web site for the book, using web-aware Stata (where there is an active connection to the Internet):

```
. insheet using http://www-personal.umich.edu/~bwest/veneer.dat
```

Next, we generate variables representing the two-way interactions between TIME and the other continuous covariates, BASE_GCF, CDA, and AGE:

```
. gen time_base_gcf = time * base_gcf
. gen time_cda = time * cda
. gen time_age = time * age
```

Step 1: Fit a model with a "loaded" mean structure (Model 7.1).

Now, we fit Model 7.1 using the `mixed` command:

```
. * Model 7.1.
. mixed gcf time base_gcf cda age time_base_gcf time_cda time_age
|| patient: time, cov(unstruct) || tooth: , variance reml
```

The first variable listed is the dependent variable, GCF. Next, we list the covariates (including the two-way interactions) with associated fixed effects in the model.

The random effects are specified after the fixed part of the model. If a multilevel data set is organized by a series of nested groups, such as patients and teeth nested within patients, then the random-effects structure of the mixed model is specified by a series of equations, each separated by ||. The nesting structure reads from left to right, with the first cluster identifier (PATIENT in this case) indicating the highest level of the data set.

For Model 7.1, we specify the random factor identifying clusters at Level 3 of the data set (PATIENT) first. We indicate that the effect of TIME is allowed to vary randomly by PATIENT. A random patient-specific intercept is included by default and is not listed. We also specify that the covariance structure for the random effects at the patient level (the $D^{(2)}$, matrix following our notation) is unstructured, with the option cov(unstruct).

Because TOOTH follows the second clustering indicator ||, Stata assumes that levels of TOOTH are nested within levels of PATIENT. We do not list any variables after TOOTH, so Stata assumes that the only random effect for each tooth is associated with the intercept. A covariance structure for the single random effect associated with each tooth is not required, because only a single variance will be estimated. Finally, the variance option requests that the estimated variances of the random patient and tooth effects, rather than their estimated standard deviations, be displayed in the output (along with the estimated variance of the residuals), and the reml option requests REML estimation.

After running the mixed command, Stata displays a summary of the clustering structure of the data set implied by this model specification:

```
-----------------------------------------------------------
              |   No. of      Observations per Group
Group Variable |   Groups   Minimum   Average    Maximum
---------------+-------------------------------------------
      patient  |     12         2        9.2         12
        tooth  |     55         2        2.0          2
-----------------------------------------------------------
```

This summary is useful to determine whether the clustering structure has been identified correctly to Stata. In this case, Stata notes that there are 12 patients and 55 teeth nested within the patients. There are from 2 to 12 observations per patient, and 2 observations per tooth.

After the command has finished running, the parameter estimates appear in the output. We can obtain information criteria associated with the fit of the model by using the estat ic command, and then save these criteria in a model fit object (model71) for later analyses:

```
. estat ic
. est store model71
```

Step 2: Select a structure for the random effects (Model 7.1 vs. Model 7.1A).

In the output associated with the fit of Model 7.1, Stata automatically reports an omnibus likelihood ratio test for all random effects at once vs. no random effects. The test statistic is calculated by subtracting the -2 REML log-likelihood for Model 7.1 from that of a simple linear regression model without any random effects. Stata reports the following note along with the test:

```
Note: LR test is conservative and provided only for reference.
```

The test is conservative because appropriate theory for the distribution of this test statistic for multiple random effects has not yet been developed (users can click on the LR test is conservative statement in the Stata Results window for an explanation of this issue). We recommend testing the variance components associated with the random effects one by one (e.g., Hypothesis 7.1), using likelihood ratio tests based on REML estimation. While this is generally appropriate for larger samples, we now illustrate this process using the Dental Veneer data.

To test Hypothesis 7.1 using an LRT, we fit Model 7.1A. We specify this model by removing the portion of the random effects specification from Model 7.1 involving teeth nested within patients (i.e., || tooth: ,):

```
. * Model 7.1A.
. mixed gcf time base_gcf cda age time_base_gcf time_cda time_age
|| patient: time, cov(unstruct) variance reml
```

To obtain a test statistic for Hypothesis 7.1, the –2 REML log-likelihood for Model 7.1 (the reference model) is subtracted from that of Model 7.1A (the nested model); both values are calculated by multiplying the reported log-restricted likelihood in the output by –2. The resulting test statistic (11.2) is referred to a mixture of χ^2 distributions with degrees of freedom of 0 and 1, and equal weights of 0.5. We calculate the appropriate p-value for this test statistic as follows:

```
. display 0.5*chiprob(1,11.2)
.00040899
```

The test is significant ($p = 0.0004$), so we retain the nested random tooth effects in the model. We retain the random patient effects without testing them to reflect the hierarchical structure of the data in the model specification.

Step 3: Select a covariance structure for the residuals (Model 7.1, and Models 7.2A through 7.2C).

We now consider alternative covariance structures for the residuals, given that we have repeated measures on the teeth nested within each patient. We begin with Model 7.2A, and specify an unstructured covariance structure for the residuals associated with each tooth within a patient:

```
. mixed gcf time base_gcf cda age time_base_gcf time_cda time_age
|| patient: time, cov(unstruct) || tooth: ,
residuals(un,t(time)) variance reml
```

We note the inclusion of the option `residuals(un,t(time))` after the random tooth effects have been specified. This option indicates that at the lowest level of the data set (the repeated measures on each tooth), the residuals have an unstructured covariance structure (indicated by `un`), and the residuals should be ordered within unique teeth by values on the variable TIME (indicated by `t(time)`).

When fitting this model, Stata does not generate any warnings or error messages, but we see red flags in the output similar to those noted when performing this analysis using the `lme()` function in R. The estimated standard errors for the estimated variance of the random tooth effects and the three residual covariance parameters (the residual variances at the two time points and the covariance of the two residuals) are all extremely large and nearly equal (about 2416), suggesting an identifiability problem with these four covariance parameters. The corresponding approximate confidence intervals for these covariance parameters are defined by entirely unrealistic limits that are either extremely large or extremely small, so we would view these results with some suspicion and consider alternative covariance structures.

We now consider Model 7.2B, with a compound symmetry covariance structure for the residuals associated with each tooth within a patient:

```
. mixed gcf time base_gcf cda age time_base_gcf time_cda time_age
|| patient: time, cov(unstruct) || tooth: ,
residuals(exchangeable) variance reml
```

The only difference between this code and the code for Model 7.2A is the specification of an *exchangeable,* or compound symmetry, covariance structure for the residuals associated with each tooth (where no time ordering is needed given this covariance structure). When fitting this model, Stata does not converge to a solution after several hundred iterations, and the estimation algorithm needs to be stopped manually. This is due to the fact that the covariance of the residuals and the variance of the random tooth effects in this structure are perfectly aliased with each other, and Stata is unable to generate parameter estimates.

We therefore consider Model 7.2C, which removes the covariance of the residuals and allows the variances of the residuals at each time point to be unique:

```
. mixed gcf time base_gcf cda age time_base_gcf time_cda time_age
|| patient: time, cov(unstruct) || tooth: ,
residuals(ind, by(time)) variance reml
```

We now specify that the residuals are independent (using the option **ind**), meaning that the residuals have a diagonal covariance structure, and allow the residuals at each time point to have unique variances (using the **by(time)** option). We do not encounter any problems or red flags when fitting this model, and we save the information criteria in a second model fit object, enabling a likelihood ratio test of Hypothesis 7.2 (that the residual variances at each time point are equal):

```
. est store model72C
```

We now perform a formal test of Hypothesis 7.2 using the **lrtest** command:

```
. lrtest model72C model71
```

The resulting test statistic is not significant ($p = 0.33$), suggesting that we would fail to reject the null hypothesis that the residual variance is constant at the two time points. We therefore proceed with Model 7.1.

Step 4: Reduce the model by removing nonsignificant fixed effects (Model 7.1 vs. Model 7.3).

We now test Hypothesis 7.3 using a likelihood ratio test based on ML estimation to decide whether we want to retain the fixed effects associated with the interactions between TIME and the other covariates in the model. To do this, we first refit Model 7.1 using ML estimation for all parameters in the model (note that the **reml** option has been removed):

```
. * Model 7.1.
. mixed gcf time base_gcf cda age time_base_gcf time_cda time_age
|| patient: time, cov(unstruct) || tooth: , variance
```

We view the model fit criteria by using the **estat ic** command and store the model estimates and fit criteria in a new object named **model7_1_ml**:

```
. estat ic
. est store model7_1_ml
```

Next, we fit a nested model (Model 7.3), again using ML estimation, by excluding the two-way interaction terms from the model specification:

```
. * Model 7.3.
. mixed gcf time base_gcf cda age
|| patient: time, cov(unstruct) || tooth: , variance
```

We display the model information criteria for this nested model, and store the model fit criteria and related estimates in another new object named `model7_3_ml`:

```
. estat ic
. est store model7_3_ml
```

The appropriate likelihood ratio test for Hypothesis 7.3 can now be carried out by applying the `lrtest` command to the two objects containing the model fit information:

```
. lrtest model7_1_ml model7_3_ml
```

The results of this test are displayed in the Stata output:

```
Likelihood-ratio test                          LR chi2(3)  =      1.84
(Assumption: model7_3_ml nested in model7_1_ml)  Prob > chi2 =    0.6060
```

Because the test is not significant ($p = 0.61$), we omit the two-way interactions and choose Model 7.3 as our preferred model.

Additional tests could be performed for the fixed effects associated with the four covariates in the model, but we stop at this point (so that we can interpret the main effects of the covariates) and refit our final model (Model 7.3) using REML estimation:

```
. * Model 7.3 (REML).
. mixed gcf time base_gcf cda age
|| patient: time, cov(unstruct) || tooth: , variance reml
```

We carry out diagnostics for Model 7.3 using informal graphical procedures in Stata in Section 7.10.

7.4.5   HLM

We use the HMLM2 (Hierarchical Multivariate Linear Model 2) procedure to fit the models for the Dental Veneer data set, because this procedure allows for specification of alternative residual covariance structures (unlike the HLM3 procedure). We note that only ML estimation is available in the HMLM2 procedure.

7.4.5.1   Data Set Preparation

To fit the models outlined in Section 7.3 using the HLM software, we need to prepare three separate data sets:

1. The **Level 1 (longitudinal measures) data set**: This data set has two observations (rows) per tooth and contains variables measured at each time point (such as GCF and the TIME variable). It also contains the mandatory TOOTH and PATIENT variables. In addition, the data set needs to include two indicator variables, one for each time point. We create two indicator variables: TIME3 has a value of 1 for all observations at 3 months, and 0 otherwise, whereas TIME6 equals 1 for all observations at 6 months, and 0 otherwise. These indicator variables must be created prior to importing the data set into HLM. The data must be sorted by PATIENT, TOOTH, and TIME.

2. The **Level 2 (tooth-level) data set**: This data set has one observation (row) per tooth and contains variables measured once for each tooth (e.g., TOOTH, CDA, and BASE_GCF). This data set must also include the PATIENT variable. The data must be sorted by PATIENT and TOOTH.

3. The **Level 3 (patient-level) data set**: This data set has one observation (row) per patient and contains variables measured once for each patient (e.g., PATIENT and AGE). The data must be sorted by PATIENT.

The Level 1, Level 2, and Level 3 data sets can easily be derived from a single data set having the "long" structure shown in Table 7.2. For this example, we assume that all three data sets are stored in SPSS for Windows format.

7.4.5.2 Preparing the Multivariate Data Matrix (MDM) File

In the main HLM window, click **File**, **Make new MDM file**, and then **Stat package input**. In the dialog box that opens, select **HMLM2** to fit a **H**ierarchical (teeth nested within patients), **M**ultivariate (repeated measures on the teeth) **L**inear **M**odel, and click **OK**. In the **Make MDM** window, choose the **Input File Type** as SPSS/Windows.

Locate the **Level 1 Specification**, **Browse** to the location of the Level 1 data set defined earlier, and open the file. Now, click on the **Choose Variables** button, and select the following variables: PATIENT (check "L3id," because this variable identifies Level 3 units), TOOTH (check "L2id," because this variable identifies the Level 2 units), TIME (check "MDM" to include this variable in the MDM file), the response variable GCF (check "MDM"), and finally, TIME3 and TIME6 (check "ind" for both, because they are indicators for the repeated measures). Click **OK** when finished selecting these six variables.

Next, locate the **Level 2 Specification** area, **Browse** to the Level 2 data set defined earlier, and open it. In the **Choose Variables** dialog box, select the following variables: PATIENT (check "L3id"), TOOTH (check "L2id"), CDA (check "MDM" to include this tooth-level variable in the MDM file), and BASE_GCF (check "MDM"). Click **OK** when finished selecting these four variables.

Now, in the **Level 3 Specification** area, **Browse** to the Level 3 data set defined earlier, and open it. Select the PATIENT variable (check "L3id") and the AGE variable (check "MDM"). Click on **OK** to continue.

Next, select "longitudinal" as the structure of the data. As the MDM template window indicates, this selection will only affect the notation used when the models are displayed in the HLM model-building window.

Once all three data sets have been identified and the variables of interest have been selected, type a name for the MDM file (with an .mdm extension), and go to the **MDM template file** portion of the window. Click on **Save mdmt file** to save this setup as an MDM template file for later use (you will be prompted to supply a file name with an .mdmt suffix). Finally, click on the **Make MDM** button to create the MDM file using the three input files. You should briefly see a screen displaying descriptive statistics and identifying the number of records processed in each of the three input files. After this screen disappears, you can click on the **Check Stats** button to view descriptive statistics for the selected MDM variables at each level of the data. Click on the **Done** button to proceed to the model specification window.

In the following model-building steps, we use notation from the HLM software. Table 7.3 shows the correspondence between the HLM notation and that used in (7.1) through (7.4).

Step 1: Fit a model with a "loaded" mean structure (Model 7.1).

We begin by specifying the **Level 1** model, i.e., the model for the longitudinal measures collected on the teeth. The variables in the Level 1 data set are displayed in a list at the left-hand side of the model specification window. Click on the outcome variable (GCF), and identify it as the **Outcome variable**. Go to the **Basic Settings** menu, and click on **Skip Unrestricted** (the "unrestricted" model in HMLM2 refers to a model with no random effects and an unstructured covariance matrix for the residuals, which will be considered in the next step), and click on **Homogeneous** (to specify that the residual variance will be constant and that the covariance of the residuals will be zero in this initial model). Choose a title for this analysis (such as "Veneer Data: Model 7.1"), and choose a location and name for the output (.html) file that will contain the results of the model fit. Click on **OK** to return to the model-building window. Click on **File** and **Preferences**, and then select **Use level subscripts**, to display subscripts in the model-building window.

Three models will now be displayed. The initial Level 1 model, as displayed in the HLM model specification window, is as follows:

Model 7.1: Level 1 Model (Initial)

$$\text{GCF*}_{tij} = \pi_{0ij} + \varepsilon_{tij}$$

In this simplest specification of the Level 1 model, the GCF_{tij} for an individual measurement on a tooth depends on the tooth-specific intercept, denoted by π_{0ij}, and a residual for the individual measure, denoted by ε_{tij}. We now add the TIME variable from the Level 1 data set to this model, by clicking on the TIME variable and then clicking on "Add variable uncentered." The Level 1 model now has the following form:

Model 7.1: Level 1 Model (Final)

$$\text{GCF*}_{tij} = \pi_{0ij} + \pi_{1ij}(\text{TIME}_{tij}) + \varepsilon_{tij}$$

The **Level 2 model** describes the equations for the tooth-specific intercept, π_{0ij}, and the tooth-specific time effect, π_{1ij}. The simplest Level 2 model is given by the following equations:

Model 7.1: Level 2 Model (Initial)

$$\pi_{0ij} = \beta_{00j} + r_{0ij}$$
$$\pi_{1ij} = \beta_{10j}$$

The tooth-specific intercept, π_{0ij}, depends on the patient-specific intercept, β_{00j}, and a random effect associated with the tooth, r_{0ij}. The tooth-specific time effect, π_{1ij}, does not vary from tooth to tooth within the same patient and is simply equal to the patient-specific time effect, β_{10j}. If we had more than two time points, this random effect could be included by clicking on the shaded r_{1ij} term in the model for π_{1ij}.

We now include the tooth-level covariates by clicking on the **Level 2** button, and then selecting the Level 2 equations for π_{0ij} and π_{1ij}, in turn. We add the uncentered versions of CDA and BASE_GCF to both equations to get the completed version of the Level 2 model:

Model 7.1: Level 2 Model (Final)

$$\pi_{0ij} = \beta_{00j} + \beta_{01j}(\text{BASE\_GCF}_{ij}) + \beta_{02j}(\text{CDA}_{ij}) + r_{0ij}$$

$$\pi_{1ij} = \beta_{10j} + \beta_{11j}(\text{BASE\_GCF}_{ij}) + \beta_{12j}(\text{CDA}_{ij})$$

After defining the Level 1 and Level 2 models, HLM displays the combined Level 1 and Level 2 model in the model specification window. It also shows how the marginal variance-covariance matrix will be calculated based on the random tooth effects and the residuals currently specified in the Level 1 and Level 2 models. By choosing the **Homogeneous** option in the **Basic Settings** menu earlier, we specified that the residuals are assumed to be independent with constant variance.

The **Level 3** portion of the model specification window shows the equations for the patient-specific intercept, β_{00j}, the patient-specific time slope, β_{10j}, and the patient-specific effects of BASE_GCF and CDA, which were defined in the Level 2 model. The simplest Level 3 equations for the patient-specific intercepts and slopes include only the overall fixed effects, γ_{000} and γ_{100}, and the patient-specific random effects for the intercept, u_{00j}, and slope, u_{10j}, respectively (one needs to click on the shaded u_{10j} term to include it in the model). We add the patient-level covariate AGE to these equations:

Model 7.1: Level 3 Model (Final)

$$\beta_{00j} = \gamma_{000} + \gamma_{001}(\text{AGE}_j) + u_{00j}$$

$$\beta_{10j} = \gamma_{100} + \gamma_{101}(\text{AGE}_j) + u_{10j}$$

The patient-specific intercept, β_{00j}, depends on the overall fixed intercept, γ_{000}, the patient-level covariate AGE, and a random effect for the intercept associated with the patient, u_{00j}. The patient-specific time slope, β_{10j}, depends on the fixed overall time slope, γ_{100}, the patient-level covariate AGE, and a random effect for TIME associated with the patient, u_{10j}. The expressions for the Level 3, Level 2, and Level 1 models defined earlier can be combined to obtain the LMM defined in (7.1). The correspondence between the HLM notation and the notation we use for (7.1) can be found in Table 7.3.

To fit Model 7.1, click on **Run Analysis**, and select **Save and Run** to save the .hlm command file. HLM will prompt you to supply a name and location for this file. After the estimation of the model has finished, click on **File** and select **View Output** to see the resulting parameter estimates and fit statistics. Note that HLM automatically displays the fixed-effect parameter estimates with model-based standard errors ("Final estimation of fixed effects"). The estimates of the covariance parameters associated with the random effects at each level of the model are also displayed.

Step 2: Select a structure for the random effects (Model 7.1 vs. Model 7.1A).

At this stage of the analysis, we wish to test Hypothesis 7.1 using a likelihood ratio test. However, HLM cannot fit models in which all random effects associated with units at a given level of a clustered data set have been removed. Because Model 7.1A has no random effects at the tooth level, we cannot consider a likelihood ratio test of Hypothesis 7.1 in HLM, and we retain all random effects in Model 7.1, as we did when using the other software procedures.

Step 3: Select a covariance structure for the residuals (Model 7.1, and Models 7.2A through 7.2C).

We are unable to specify Model 7.2A, having random effects at the patient and tooth levels and an unstructured covariance structure for the residuals, using the HMLM2 procedure. However, we can fit Model 7.2B, which has random effects at the patient- and tooth-level and a **compound symmetry** covariance structure for the residuals. To do this, we use a first-order autoregressive, or AR(1), covariance structure for the Level 1 (or residual) variance. In this case, the AR(1) covariance structure is equivalent to the compound symmetry structure, because there are only two time points for each tooth in the Dental Veneer data set.

Open the .hlm file saved in the process of fitting Model 7.1, and click **Basic Settings**. Choose the **1st order autoregressive** covariance structure option, make sure that the "unrestricted" model is still being skipped, enter a new title for the analysis and a different name for the output (.html) file to save, and click on **OK** to continue. We recommend saving the .hlm file under a different name as well when making these changes. Next, click on **Run Analysis** to fit Model 7.2B. In the process of fitting the model, HLM displays the following message:

> Invalid info, score, or likelihood

This warning message arises because the residual covariance parameters are aliased with the variance of the nested random tooth effects in this model.

The output for Model 7.2B has two parts. The first part is essentially a repeat of the output for Model 7.1, which had a homogeneous residual variance structure (under the header OUTPUT FOR RANDOM EFFECTS MODEL WITH HOMOGENEOUS LEVEL-1 VARIANCE). The second part (under the header OUTPUT FOR RANDOM EFFECTS MODEL FIRST-ORDER AUTOREGRESSIVE MODEL FOR LEVEL-1 VARIANCE) does not include any estimates of the fixed-effect parameters, due to the warning message indicated above. Because of the problems encountered in fitting Model 7.2B, we do not consider these results.

Next, we fit Model 7.2C, which has a **heterogeneous residual variance** structure. In this model, the residuals at time 1 and time 2 are allowed to have different residual variances, but they are assumed to be uncorrelated. Click on the **Basic Settings** menu in the model-building window, and then click on the **Heterogeneous** option for the residual variance (make sure that "Skip unrestricted" is still checked). Enter a different title for this analysis (e.g., "Veneer Data: Model 7.2C") and a new name for the output file, and then click on **OK** to proceed with the analysis. Next, save the .hlm file under a different name, and click on **Run Analysis** to fit this model and investigate the output.

No warning messages are generated when Model 7.2C is fitted. However, because HLM only fits models of this type using ML estimation, we cannot carry out the REML-based likelihood ratio test of Hypothesis 7.2. The HMLM2 procedure by default performs an ML-based likelihood ratio test, calculating the difference in the deviance (or -2 ML log-likelihood) statistics from Model 7.1 and Model 7.2C, and displays the result of this test at the bottom of the output for the heterogeneous variances model (Model 7.2C). This nonsignificant likelihood ratio test ($p = 0.31$) suggests that the simpler nested model (Model 7.1) is preferable at this stage of the analysis.

Step 4: Reduce the model by removing nonsignificant fixed effects (Model 7.1 vs. Model 7.3).

We now fit Model 7.3, which omits the fixed effects associated with the two-way interactions between TIME and the other covariates from Model 7.1. In HLM, this is accomplished by removing the effects of the covariates in question from the Level 2 and Level 3 equations for the effects of TIME.

First, the fixed effects associated with the tooth-level covariates, BASE_GCF and CDA, are removed from the Level 2 equation for the tooth-specific effect of TIME, π_{1ij}, as follows:

Model 7.1: Level 2 Equation for the Effect of Time

$$\pi_{1ij} = \beta_{10j} + \beta_{11j}(\text{BASE\_GCF}_{ij}) + \beta_{12j}(\text{CDA}_{ij})$$

Model 7.3: Level 2 Equation for the Effect of Time with Covariates Removed

$$\pi_{1ij} = \beta_{10j}$$

We also remove the patient-level covariate, AGE, from the Level 3 model for the patient-specific effect of TIME, β_{10j}:

Model 7.1: Level 3 Equation for the Effect of Time

$$\beta_{10j} = \gamma_{100} + \gamma_{101}(\text{AGE}_j) + u_{10j}$$

Model 7.3: Level 3 Equation for the Effect of Time with Covariates Removed

$$\beta_{10j} = \gamma_{100} + u_{10j}$$

To accomplish this, open the .hlm file defining Model 7.1. In the HLM model specification window, click on the Level 2 equation for the effect of TIME, click on the BASE_GCF covariate in the list of covariates at the left of the window, and click on **Delete variable from model**. Repeat this process for the CDA variable in the Level 2 model. Then, click on the Level 3 equation for the effect of TIME, and delete the AGE variable.

After making these changes, click on **Basic Settings** to change the title for this analysis and the name of the text output file, and click **OK**. To set up a likelihood ratio test of Hypothesis 7.3, click **Other Settings** and **Hypothesis Testing**. Enter the deviance reported for Model 7.1 (843.65045) and the number of parameters in Model 7.1 (13), and click **OK**. Save the .hlm file associated with the Model 7.3 specification under a different name, and then click **Run Analysis**. The results of the likelihood ratio test for the fixed-effect parameters that have been removed from this model (Hypothesis 7.3) can be viewed at the bottom of the output for Model 7.3.

7.5   Results of Hypothesis Tests

The results of the hypothesis tests reported in this section were based on the analysis of the Dental Veneer data using Stata, and are summarized in Table 7.5.

7.5.1   Likelihood Ratio Tests for Random Effects

Hypothesis 7.1.   The nested random effects, $u_{0i|j}$, associated with teeth within the same patient can be omitted from Model 7.1.

The likelihood ratio test statistic for Hypothesis 7.1 is calculated by subtracting the value of the -2 REML log-likelihood associated with Model 7.1 (the reference model including the random tooth-specific intercepts) from that of Model 7.1A (the nested model excluding the random tooth effects). Because the null hypothesis value of the variance of the random tooth-specific intercepts is at the boundary of the parameter space ($H_0 : \sigma^2_{int\,:\,tooth(patient)} = 0$), the null distribution of the test statistic is a mixture of χ^2_0 and χ^2_1 distributions, each with equal weight 0.5 (Verbeke & Molenberghs, 2000). To evaluate the significance of the test statistic, we calculate the p-value as follows:

$$p\text{-value} = 0.5 \times \text{P}(\chi^2_0 > 11.2) + 0.5 \times \text{P}(\chi^2_1 > 11.2) < 0.001$$

We reject the null hypothesis and retain the random effects associated with teeth nested within patients in Model 7.1 and all subsequent models.

TABLE 7.5: Summary of Hypothesis Test Results for the Dental Veneer Analysis

Hypo-thesis Label	Test	Estima-tion Method	Models Compared (Nested vs. Reference)	Test Statistic Values (Calculation)	p-Value
7.1	LRT	REML	7.1A vs. 7.1	$\chi^2(0:1) = 11.2$ $(858.3 - 847.1)$	$< .001$
7.2	LRT	REML	7.1 vs. 7.2C	$\chi^2(1) = 0.9$ $(847.1 - 846.2)$	0.34
7.3	LRT	ML	7.3 vs. 7.1	$\chi^2(3) = 1.8$ $(845.5 - 843.7)$	0.61

Note: See Table 7.4 for null and alternative hypotheses, and distributions of test statistics under H_0.

The presence of the random tooth-specific intercepts implies that different teeth within the same patient tend to have consistently different GCF values over time, which is in keeping with what we observed in the initial data summary. To preserve the hierarchical nature of the model, we do not consider fitting a model without the random patient-specific effects, u_{0j} and u_{1j}.

7.5.2 Likelihood Ratio Tests for Residual Variance

Hypothesis 7.2. The variance of the residuals is constant (homogeneous) across the time points in Model 7.2C.

To test Hypothesis 7.2, we use a REML-based likelihood ratio test. The test statistic is calculated by subtracting the –2 REML log-likelihood value for Model 7.2C, the reference model with heterogeneous residual variances, from that for Model 7.1, the nested model. Because Model 7.2C has one additional variance parameter compared to Model 7.1, the asymptotic null distribution of the test statistic is a χ_1^2 distribution. We do not reject the null hypothesis in this case ($p = 0.34$) and decide to keep a homogeneous residual variance structure in Model 7.1 and all of the subsequent models.

7.5.3 Likelihood Ratio Tests for Fixed Effects

Hypothesis 7.3. The fixed effects associated with the two-way interactions between TIME and the patient- and tooth-level covariates can be omitted from Model 7.1.

To test Hypothesis 7.3, we use an ML-based likelihood ratio test. We calculate the test statistic by subtracting the –2 ML log-likelihood for Model 7.1 from that for Model 7.3. The asymptotic null distribution of the test statistic is a χ^2 with 3 degrees of freedom, corresponding to the 3 fixed-effect parameters that are omitted in the nested model (Model 7.3) compared to the reference model (Model 7.1). There is not enough evidence to reject the null hypothesis for this test ($p = 0.61$), so we remove the two-way interactions involving TIME from the model.

We do not attempt to reduce the model further, because the research is focused on the effects of the covariates on GCF. Model 7.3 is the final model that we consider for the analysis of the Dental Veneer data.

7.6   Comparing Results across the Software Procedures

7.6.1   Comparing Model 7.1 Results

Table 7.6 shows a comparison of selected results obtained using the six software procedures to fit the initial three-level model, Model 7.1, to the Dental Veneer data. This model is "loaded" with fixed effects, has two patient-specific random effects (associated with the intercept and with the effect of time), has a random effect associated with each tooth nested within a patient, and has residuals that are independent and identically distributed. Model 7.1 was fitted using REML estimation in SAS, SPSS, R, and Stata, and was fitted using ML estimation in HLM (the current version of the HMLM2 procedure does not allow models to be fitted using REML estimation).

Table 7.6 demonstrates that the procedures in SAS, SPSS, R, and Stata agree in terms of the estimated fixed-effect parameters and their standard errors for Model 7.1. The procedures in all of these software packages use REML estimation by default. REML estimation is not available in HMLM2, so HMLM2 uses ML estimation instead. Consequently, the fixed-effect estimates from HMLM2 are not comparable to those from the other software procedures. As expected, the fixed-effect estimates from HMLM2 differ somewhat from those for the other software procedures. Most notably, the estimated standard errors reported by HMLM2 are smaller in almost all cases than those reported in the other software procedures. We expect this because of the bias in the estimated covariance parameters when ML estimation is used instead of REML (see Subsection 2.4.1).

The estimated covariance parameters generated by HMLM2 differ more markedly from those in the other five software procedures. Although the results generated by the procedures in SAS, SPSS, R, and Stata are the same, the covariance parameters estimated by HMLM2 tend to be smaller and to have smaller standard errors than those reported by the other software procedures. Again, this is anticipated, in view of the bias in the ML estimates of the covariance parameters. The difference is most apparent in the variance of the random patient-specific intercepts, $\sigma^2_{int:patient}$, which is estimated to be 555.39 with a standard error of 279.75 by the `mixed` procedure in Stata, and is estimated to be 447.13 with a standard error of 212.85 by HMLM2.

There are also differences in the information criteria reported across the software procedures. The programs that use REML estimation agree in terms of the –2 REML log-likelihoods, but disagree in terms of the other information criteria, because of different calculation formulas that are used (see Section 3.6 for a discussion of these differences). The –2 log-likelihood reported by HMLM2 (referred to as the *deviance*) is not comparable with the other software procedures, because it is calculated using ML estimation.

7.6.2   Comparing Results for Models 7.2A, 7.2B, and 7.2C

Table 7.7 presents a comparison of selected results across the procedures in SAS, SPSS, R, Stata, and HLM for Models 7.2A, 7.2B, and 7.2C. Recall that each of these models has a different residual covariance structure, and that there were problems with aliasing of the covariance parameters in Models 7.2A and 7.2B. We do not display results for the `lmer()` procedure in R, given that one cannot currently fit models with conditional errors that are correlated and/or have nonconstant variance when using this procedure.

In Table 7.7 we present the information criteria calculated by the procedures in SAS, SPSS, R, and Stata for Model 7.2A, Model 7.2B, and Model 7.2C. Because the covariance parameters in Model 7.2A and Model 7.2B are aliased, we do not compare their results with those for Model 7.2C, but present brief descriptions of how the problem might be detected in the software procedures. We note that the −2 REML log-likelihoods are virtually the same for a given model across the procedures. The other information criteria (AIC and BIC) differ because of different calculation formulas.

We report the model information criteria and covariance parameter estimates for Model 7.2C, which has a heterogeneous residual variance structure and is the only model in Table 7.7 that does not have an aliasing problem. The estimated covariance parameters and their respective standard errors are comparable across the procedures in SAS, SPSS, R, and Stata, each of which use REML estimation. The covariance parameter estimates reported by the HMLM2 procedure, which are calculated using ML estimation, are in general smaller than those reported by the other procedures, and their estimated standard errors are smaller as well. We expect this because of the bias in the ML estimation of the covariance parameters. We do not present the estimates of the fixed-effect parameters for Model 7.2C in Table 7.7.

7.6.3 Comparing Model 7.3 Results

Table 7.8 shows results from fitting the final model, Model 7.3, using REML estimation in SAS, SPSS, R, and Stata, and using ML estimation in HLM. This model has the same random effects and residual covariance structure as in Model 7.1, but omits the fixed effects associated with the two-way interactions between TIME and the other covariates from the model.

The fixed-effect parameter estimates and their estimated standard errors are nearly identical across the five procedures (`proc mixed` in SAS, `MIXED` in SPSS, `lme()` in R, `lmer()` in R, and `mixed` in Stata) that use REML estimation. Results from the HMLM2 procedure again differ because HMLM2 uses ML estimation. In general, the estimated standard errors of the fixed-effect parameter estimates are smaller in HMLM2 than in the other software procedures.

As noted in the comparison of results for Model 7.1, the −2 REML log-likelihood values agree very well across the procedures. The AIC and BIC differ because of different computational formulas. The information criteria are not computed by the HMLM2 procedure.

The estimated covariance parameters and their estimated standard errors are also very similar across the procedures in SAS, SPSS, R, and Stata. Again, we note that these estimated parameters and their standard errors are consistently smaller in HMLM2, which uses ML estimation.

TABLE 7.6: Comparison of Results for Model 7.1

Estimation Method	SAS: proc mixed REML	SPSS: MIXED REML	R: lme() function REML	R: lmer() function REML	Stata: mixed REML	HMLM2 ML
Fixed-Effect Parameter	*Estimate (SE)*	*Estimate (SE)*	*Estimate (SE)*	*Estimate (SE)*	*Estimate (SE)*	*Estimate (SE)*
β_0(Intercept)	69.92(28.40)	69.92(28.40)	69.92(28.40)	69.92(28.40)	69.92(28.40)	70.47(26.11)
β_1(Time)	−6.02(7.45)	−6.02(7.45)	−6.02(7.45)	−6.02(7.45)	−6.02(7.45)	−6.11(6.83)
β_2(Baseline GCF)	−0.32(0.29)	−0.32(0.29)	−0.32(0.29)	−0.32(0.29)	−0.32(0.29)	−0.32(0.28)
β_3(CDA)	−0.88(1.08)	−0.88(1.08)	−0.88(1.08)	−0.88(1.08)	−0.88(1.08)	−0.88(1.05)
β_4(Age)	−0.97(0.61)	−0.97(0.61)	−0.97(0.61)	−0.97(0.61)	−0.97(0.61)	−0.98(0.55)
β_5(Time × Base_GCF)	0.07(0.06)	0.07(0.06)	0.07(0.06)	0.07(0.06)	0.07(0.06)	0.07(0.06)
β_6(Time × CDA)	0.13(0.22)	0.13(0.22)	0.13(0.22)	0.13(0.22)	0.13(0.22)	0.13(0.21)
β_7(Time × Age)	0.11(0.17)	0.11(0.17)	0.11(0.17)	0.11(0.17)	0.11(0.17)	0.11(0.15)
Covariance Parameter	*Estimate (SE)*	*Estimate (SE)*	*Estimate (n.c.)*	*Estimate (n.c.)*	*Estimate (SE)*	*Estimate (SE)*
$\sigma^2_{int:patient}$	555.39[a](279.75)	555.39[a](279.75)	555.39[b,c]	555.39[c]	555.39(279.75)	447.13[d](212.85)
$\sigma_{int,time:patient}$	−149.76(74.55)[e]	−149.76(74.55)[e]	−0.95(corr.)	−0.95(corr.)	−149.77(74.55)	−122.23(57.01)
$\sigma^2_{time:patient}$	44.72(21.15)[f]	44.72(21.15)[f]	44.72[c]	44.72[c]	44.72(21.15)	36.71(16.21)
$\sigma^2_{int:tooth(patient)}$	46.96(16.67)	46.96(16.67)	46.96[c]	46.96	46.96(16.67)	45.14(15.66)
σ^2	49.69(10.92)	49.69(10.92)	49.69	49.69	49.69(10.92)	47.49(10.23)
Model Information Criteria						
−2 RE/ML log-likelihood	847.1	847.1	847.1	847.1	847.1	843.7[g]
AIC	857.1	857.1	873.1	873.1	873.1	n.c.
BIC	859.5	870.2	907.2	908.2	908.2	n.c.

Note: (n.c.) = not computed

Note: 110 Longitudinal Measures at Level 1; 55 Teeth at Level 2; 12 Patients at Level 3.

[a]Reported as UN(1,1) in SAS and SPSS.

[b]The nlme version of the lme() function reports the estimated standard deviations of the random effects and residuals by default; these estimates have been squared in Tables 7.6, 7.7, and 7.8. The intervals() function can be applied to obtain CIs for the parameters.

[c]Standard errors are not reported.

[d]HLM reports the four covariance parameters associated with the random effects in the 2×2 Tau(beta) matrix and the scalar Tau(pi), respectively.

[e]Reported as UN(2,1) in SAS and SPSS.

[f]Reported as UN(2,2) in SAS and SPSS.

[g]The -2 ML log-likelihood associated with the model fit is referred to in the HLM output as the model deviance.

TABLE 7.7: Comparison of Results for Models 7.2A, 7.2B (Both with Aliased Covariance Parameters), and 7.2C

	SAS: proc mixed	SPSS: MIXED	R: lme() function	Stata: mixed	HMLM2
Estimation Method	**REML**	**REML**	**REML**	**REML**	**ML**
Model 7.2A (Unstructured)					
Software Notes	Warning: Hessian not positive-definite	Warning: validity of the fit uncertain	Wide intervals for cov. parms.	Extremely large SEs for cov. parms.	Cannot be fitted
−2 REML log-likelihood	846.2	846.7	846.2	846.2	
AIC (smaller the better)	860.2	860.7	876.2	876.2	
BIC (smaller the better)	863.6	879.0	915.5	916.7	
Model 7.2B (Comp. Symm.)					
Software Notes	Stopped: Too many likelihood evaluations	Warning: validity of the fit uncertain	Wide intervals for cov. parms.	Failure to converge	Invalid likelihood
−2 REML/ML log-likelihood	N/A	847.1	847.1	N/A	N/A
AIC	N/A	859.1	875.1	N/A	N/A
BIC	N/A	874.9	911.9	N/A	N/A
Model 7.2C (Heterogeneous)					
−2 REML/ML log-likelihood	846.2	846.2	846.2	846.2	842.6
AIC	858.2	858.2	874.2	874.2	*n.c.*
BIC	861.1	873.9	910.9	912.0	*n.c.*
Covariance Parameters (Model 7.2C)					
	Estimate (SE)	*Estimate (SE)*	*Estimate (SE)*	*Estimate (SE)*	*Estimate (SE)*
$\sigma^2_{int:patient}$	546.60(279.33)	546.61(279.34)	$546.61^{a,b}$	546.61(279.34)	438.18(212.68)
$\sigma_{int,time:patient}$	−148.64(74.38)	−148.64(74.38)	−0.95(*corr.*)	−148.64(74.38)	−121.14(56.90)

TABLE 7.7: (Continued)

| Estimation Method | SAS: proc mixed | SPSS: MIXED | R: lme() function | Stata: mixed | HMLM2 |
	REML	REML	REML	REML	ML
$\sigma^2_{time:patient}$	44.64(21.10)	44.64(21.10)	44.64[b]	44.64(21.10)	36.65(16.16)
$\sigma^2_{int:tooth(patient)}$	46.92(16.53)	46.92(16.53)	46.92[b]	46.92(16.53)	45.12(15.54)
σ^2_{t1}	62.38(18.81)	62.38(18.81)	62.38[b]	62.38(18.81)	59.93(17.70)
σ^2_{t2}	36.95(15.30)	36.95(15.30)	36.95[b,c]	36.95(15.30)	35.06(14.32)

Note: SE = Standard error, (n.c.) = not computed

Note: 110 Longitudinal Measures at Level 1; 55 Teeth at Level 2; 12 Patients at Level 3.

[a]Users of R can employ the function intervals(model7.2c.fit) to obtain approximate 95% confidence intervals for the covariance parameters.

[b]Standard errors are not reported.

[c]See Subsection 3.4.3 for a discussion of the lme() function output from models with heterogeneous residual variance.

TABLE 7.8: Comparison of Results for Model 7.3

Estimation Method	SAS: proc mixed REML	SPSS: MIXED REML	R: lme() function REML	R: lmer() function REML	Stata: mixed REML	HMLM2 ML
Fixed-Effect Parameter	*Estimate (SE)*	*Estimate (SE)*	*Estimate (SE)*	*Estimate (SE)*	*Estimate (SE)*	*Estimate (SE)*
β_0 (Intercept)	45.74(12.55)	45.74(12.55)	45.74(12.55)	45.74(12.55)	45.74(12.55)	46.02(11.70)
β_1 (Time)	0.30(1.94)	0.30(1.94)	0.30(1.94)	0.30(1.94)	0.30(1.94)	0.29(1.86)
β_2 (Baseline GCF)	−0.02(0.14)	−0.02(0.14)	−0.02(0.14)	−0.02(0.14)	−0.02(0.14)	−0.02(0.14)
β_3 (CDA)	−0.33(0.53)	−0.33(0.53)	−0.33(0.53)	−0.33(0.53)	−0.33(0.53)	−0.31(0.51)
β_4 (Age)	−0.58(0.21)	−0.58(0.21)	−0.58(0.21)	−0.58(0.21)	−0.58(0.21)	−0.58(0.19)
Covariance Parameter	*Estimate (SE)*	*Estimate (SE)*	*Estimate (n.c.)*	*Estimate (n.c.)*	*Estimate (SE)*	*Estimate (SE)*
$\sigma^2_{int:patient}$	524.95(252.99)	524.98(253.02)	524.99	524.98	524.99(253.02)	467.74(221.98)
$\sigma_{int,time:patient}$	−140.42(66.57)	−140.42(66.58)	−0.95(*corr.*)	−0.95(*corr.*)	−140.42(66.58)	−127.80(59.45)
$\sigma^2_{time:patient}$	41.89(18.80)	41.89(18.80)	41.89	41.89	41.89(18.80)	38.23(16.86)
$\sigma^2_{int:tooth(patient)}$	47.45(16.63)	47.46(16.63)	47.46	47.46	47.46(16.63)	44.57(15.73)
σ^2	48.87(10.51)	48.87(10.51)	48.87	48.87	48.87(10.51)	48.85(10.52)
Model Information Criteria						
−2 log-likelihood	841.9	841.9	841.9	841.9	841.9	845.5
AIC	851.9	851.9	861.9	861.9	861.9	n.c.
BIC	854.3	865.1	888.4	888.9	888.9	n.c.

Note: SE = Standard error, (n.c.) = not computed

Note: 110 Longitudinal Measures at Level 1; 55 Teeth at Level 2; 12 Patients at Level 3.

7.7 Interpreting Parameter Estimates in the Final Model

Results in this section were obtained by fitting Model 7.3 to the Dental Veneer data using the `mixed` command in Stata.

7.7.1 Fixed-Effect Parameter Estimates

The Stata output for the fixed-effect parameter estimates and their estimated standard errors is shown below.

```
                                              Wald chi2(4) = 7.48
      Log restricted-likelihood = -420.92761   Prob > chi2 = 0.1128


 gcf          Coef.      Std. Err.    z    P > |z|      [95% Conf. Interval]
 -------------------------------------------------------------------------------
 time        0.3009815   1.9368630   0.16   0.877     -3.4952000    4.0971630
 base_gcf   -0.0183127   0.1433094  -0.13   0.898     -0.2991940    0.2625685
 cda        -0.3293040   0.5292525  -0.62   0.534     -1.3666190    0.7080128
 age        -0.5773932   0.2139656  -2.70   0.007     -0.9967582   -0.1580283
 _cons      45.7386200  12.5549700   3.64   0.000     21.1313300   70.3459100
```

The first part of the output above shows the value of the REML log-likelihood for Model 7.3. Note that in Table 7.8 we report the –2 REML log-likelihood value for this model (841.9). The Wald chi-square statistic and corresponding p-value reported at the top of the output represent an omnibus test of all fixed effects (with the exception of the intercept). The null distribution of the test statistic is a χ^2 with 4 degrees of freedom, corresponding to the 4 fixed effects in the model. This test statistic is not significant ($p = 0.11$), suggesting that these covariates do not explain a significant amount of variation in the GCF measures.

The `mixed` command reports z-tests for the fixed-effect parameters, which are asymptotic (i.e., they assume large sample sizes). The z-tests suggest that the only fixed-effect parameter significantly different from zero is the one associated with AGE ($p = 0.007$). There appears to be a negative effect of AGE on GCF, after controlling for the effects of time, baseline GCF, and CDA. Patients who are one year older are predicted to have an average value of GCF that is 0.58 units lower than similar patients who are one year younger. There is no significant fixed effect of TIME on GCF overall. This result is not surprising, given the initial data summary in Figure 7.2, in which we saw that the GCF for some patients went up over time, whereas that for other patients decreased over time.

The effect of contour difference (CDA) is also not significant, indicating that a greater discrepancy in tooth contour after veneer placement is not necessarily associated with a higher mean value of GCF.

Earlier, when we tested the two-way interactions between TIME and the other covariates (Hypothesis 7.3), we found that none of the fixed effects associated with the two-way interactions were significant ($p = 0.61$; see Table 7.5). As a result, all two-way interactions between TIME and the other covariates were dropped from the model. The fact that there were no significant interactions between TIME and the other covariates suggests that the effect of TIME on GCF does not tend to differ for different values of AGE, baseline GCF, or contour difference.

7.7.2   Covariance Parameter Estimates

The Stata output below displays the estimates of the covariance parameters associated with the random effects in Model 7.3, reported by the `mixed` command.

```
Random-effects Parameters     Estimate    Std. Err.     [95% Conf. Interval]
-----------------------------------------------------------------------------
patient: Unstructured
  var(time)                    41.88772    524.98510    -140.42290   18.799970
  var(_cons)                  253.02050     66.57623      17.38009  204.128700
  cov(time, _cons)           -270.90990    100.95350     1350.17500  -9.935907

tooth: Identity
  var(_cons)                   47.45738     16.63034      23.87920   94.31650

  var(Residual)                48.86704     10.50523      32.06479   74.47382

LR test vs. linear regression: chi2(4) = 91.12     Prob > chi2 = 0.0000
Note: LR test is conservative and provided only for reference.
```

Stata reports the covariance parameter estimates, their standard errors, and approximate 95% confidence intervals. The output table divides the parameter estimates into three groups. The top group corresponds to the patient level (Level 3) of the model, where the `Unstructured` covariance structure produces three covariance parameter estimates. These include the variance of the random patient-specific time effects, `var(time)`, the variance of the random patient-specific intercepts, `var(_cons)`, and the covariance between these two random effects, `cov(time, _cons)`. We note that the covariance between the two patient-specific random effects is negative. This means that patients with a higher (lower) time 1 value for their GCF tend to have a lower (higher) time 2 value.

Because there is only a single random effect at the tooth level of the model (Level 2), the variance-covariance matrix for the nested random tooth effects (which only has one element) has an `Identity` covariance structure. The single random tooth effect is associated with the intercept, and the estimated covariance parameter at the tooth level represents the estimated variance of these nested random tooth effects.

At the lowest level of the data (Level 1), there is a single covariance parameter associated with the variance of the residuals, labeled `var(Residual)`.

Stata also displays approximate 95% confidence intervals for the covariance parameters based on their standard errors, which can be used to get an impression of whether the true covariance parameters in the population of patients and teeth are equal to zero. We note that none of the reported confidence intervals cover zero. However, Stata does not automatically generate formal tests for any of these covariance parameters. Readers should note that interpreting these 95% confidence intervals for covariance parameters can be problematic, especially when estimates of variances are small. See Bottai & Orsini (2004) for more details.

Finally, we note an omnibus likelihood ratio test for all covariance parameters. Stata generates a test statistic by calculating the difference in the –2 REML log-likelihood of Model 7.3 (the reference model) and that of a linear regression model with the same fixed effects but without any random effects (the nested model, which has four fewer covariance parameters). The result of this conservative test suggests that some of the covariance parameters are significantly different from zero, which is in concordance with the approximate 95% confidence intervals for the parameters. Stata allows users to click on the note about the likelihood ratio test in the output, for additional information about the reason why this test should be considered conservative (see Subsection 5.4.4 for more details on this type of test).

Based on these results and the formal test of Hypothesis 7.1, we have evidence of between-patient variance and between-tooth variance within the same patient that is not being explained by the fixed effects of the covariates included in Model 7.3.

7.8 The Implied Marginal Variance-Covariance Matrix for the Final Model

In this section, we present the estimated V_j matrix for the marginal model implied by Model 7.3 for the first patient in the Dental Veneer data set. We use SAS `proc mixed` to generate this output, because the post-estimation commands associated with the current implementation of the `mixed` command in Stata 13 do not allow one to display blocks of the estimated V_j matrix in the output.

Recall that prior to fitting the models in SAS, we first sorted the data by PATIENT, TOOTH, and TIME, as shown in the following syntax. This was done to facilitate reading the output for the marginal variance-covariance and correlation matrices.

```
proc sort data = veneer;
by patient tooth time;
run;
```

The estimated V_j matrix for patient 1 shown in the SAS output that follows can be generated by using the v = 1 option in either of the `random` statements in the `proc mixed` syntax for Model 7.3 (see Subsection 7.4.1). The following output displays an 8×8 matrix, because there are eight observations for the first patient, corresponding to measurements at 3 months and at 6 months for each of the patient's four treated teeth. If another patient had three teeth, we would have had a 6×6 marginal covariance matrix.

The estimated marginal variances and covariances for each tooth are represented by 2×2 blocks, along the diagonal of the V_j matrix. Note that the 2×2 tooth-specific covariance matrix has the same values across all teeth. The estimated marginal variance for a given observation on a tooth at 3 months is 155.74, whereas that at 6 months is 444.19. The estimated marginal covariance between observations on the same tooth at 3 and at 6 months is 62.60.

```
                    Estimated V Matrix for PATIENT 1

Row    Col1       Col2       Col3       Col4       Col5       Col6       Col7       Col8
--------------------------------------------------------------------------------------------
 1    155.7400    62.6020    59.4182    15.1476    59.4182    15.1476    59.4182    15.1476
 2     62.6020   444.1900    15.1476   347.8600    15.1476   347.8600    15.1476   347.8600
 3     59.4182    15.1476   155.7400    62.6020    59.4182    15.1476    59.4182    15.1476
 4     15.1476   347.8600    62.6020   444.1900    15.1476   347.8600    15.1476   347.8600
 5     59.4182    15.1476    59.4182    15.1476   155.7400    62.6020    59.4182    15.1476
 6     15.1476   347.8600    15.1476   347.8600    62.6020   444.1900    15.1476   347.8600
 7     59.4182    15.1476    59.4182    15.1476    59.4182    15.1476   155.7400    62.6020
 8     15.1476   347.8600    15.1476   347.8600    15.1476   347.8600    62.6020   444.1900
```

The following SAS output shows the estimated marginal *correlation* matrix for patient 1 generated by `proc mixed`, obtained by using the `vcorr = 1` option in either of the `random` statements in the syntax for Model 7.3. The 2×2 submatrices along the diagonal represent the marginal correlations between the two measurements on any given tooth at 3 and at 6 months.

```
              Estimated V Correlation Matrix for PATIENT 1

   Row   Col1     Col2     Col3     Col4     Col5     Col6     Col7     Col8
   -----------------------------------------------------------------------
    1   1.00000  0.23800  0.38150  0.05759  0.38150  0.05759  0.38150  0.05759
    2   0.23800  1.00000  0.05759  0.78310  0.05759  0.78310  0.05759  0.78310
    3   0.38150  0.05759  1.00000  0.23800  0.38150  0.05759  0.38150  0.05759
    4   0.05759  0.78310  0.23800  1.00000  0.05759  0.78310  0.05759  0.78310
    5   0.38150  0.05759  0.38150  0.05759  1.00000  0.23800  0.38150  0.05759
    6   0.05759  0.78310  0.05759  0.78310  0.23800  1.00000  0.05759  0.78310
    7   0.38150  0.05759  0.38150  0.05759  0.38150  0.05759  1.00000  0.23800
    8   0.05759  0.78310  0.05759  0.78310  0.05759  0.78310  0.23800  1.00000
```

We note in this matrix that the estimated covariance parameters for Model 7.3 imply that observations on the same tooth are estimated to have a rather small marginal correlation of approximately 0.24.

If we re-sort the data by PATIENT, TIME, and TOOTH and refit Model 7.3 using SAS `proc mixed`, the rows and columns in the correlation matrix are reordered correspondingly, and we can more readily view the blocks of marginal correlations among observations on all teeth at each time point.

```
proc sort data = veneer;
by patient time tooth;
run;
```

We run identical syntax to fit Model 7.3 and display the resulting marginal correlation matrix for patient 1:

```
              Estimated V Correlation Matrix for PATIENT 1

   Row   Col1     Col2     Col3     Col4     Col5     Col6     Col7     Col8
   -----------------------------------------------------------------------
    1   1.00000  0.38150  0.38150  0.38150  0.23800  0.05759  0.05759  0.05759
    2   0.38150  1.00000  0.38150  0.38150  0.05759  0.23800  0.05759  0.05759
    3   0.38150  0.38150  1.00000  0.38150  0.05759  0.05759  0.23800  0.05759
    4   0.38150  0.38150  0.38150  1.00000  0.05759  0.05759  0.05759  0.23800
    5   0.23800  0.05759  0.05759  0.05759  1.00000  0.78310  0.78310  0.78310
    6   0.05759  0.23800  0.05759  0.05759  0.78310  1.00000  0.78310  0.78310
    7   0.05759  0.05759  0.23800  0.05759  0.78310  0.78310  1.00000  0.78310
    8   0.05759  0.05759  0.05759  0.23800  0.78310  0.78310  0.78310  1.00000
```

In this output, we focus on the 4×4 blocks that represent the correlations among the four teeth for patient 1 at time 1 and at time 2. It is readily apparent that the marginal correlation at time 1 is estimated to have a constant value of 0.38, whereas the marginal correlation among observations on the four teeth at time 2 is estimated to have a higher value of 0.78.

As noted earlier, the estimated marginal correlation between the observations on tooth 1 at time 1 and time 2 (displayed in this output in row 1, column 5) is 0.24. We also note that the estimated marginal correlation of observations on tooth 1 at time 1 and the other three teeth at time 2 (displayed in this output in row 1, columns 6 through 8) is rather low, not surprisingly, and is estimated to be 0.06.

7.9 Diagnostics for the Final Model

In this section, we check the assumptions for the REML-based fit of Model 7.3, using informal graphical procedures available in Stata. Similar plots can be generated using the other four software packages by saving the conditional residuals, conditional predicted values, and EBLUPs of the random effects based on the fit of Model 7.3. We include syntax for performing these diagnostics in the other software packages on the book's web page (see Appendix A).

7.9.1 Residual Diagnostics

We first assess the assumption of constant variance for the residuals in Model 7.3. Figure 7.4 presents a plot of the standardized conditional residuals vs. the conditional predicted values (based on the fit of Model 7.3) to assess whether the variance of the residuals is constant. The final command used to fit Model 7.3 in Stata is repeated from Subsection 7.4.4 as follows:

```
. * Model 7.3 (REML).
. mixed gcf time base_gcf cda age
|| patient: time, cov(unstruct) || tooth: , variance reml
```

After fitting Model 7.3, we save the standardized residuals in a new variable named ST_RESID, by using the `predict` post-estimation command in conjunction with the `rstandard` option (this option requests that standardized residuals be saved in the data set):

```
. predict st_resid, rstandard
```

We also save the conditional predicted GCF values (including the EBLUPs) in a new variable named PREDVALS, by using the `fitted` option:

```
. predict predvals, fitted
```

We then use the two new variables to generate the fitted-residual scatter plot in Figure 7.4, with a reference line at zero on the y-axis:

```
. twoway (scatter st_resid predvals), yline(0)
```

The plot in Figure 7.4 suggests nonconstant variance in the residuals as a function of the predicted values, and that a variance-stabilizing transformation of the GCF response variable (such as the square-root transformation) may be needed, provided that no important fixed effects of relevant covariates have been omitted from the model.

Analysts interested in generating marginal predicted values for various subgroups defined by the covariates and factors included in a given model, in addition to plots of those marginal predicted values, can make use of the post-estimation commands `margins` and `marginsplot` (where the latter command needs to directly follow the former command) to visualize these

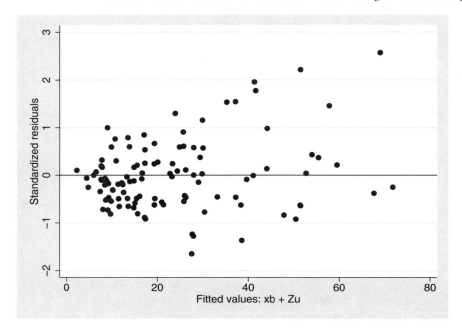

FIGURE 7.4: Residual vs. fitted plot based on the fit of Model 7.3.

predicted values. For example, to plot marginal predicted values of GCF at TIME = 0, TIME = 3, and TIME = 6 in addition to 95% confidence intervals for the predicted values, the following two commands can be used (plot not shown):

```
. margins, at(time=(0,3,6))
. marginsplot
```

These plots can be useful for illustrating predicted values of the dependent variable based on a given model, and can be especially useful for interpreting more complex interactions. Stata users can submit the command **help margins** for more details.

The assumption of normality for the conditional residuals can be checked by using the **qnorm** command to generate a normal Q–Q plot:

```
. qnorm st_resid
```

The resulting plot in Figure 7.5 suggests that the distribution of the conditional residuals deviates from a normal distribution. This further suggests that a transformation of the response variable may be warranted, provided that no fixed effects of important covariates or interaction terms have been omitted from the final model. In this example, a square-root transformation of the response variable (GCF) prior to model fitting was found to improve the appearance of both of these diagnostic plots, suggesting that such a transformation would be recommended before making any final inferences about the parameters in this model.

7.9.2 Diagnostics for the Random Effects

We now check the distributions of the predicted values (EBLUPs) for the three random effects in Model 7.3. After refitting Model 7.3, we save the EBLUPs of the two random

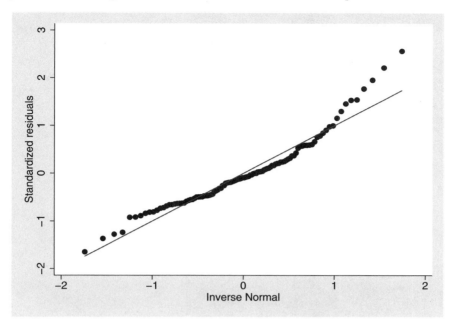

FIGURE 7.5: Normal Q–Q plot of the standardized residuals based on the fit of Model 7.3.

patient effects and the nested random tooth effects in three new variables, again using the `predict` command:

```
. predict pat_eblups*, reffects level(patient)
. predict tooth_eblups, reffects level(tooth)
```

The first command saves the predicted random effects (EBLUPs) for each level of PATIENT in new variables named PAT_EBLUPS1 (for the random TIME effects) and PAT_EBLUPS2 (for the random effects associated with the intercept) in the data set. The asterisk (*) requests that a single new variable be created for each random effect associated with the levels of PATIENT. In this case, two new variables are created, because there are two random effects in Model 7.3 associated with each patient. The second command saves the EBLUPs of the random effects associated with the intercept for each tooth in a new variable named TOOTH_EBLUPS.

After saving these three new variables, we generate a new data set containing a single case per patient, and including the individual patient EBLUPs (the original data set should be saved before creating this collapsed data set):

```
. save "C:\temp\veneer.dta", replace
. collapse pat_eblups1 pat_eblups2, by(patient)
```

We then generate normal Q–Q plots for each set of patient-specific EBLUPs and check for outliers:

```
. qnorm pat_eblups1, ytitle(EBLUPs of Random Patient TIME Effects)
. graph save "C:\temp\figure76_part1.gph"
```

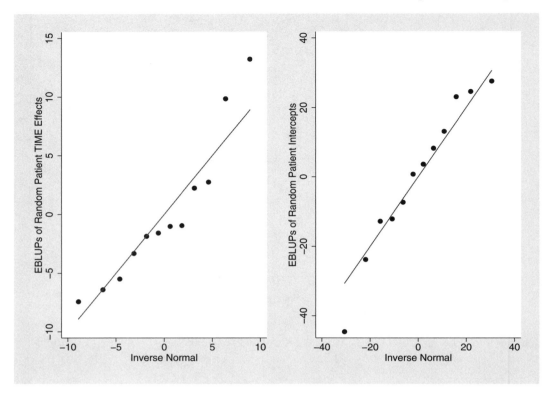

FIGURE 7.6: Normal Q–Q plots for the EBLUPs of the random patient effects.

```
. qnorm pat_eblups2, ytitle(EBLUPs of Random Patient Intercepts)
. graph save "C:\temp\figure76_part2.gph"

. graph combine "C:\temp\figure76_part1.gph"
"C:\temp\figure76_part2.gph"
```

Note that we make use of the **graph save** and **graph combine** commands to save the individual plots and then combine them into a single figure.

Figure 7.6 suggests that there are two positive outliers in terms of the random TIME effects (EBLUPs greater than 5) associated with the patients (left panel). We can investigate selected variables for these patients after opening the original data set (including the saved EBLUPs) once again:

```
. use "C:\temp\veneer.dta", clear
. list patient tooth gcf time age cda base_gcf if pat_eblups1 > 5
```

```
+--------------------------------------------------+
|    patient tooth gcf time age cda     base_gcf   |
|--------------------------------------------------|
|  1.  1       6    11   3    46  4.666667  17      |
|  2.  1       6    68   6    46  4.666667  17      |
|  3.  1       7    13   3    46  4.666667  22      |
|  4.  1       7    47   6    46  4.666667  22      |
|  5.  1       8    14   3    46  5.000000  18      |
|--------------------------------------------------|
|  6.  1       8    58   6    46  5.000000  18      |
|  7.  1       9    10   3    46  3.333333  12      |
|  8.  1       9    57   6    46  3.333333  12      |
|  9.  1      10    14   3    46  8.666667  10      |
| 10.  1      10    44   6    46  8.666667  10      |
|--------------------------------------------------|
| 11.  1      11    11   3    46  5.666667  17      |
| 12.  1      11    53   6    46  5.666667  17      |
| 85. 10       6    29   3    36  2.333333  27      |
| 86. 10       6    87   6    36  2.333333  27      |
| 87. 10       7    28   3    36  8.666667  10      |
|--------------------------------------------------|
| 88. 10       7    61   6    36  8.666667  10      |
| 89. 10       8    17   3    36  6.333333  25      |
| 90. 10       8    47   6    36  6.333333  25      |
| 91. 10       9    11   3    36  6.666667   7      |
| 92. 10       9    42   6    36  6.666667   7      |
|--------------------------------------------------|
| 93. 10      10    48   3    36  6.333333  15      |
| 94. 10      10    65   6    36  6.333333  15      |
| 95. 10      11    55   3    36  4.666667  19      |
| 96. 10      11    70   6    36  4.666667  19      |
+--------------------------------------------------+
```

As expected, these two patients (PATIENT = 1 and 10) consistently have large increases in GCF as a function of TIME for all of their teeth, and their data should be checked for validity.

The tooth-specific random effects can be assessed in a similar manner by first creating a tooth-specific data set containing only the PATIENT, TOOTH, and TOOTH_EBLUP variables (after generating predicted values of the random effects in the original data set, as shown above), and then generating a normal Q–Q plot:

```
. collapse tooth_eblups, by(patient tooth)
. qnorm tooth_eblups, ytitle(EBLUPs of Random Tooth Effects)
```

The resulting plot (not displayed) does not provide any evidence of extremely unusual random tooth effects.

7.10   Software Notes and Recommendations

7.10.1   ML vs. REML Estimation

In this chapter, we introduce for the first time the HMLM2 (hierarchical multivariate linear models) procedure, which was designed for analyses of clustered longitudinal data sets in

HLM. Unlike the LMM procedures in SAS, SPSS, R, and Stata, this procedure only uses ML estimation. The procedures in SAS, SPSS, R, and Stata provide users with a choice of either REML or ML estimation when fitting these models.

This difference could have important consequences when developing a model. We recommend using likelihood ratio tests based on REML estimation when testing hypotheses, such as Hypotheses 7.1 and 7.2, involving covariance parameters. This is not possible if the models are fitted using ML estimation.

A second important consequence of using ML estimation is that the covariance parameter estimates are known to be biased. This can result in smaller estimated standard errors for the estimates of the fixed effects in the model and also has implications for the fixed-effect parameters that are estimated. Some of these differences were apparent in Tables 7.6 through 7.8.

7.10.2   The Ability to Remove Random Effects from a Model

The HMLM2 procedure requires that at least one random effect be specified in the model at Level 3 and at Level 2 of the data. The procedures in SAS, SPSS, R, and Stata all allow more flexibility in specifying and testing which levels of the data (e.g., patient or teeth nested within patients) should have random effects included in the model.

Although the HMLM2 procedure is more restrictive than some of the other software procedures in this sense, it also ensures that the hierarchy of the data is maintained in the analysis. Users of SAS, SPSS, R, and Stata must think carefully about how the hierarchy of the data is specified in the model, and then correctly specify the appropriate random effects in the syntax. HMLM2 forces the hierarchical structure of these data sets to be taken into consideration.

7.10.3   Considering Alternative Residual Covariance Structures

With the exception of the `lmer()` function in R, all of the other LMM procedures considered in this chapter allow users to fit models with nonidentity residual covariance (R_{ij}) matrices. The **unstructured** residual covariance matrix (Model 7.2A) is not available in HMLM2 when random effects are also considered simultaneously, and we had to use an alternative setup of the model to allow us to fit the compound symmetry structure (Model 7.2B) using HMLM2 (see Subsection 7.4.5).

In this analysis, we found that the identity residual covariance structure was the better and more parsimonious choice for our models, but this would not necessarily be the case in analyses of other data sets. Heterogeneity of variances and correlation of residuals is a common feature in longitudinal data sets, and the ability to accommodate a wide range of residual covariance structures is very important. We recommend that analysts of clustered longitudinal data sets with more than two longitudinal observations consider alternative covariance structures for the residuals, and attempt to identify the structures that provide the best fit to a given set of data. The information criteria (AIC, BIC, etc.) can be useful in this regard.

In the Dental Veneer example, there were only a small number of residual covariance structures that could be considered for the 2×2 R_{ij} matrix, because it contained only three parameters, at the most, and aliasing with other covariance parameters was involved. In other data sets with more longitudinal observations, a wider variety of residual covariance structures could (and should) be considered. The procedures in SAS, SPSS, Stata, HLM, and the `lme()` function in R offer flexibility in this regard, with **proc mixed** in SAS having the largest list of available residual covariance structures.

7.10.4 Aliasing of Covariance Parameters

We had difficulties when fitting Models 7.2A and 7.2B because of aliasing (nonidentifiability) of the covariance parameters. The problems with these models arose because we were specifying random effects at two levels of the data (patient and teeth within patients), as well as an additional residual covariance at the tooth level. If we had more than two observations per tooth, this would have been a problem for Model 7.2B only.

The symptoms of aliasing of covariance parameters manifest themselves in different fashions in the different software programs. For Model 7.2A, SAS complained in a `NOTE` in the log that the estimated Hessian matrix (which is used to compute the standard errors of the estimated covariance parameters) was not positive-definite. Users of SAS need to be aware of these types of messages in the log file. SAS also reported a value of zero for the UN(2,2) covariance parameter (i.e., the residual variance at time 2) and did not report a standard error for this parameter estimate in the output. For Model 7.2B, SAS did not provide estimates due to too many likelihood evaluations.

SPSS produced a warning message in the output window about lack of convergence for both Models 7.2A and 7.2B. In this case, results from SPSS should not be interpreted, because the estimation algorithm has not converged to a valid solution for the parameter estimates.

After fitting Models 7.2A and 7.2B with the `lme()` function in R, attempts to use the `intervals()` function to obtain confidence intervals for the estimated covariance parameters resulted in extremely wide and unrealistic intervals. Simply fitting these two models in R did not indicate any problems with the model specification, but the intervals for these parameters provide an indication of instability in the estimates of the standard errors for these covariance parameter estimates. Similar problems were apparent when fitting Model 7.2A in Stata, and Model 7.2B could not be fitted in Stata due to a lack of convergence (similar to the procedures in SAS and SPSS).

We were not able to fit Model 7.2A using HMLM2, because the **unstructured** residual covariance matrix is not available as an option in a model that also includes random effects. In addition, HMLM2 reported a generic message for Model 7.2B that stated "Invalid info, score, or likelihood" and did not report parameter estimates for this model.

In general, users of these software procedures need to be very cautious about interpreting the output for covariance parameters. We recommend always examining the estimated covariance parameters and their standard errors to see if they are reasonable. The procedures in SAS, SPSS and Stata make this relatively easy to do. In R, the `intervals()` function is helpful (when using the `lme()` function). HMLM2 is fairly direct and obvious about problems that occur, but it is not very helpful in diagnosing this particular problem.

Readers should be aware of potential problems when fitting models to clustered longitudinal data, pay attention to warnings and notes produced by the software, and check model specification carefully. We considered three possible structures for the residual covariance matrix in this example to illustrate potential problems with aliasing. We advise exercising caution when fitting these models so as not to overspecify the covariance structure.

7.10.5 Displaying the Marginal Covariance and Correlation Matrices

The ability to examine implied marginal covariance matrices and their associated correlation matrices can be very helpful in understanding an LMM that has been fitted (see Section 7.8). SAS makes it easy to do this for any subject desired, by using the `v =` and `vcorr =` options in the `random` statement. In fact, `proc mixed` in SAS is currently the only procedure that allows users to examine the marginal covariance matrix implied by a LMM fitted to a clustered longitudinal data set with three levels.

7.10.6   Miscellaneous Software Notes

1. SPSS: The syntax to set up the subject in the `RANDOM` subcommand for TOOTH nested within PATIENT is (TOOTH*PATIENT), which appears to be specifying TOOTH crossed with PATIENT, but is actually the syntax used for nesting. Alternatively, one could use a `RANDOM` subcommand of the form `/RANDOM tooth(patient)`, without any SUBJECT variable(s), to include nested random tooth effects in the model; however, this would not allow one to specify multiple random effects at the tooth level.

2. HMLM2: This procedure requires that the Level 1 data set include an indicator variable for each time point. For instance, in the Dental Veneer example, the Level 1 data set needs to include two indicator variables: one for observations at 3 months, and a second for observations at 6 months. These indicator variables are not necessary when using the procedures in SAS, SPSS, R, and Stata.

7.11   Other Analytic Approaches

7.11.1   Modeling the Covariance Structure

In Section 7.8 we examined the marginal covariance of observations on patient 1 implied by the random effects specified for Model 7.3. As discussed in Chapter 2, we can model the marginal covariance structure directly by allowing the residuals for observations on the same tooth to be correlated.

For the Dental Veneer data, we can model the tooth-level marginal covariance structure implied by Model 7.3 by removing the random tooth-level effects from the model and specifying a compound symmetry covariance structure for the residuals, as shown in the following SAS syntax for Model 7.3A:

```
title "Alternative Model 7.3A";
proc mixed data = veneer noclprint covtest;
class patient tooth cattime;
model gcf = time base_gcf cda age / solution outpred = resids;
random intercept time / subject = patient type = un solution
v = 1 vcorr = 1;
repeated cattime / subject = tooth(patient) type=cs;
run;
```

We can view the estimated covariance parameters for Model 7.3A in the following output:

```
          Covariance Parameter Estimates (Model 7.3A)

                                      Standard    Z
Cov Parm    Subject         Estimate    Error    Value  Pr Z
UN(1,1)     PATIENT         524.9700   253.0100   2.07  0.0190
UN(2,1)     PATIENT        -140.4200    66.5737  -2.11  0.0349
UN(2,2)     PATIENT          41.8869    18.7993   2.23  0.0129
CS          TOOTH (PATIENT)  47.4573    16.6304   2.85  0.0043
Residual                     48.8675    10.5053   4.65 <0.0001
```

The analogous syntax and output for Model 7.3 are shown below for comparison. Note that the output for the models is nearly identical, except for the labels assigned to the covariance parameters in the output. The −2 REML log-likelihood values are the same for the two models, as are the AIC and BIC.

```
title "Model 7.3";
proc mixed data = data.veneer noclprint covtest;
class patient tooth cattime;
model gcf = time base_gcf cda age / solution outpred = resids;
random intercept time / subject = patient type = un
v = 1 vcorr = 1;
random intercept / subject = tooth(patient) solution;
run;
```

```
             Covariance Parameter Estimates (Model 7.3)

                                       Standard      Z
   Cov Parm    Subject         Estimate    Error    Value   Pr Z
   UN(1,1)     PATIENT         524.9500   252.9900   2.07   0.0190
   UN(2,1)     PATIENT        -140.4200    66.5725  -2.11   0.0349
   UN(2,2)     PATIENT          41.8874    18.7998   2.23   0.0129
   Intercept   TOOTH (PATIENT)  47.4544    16.6298   2.85   0.0022
   Residual                     48.8703    10.5059   4.65  <0.0001
```

It is important to note that the model setup used for Model 7.3 only allows for positive marginal correlations among observations on the same tooth over time, because the implied marginal correlations are a result of the variance of the random intercepts associated with each tooth. The specification of Model 7.3A allows for *negative* correlations among observations on the same tooth.

7.11.2   The Step-Up vs. Step-Down Approach to Model Building

The step-up approach to model building commonly used in the HLM literature (Raudenbush & Bryk, 2002) begins with an "unconditional" model, containing only the intercept and random effects. The reduction in the estimated variance components at each level of the data is then monitored as fixed effects are added to the model. The mean structure is considered complete when adding fixed-effect terms provides no further reduction in the variance components. This step-up approach to model building (see Chapter 4, or Subsection 2.7.2) could also be considered for the Dental Veneer data.

The step-down (or top-down) approach involves starting the analysis with a "loaded" mean structure and then working on the covariance structure. One advantage of this approach is that the covariances can then be truly thought of as measuring "variance" and not simply variation due to fixed effects that have been omitted from the model. An advantage of using the step-up approach is that the effect of each covariate on reducing the model "variance" can be viewed for each level of the data. If we had used the step-up approach and adopted a strategy of only including significant main effects in the model, our final model for the Dental Veneer data might have been different from Model 7.3.

7.11.3   Alternative Uses of Baseline Values for the Dependent Variable

The baseline (first) value of the dependent variable in a series of longitudinal measures may be modeled as simply one of the repeated outcome measures, or it can be considered as a baseline covariate, as we have done in the Dental Veneer example.

There are strong theoretical reasons for treating the baseline value as another measure of the outcome. If the subsequent measures represent values on the dependent variable, measured with error, then it is difficult to argue that the first of the series is "fixed," as required for covariates. In this sense it is more natural to consider the entire sequence, including the baseline values, as having a multivariate normal distribution. However, when using this approach, if a treatment is administered after the baseline measurement, the treatment effect must be modeled as a treatment by time interaction if treatment groups are similar at baseline. A changing treatment effect over time may lead to a complex interaction between treatment and a function of time.

Those who consider the baseline value as a covariate argue that the baseline value is inherently different from other values in the series. The baseline value is often taken prior to a treatment or intervention, as in the Dental Veneer data. There is a history of including baseline values as covariates, particularly in clinical trials. The inclusion of baseline covariates in a model may substantially reduce the residual variance (because of strong correlations with the subsequent values), thus increasing the power of tests for other covariates. The inclusion of baseline covariates also allows an appropriate adjustment for baseline imbalance between groups.

Finally, the values in the subsequent series of response measurements may be a function of the initial value. This can happen in instances when there is large room for improvement when the baseline level is poor, but little room for improvement when the baseline level is already good. This situation is easily modeled with an interaction between time and the baseline covariate, but more difficult to handle in the model considering the baseline value as one of the outcome measures.

In summary, we find both model frameworks to be useful in different settings. The longitudinal model, which includes baseline values as measures on the dependent variable, is more elegant; the model considering the first outcome measurement as a baseline covariate is often more practical.

8

**Models for Data with Crossed Random Factors:
The SAT Score Example**

8.1 Introduction

This chapter introduces the analysis of data sets with **crossed random factors,** where there are multiple random factors with levels that are *crossed* with each other, rather than having an explicit nesting structure. For example, in Chapter 4, we analyzed a data set where students were nested in classrooms, and classrooms were nested within schools. The classroom and school ID variables were both random factors, where the levels of these variables were randomly selected from larger populations of classrooms and schools. Further, the levels of the classroom factor were *nested* within levels of the school factor; a given classroom could not exist in multiple schools. In this chapter, we consider an example data set where there are repeated measurements of math scores on an SAT test (Student Aptitude Test) collected on randomly sampled students within a given school, and those students have multiple teachers from the school over time. As a result, both students and teachers have multiple measures on the dependent variable associated with them, but the levels of these two random factors (student ID and teacher ID) are *crossed* with each other.

Linear mixed models with **crossed random effects** enable the potential correlations of the repeated observations associated with each level of these crossed random factors to be modeled simultaneously. For example, we might expect between-student variance in math performance over time; at the same time, we might expect that some teachers are better math instructors, resulting in between-teacher variance in the math scores. These models enable simultaneous estimation of the components of variance associated with the levels of the crossed random factors, and assessment of which random factor tends to contribute the most to variability in measures on the dependent variable.

As discussed in Chapter 2, models with *crossed* random effects tend to be more difficult to estimate. Estimation is facilitated by the use of **sparse matrices** in model specification and maximum likelihood estimation, but different software procedures will tend to use different algorithms when fitting these types of models. We do not highlight one package in particular in this chapter, and discuss some of the notable differences between the software procedures in Section 8.10.

8.2 The SAT Score Study

8.2.1 Study Description

The data used in this example have been borrowed from the example data sets provided by the developers of the HLM software. Specifically, we analyze a subset of the data from an

TABLE 8.1: Sample of the SAT Score Data Set in the "Long" Format

STUDID	TCHRID	MATH	YEAR
13099	14433	631	−1
13100	14433	596	−1
13100	14484	575	0
13100	14755	591	1
13101	14433	615	−1
13101	14494	590	0
13102	14433	621	−1
13102	14494	624	0
13102	14545	611	1
. . .			

Note: "..." indicates portion of the data not displayed.

educational study, focusing on one of the original 67 schools in the study, and the teachers and students within that school. In this study, students were given a math test in grades 3, 4, and 5 that was similar to the SAT math test used to evaluate college applicants. While different treatments were applied in some years for randomly sampled students, we do not consider treatment effects in the analysis in this chapter. We focus mainly on the components of variance in the math scores associated with students and teachers, in addition to change over time in the scores. The subset of the original data set that we work with in this chapter features 234 repeated measures collected from 122 students who have been instructed by a total of 12 teachers.

Before we carry out an analysis of the SAT score data using the procedures in SAS, SPSS, R, Stata, or HLM, we need to make sure that the data set has been restructured into the "long" format (similar to the other case studies presented in the book). A portion of the SAT score data in the "long" format is shown in Table 8.1.

The portion of the SAT score data set presented in Table 8.1 demonstrates some key features of both this specific data set and data sets with crossed random factors more generally. First, note that the YEAR variable represents a centered version of the grade in which the student was measured (centered at Grade 4). Second, note that there are repeated measures on both students and teachers; students are measured repeatedly over time (with not all students measured in each of the three years), and different students might have had the same teacher in a given year. For example, teacher ID 14433 instructed all four of the students in Table 8.1 in grade 3. Students are therefore not nested within teachers. This crossed structure, illustrated in the matrix below, introduces multiple levels of potentially correlated observations in the data set, and linear mixed models including random effects for both students and teachers enable decomposition of the components of variance due to each of these crossed random factors.

Student ID	Teacher ID 14433	14484	14494	14545	14755	...
13099	X					
13100	X	X			X	
13101	X		X			
13102	X		X	X		
...						

We note below how this crossed structure, where an "X" indicates an observation on the dependent variable measuring math achievement on the SAT for a given student–teacher pairing, results in a matrix with several empty cells, where a particular student was not instructed by a particular teacher. This is why methods using *sparse matrices* for estimation are the most efficient when fitting these types of models.

In the analysis in this chapter, we consider models including a fixed effect of YEAR (enabling assessment of change over time in the mean math score) and random effects for the levels of TCHRID (teachers) and STUDID (students), to see whether the variation in math scores is being driven by students or teachers.

To summarize, the following variables are included in the SAT score data set:

> - **STUDID** = Unique Student ID
>
> - **TCHRID** = Unique Teacher ID
>
> - **MATH** = Score on SAT Math Test
>
> - **YEAR** = Year of Measurement (-1 = Grade 3, 0 = Grade 4, 1 = Grade 5)

Sorting the data set by STUDID or TCHRID is not required for using the software procedures that enable fitting models with crossed random effects.

8.2.2 Data Summary

In this section, we consider some exploratory graphical analyses of the SAT score data, using the R software. These plots can easily be generated using the other four software packages as well.

We first consider the distribution of the SAT scores as a function of the year (or grade) in which the data were collected. The following R syntax can be used to read in the data (with variable names in the first row) from the `C:\temp` directory, and then generate the side-by-side box plots in Figure 8.1.

```
> sat <- read.csv("C:\\temp\\school_data_final.csv", h=T)
> attach(sat)
> plot(math ~ factor(year), xlab = "Year (Centered at Grade 4)",
ylab = "SAT Math Score")
```

Figure 8.1 shows evidence of a fairly clear linear increase in performance on the math portion of the SAT among these students as a function of year of measurement. These plots result in an expectation that the fixed effect of YEAR (representing a linear rate of change in the SAT math scores) will likely be positive and significant in our analysis. We also see evidence of fairly constant variance in the scores as a function of YEAR.

We now consider variability in the math scores among the teachers and students. First, we examine side-by-side box plots for the teachers, using the following R syntax to generate Figure 8.2.

```
> plot(math ~ factor(tchrid), xlab = "Teacher ID",
ylab = "SAT Math Score")
```

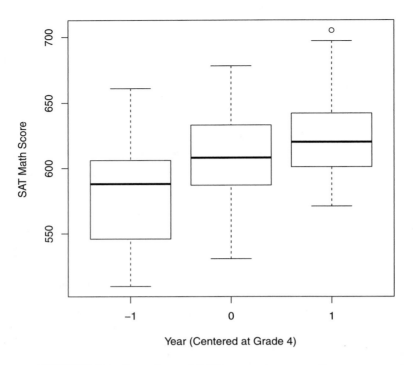

FIGURE 8.1: Box plots of SAT scores by year of measurement.

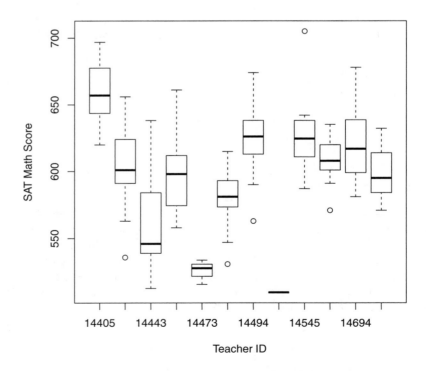

FIGURE 8.2: Box plots of SAT scores for each of the 13 teachers in the SAT score data set.

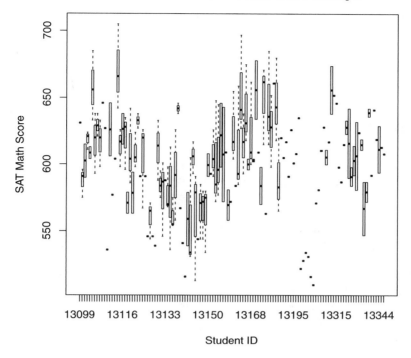

FIGURE 8.3: Box plots of SAT scores for each of the 122 students in the SAT score data set.

Figure 8.2 provides evidence of substantial variability in performance on the math portion of the SAT among these 13 teachers, suggesting that a model for these data should include random teacher effects.

Finally, we consider variability among the students, using the following R syntax to generate Figure 8.3.

```
> plot(math ~ factor(studid), xlab = "Student ID",
ylab = "SAT Math Score")
```

We also note a fair amount of variability among the students in terms of the SAT scores in Figure 8.3, and that several students have only been measured once. These results suggest that random effects associated with the students should also be included in a model for these data (meaning that the random effects of students would be crossed with the random effects of teachers, given the structure of these data).

We now consider the model that we will fit to the SAT score data.

8.3   Overview of the SAT Score Data Analysis

In this chapter, we do not consider explicit model-building steps for the SAT score data. Instead, we simply fit a single model in each of the different software procedures that includes all of the fixed effects and crossed random effects that we are interested in evaluating. We then compare the methods and syntax used to fit the model across the software procedures, in addition to the resulting estimates produced by the procedures.

8.3.1 Model Specification

8.3.1.1 General Model Specification

We specify Model 8.1 in this subsection. The general specification of Model 8.1 corresponds closely to the syntax used to fit this model when using the procedures in SAS, SPSS, Stata, and R.

The value of MATH_{tij} in a given year indexed by t ($t = 1, 2, 3$) for the i-th student ($i = 1, 2, ..., 122$) being instructed by the j-th teacher ($j = 1, 2, ..., 13$) can be written as follows:

$$\begin{aligned} \text{MATH}_{tij} = &\quad \beta_0 + \beta_1 \times \text{YEAR}_{tij} \quad \} \text{ fixed} \\ &+ u_i + v_j + \varepsilon_{tij} \qquad \} \text{ random} \end{aligned} \tag{8.1}$$

The fixed-effect parameters are represented by β_0 and β_1. The fixed intercept β_0 represents the expected value of MATH_{tij} when YEAR_{tij} in equal to zero (or Grade 4, given the centering). The parameter β_1 represents the fixed effect of the YEAR variable, which can be interpreted as the linear rate of change in the expected math score associated with a one-year increase.

The u_i term represents the random effect associated with student i, and v_j represents the random effect associated with teacher j. We assume that the random effects arise from two independent normal distributions:

$$u_i \sim \mathcal{N}(0, \sigma_i^2), v_j \sim \mathcal{N}(0, \sigma_j^2)$$

We have a total of 135 random effects in this model, corresponding to the 122 students and the 13 teachers. The overall resulting $\boldsymbol{D}$ matrix corresponding to this type of model with crossed random effects is a diagonal matrix with $122 + 13 = 135$ columns and 135 rows, corresponding to the 135 random effects of the students and teachers. The first 122×122 block-diagonal portion of the matrix will have the variance of the random student effects, σ_i^2, on the diagonal, and zeroes off the diagonal. The remaining 13×13 block-diagonal portion of this matrix will have the variance of the random teacher effects, σ_j^2, on the diagonal, and zeroes off the diagonal.

The residuals associated with the math score observations are assumed to be independent of the two random effects, and follow a normal distribution:

$$\varepsilon_{tij} \sim \mathcal{N}(0, \sigma^2)$$

We next consider the hierarchical specification of a model with crossed random effects.

8.3.1.2 Hierarchical Model Specification

We now present an equivalent hierarchical specification of Model 8.1, using the same notation as in Subsection 8.3.1.1. The hierarchical model has two components, reflecting contributions from the two levels of the data: the repeated measures at Level 1, and the crossing of students and teachers at Level 2 (i.e., the repeated measures of the dependent variable are associated with both students and teachers simultaneously). We write the **Level 1** component as

Level 1 Model (Repeated Measures)

$$\text{MATH}_{tij} = b_{0ij} + b_{1ij} \times \text{YEAR}_{tij} + \varepsilon_{tij} \tag{8.2}$$

where the residuals (ε_{tij}) have the distribution defined in the general specification of Model 8.1 in Subsection 8.3.1.1, with constant variance.

In the **Level 1** model, we assume that MATH_{tij}, the math SAT score for an individual combination of student i and teacher j at time t, follows a linear model, defined by the intercept specific to the student–teacher combination, b_{0ij}, and the effect of YEAR specific to the student–teacher combination, b_{1ij}.

The **Level 2** model then describes variation between the various student–teacher combinations in terms of the random intercepts and time effects, using the crossed random effects:

Level 2 Model (Student–Teacher Combination)

$$b_{0ij} = \beta_0 + u_i + v_j$$
$$b_{1ij} = \beta_1 \tag{8.3}$$

where

$$u_i \sim \mathcal{N}(0, \sigma_i^2), v_j \sim \mathcal{N}(0, \sigma_j^2)$$

In this **Level 2** model, the intercept b_{0ij} for student i and teacher j depends on the overall fixed intercept, β_0, the random effect associated with student i, u_i, and the random effect associated with teacher j, v_j. We note that crossed random effects associated with the students and teachers are not included in the Level 2 equation for the effect of YEAR in this simple example (although they could be more generally, if one wished to test hypotheses about variance among students or teachers in terms of the YEAR effects). As a result, the YEAR effect specific to a student–teacher combination is simply defined by the overall fixed effect of YEAR, β_1.

By substituting the expressions for b_{0ij} and b_{1ij} from the Level 2 model into the Level 1 model, we obtain the general linear mixed model (LMM) with crossed random effects that was specified in (8.1).

8.3.2   Hypothesis Tests

We test a simple set of three hypotheses in this case study, related to the two crossed random effects and the overall fixed effect of YEAR on the SAT math scores for these students.

Hypothesis 8.1. The random effects associated with students (u_i) can be omitted from Model 8.1.

Model 8.1 has a single random effect, u_i, associated with the intercept for each student. To test Hypothesis 8.1, we fit a model (Model 8.2) excluding the random student effects (i.e., a two-level model with repeated measures nested within teachers), and use a REML-based likelihood ratio test. The test statistic is calculated by subtracting the –2 REML log-likelihood value for Model 8.1 (the reference model) from that for Model 8.2 (the nested model). The asymptotic null distribution of the test statistic is a mixture of χ_0^2 and χ_1^2 distributions, with equal weights of 0.5 (see Subsection 2.6.2.2).

We once again remind readers that likelihood ratio tests, such as the one used for Hypothesis 8.1, rely on asymptotic (large-sample) theory, so we would not usually carry out this type of test for such a small data set. Rather, in practice, the random effects would probably be retained without testing, so that the appropriate marginal variance-covariance structure would be obtained for the data set. We present the calculation of this likelihood ratio test for the random effects (and those that follow in this chapter) strictly for illustrative purposes.

Hypothesis 8.2. The random effects associated with teachers (v_j) can be omitted from Model 8.1.

We test Hypothesis 8.2 using a similar REML-based likelihood ratio test. The test statistic is calculated by subtracting the -2 REML log-likelihood value for Model 8.1 from that for a new nested model, Model 8.3, excluding the random teacher effects (i.e., a two-level model with repeated measures nested within students). The asymptotic distribution of the test statistic under the null hypothesis is once again a mixture of χ_0^2 and χ_1^2 distributions, with equal weights of 0.5.

Hypothesis 8.3. The fixed effects associated with the YEAR variable can be omitted from Model 8.1.

The null and alternative hypotheses are

$$H_0\colon \beta_1 = 0$$
$$H_A\colon \beta_1 \neq 0$$

We test Hypothesis 8.3 using the standard test statistics for single fixed-effect parameters that are computed automatically by the various software procedures. We talk about differences in the test statistics reported and the test results in the next section (Section 8.4). For more detail on the results of these hypothesis tests, see Section 8.5.

8.4 Analysis Steps in the Software Procedures

The modeling results for all software procedures are presented and compared in Section 8.6.

8.4.1 SAS

We first import the comma-separated data file (`school_data_final.csv`, assumed to be located in the `C:\temp` directory) into SAS, and create a temporary SAS data set named `satmath`.

```
PROC IMPORT OUT = WORK.satmath
DATAFILE="C:\temp\school_data_final.csv"
DBMS=CSV REPLACE;
GETNAMES=YES;
DATAROW=2;
RUN;
```

We now proceed with fitting Model 8.1 and testing the hypotheses outlined in Section 8.3.2. The SAS syntax to fit Model 8.1 using `proc mixed` is as follows:

```
title "Model 8.1";
proc mixed data = satmath covtest;
class studid tchrid;
model math = year / solution;
random int / subject = studid;
random int / subject = tchrid;
run;
```

We have specified the `covtest` option in the `proc mixed` statement to obtain the standard errors of the estimated variance components in the output for comparison with the other software procedures. Recall that this option also causes SAS to display a Wald test for the variance of the random effects associated with the teachers and students, which we do not recommend for use in testing whether to include random effects in a model (see Subsection 2.6.3.2).

The `class` statement identifies the categorical variables that are required to specify the model. We include the two crossed random factors, STUDID and TCHRID, in the `class` statement; as shown in earlier chapters, random factors (crossed or nested) need to be specified here. The dependent variable, MATH, and the predictor variable, YEAR, are both treated as continuous variables in this model.

The `model` statement sets up the fixed-effects portion of Model 8.1. We specify that the dependent variable, MATH, is a linear function of a fixed intercept (included by default) and the fixed effect of the YEAR variable. The `solution` option requests that the estimates of the two fixed-effect parameters be displayed in the output, along with their standard errors and a t-test for each parameter.

The two `random` statements set up the crossed random effects structure for this model. In this case, STUDID is identified as the `subject` in the first `random` statement, indicating that it is a random factor. By specifying `random int`, we include a random effect associated with the intercept for each unique student. The second `random` statement with a different `subject` variable specified (TCHRID) also includes a random effect associated with the intercept for each teacher, and indicates that the levels of these two subject variables may potentially cross with each other. We note the difference in these two random statements from those used in Chapter 4, for example; there is no nesting relationship indicated for the random factors.

We now test Hypothesis 8.1 by removing the random student effects from Model 8.1, and performing a likelihood ratio test. We first refit Model 8.1 without the first random statement:

```
title "Hypothesis 8.1";
proc mixed data = satmath covtest;
class studid tchrid;
model math = year / solution;
random int / subject = tchrid;
run;
```

The –2 REML log-likelihood value for this reduced two-level model is 2170.3, and the corresponding value for Model 8.1 was 2123.6. We compute the p-value for the likelihood ratio test using the following syntax:

```
title "p-value for Hypothesis 8.1";
data _null_;
lrtstat = 2170.3 - 2123.6;
df = 1;
pvalue = 0.5 * (1 - probchi(lrtstat, df));
format pvalue 10.8;
put lrtstat = df = pvalue = ;
run;
```

We have very strong evidence ($p < 0.001$) against the null hypothesis in this case, and would choose to retain the random student effects in the model; there is clear evidence of substantial between-student variance in performance on the math test, as was apparent in the initial data summary.

We test Hypothesis 8.2 using a similar approach. We first fit a reduced model without the random teacher effects, and then compute the likelihood ratio test statistic and p-value:

```
title "Hypothesis 8.2";
proc mixed data = satmath covtest;
class studid tchrid;
model math = year / solution;
random int / subject = studid;
run;

title "p-value for Hypothesis 8.2";
data _null_;
lrtstat = 2203.1 - 2123.6;
df = 1;
pvalue = 0.5 * (1 - probchi(lrtstat, df));
format pvalue 10.8;
put lrtstat = df = pvalue = ;
run;
```

We have even stronger evidence against the null hypothesis in this case, and would also choose to retain the random teacher effects in this model. Collectively, we can conclude that there is substantial variance among both students and teachers in performance on the math test. The **solution** option can be added to both **random** statements for Model 8.1 to examine predicted values of the random effects (EBLUPs) for individual students or teachers:

```
title "Model 8.1, EBLUPs";
proc mixed data = satmath covtest;
class studid tchrid;
model math = year / solution;
random int / subject = studid solution;
random int / subject = tchrid solution;
run;
```

Finally, we test Hypothesis 8.3 with regard to the fixed effect of YEAR by examining the t-test for the fixed year effect included in the SAS output for Model 8.1. The resulting p-value ($p = 0.0032$) suggests that the fixed effect of YEAR is significantly different from zero, and the positive estimated coefficient suggests that the average performance of the students is increasing significantly over time. At this point, the model diagnostics examined in other chapters could also be examined for Model 8.1.

8.4.2   SPSS

We assume that the .csv data set used to carry out the data summary in Subsection 8.2.2 has been imported into SPSS. We begin the SPSS analysis by setting up the syntax to fit Model 8.1 using the **MIXED** command:

```
* Model 8.1 .
MIXED math WITH year
/CRITERIA=CIN(95) MXITER(100) MXSTEP(10) SCORING(1)
SINGULAR(0.000000000001) HCONVERGE(0, ABSOLUTE)
LCONVERGE(0, ABSOLUTE) PCONVERGE(0.000001, ABSOLUTE)
/FIXED=year | SSTYPE(3)
/METHOD=REML
/PRINT=SOLUTION TESTCOV
```

```
/RANDOM=INTERCEPT | SUBJECT(studid) COVTYPE(VC)
/RANDOM=INTERCEPT | SUBJECT(tchrid) COVTYPE(VC).
```

In this syntax, MATH is listed as the dependent variable. The /FIXED subcommand then lists the variable that has an associated fixed effect in Model 8.1 (YEAR). Note that YEAR is being treated as a continuous covariate, given that it is specified following the WITH keyword in the MIXED command. A fixed intercept term is included in the model by default.

The /METHOD subcommand then specifies the REML estimation method, which is the default. The /PRINT subcommand requests that the SOLUTION for the estimated parameters in the model be displayed in the output. Furthermore, we request simple Wald tests of the two variance components with the TESTCOV option, to get an initial sense of the importance of the between-student variance and the between-teacher variance.

We then include two separate /RANDOM subcommands, indicating that crossed random effects associated with each level of STUDID and TCHRID should be included in Model 8.1. The covariance structure for the random effects is specified as variance components (the default), using the COVTYPE(VC) syntax. We note that this specification of the two /RANDOM subcommands does not identify any type of nesting relationship between the two random effects.

After fitting Model 8.1 and generating the parameter estimates and model information criteria in the SPSS output viewer, we now test Hypothesis 8.1 by removing the random student effects from Model 8.1, and performing a likelihood ratio test. We first refit Model 8.1 without the first /RANDOM subcommand:

```
* Hypothesis 8.1 .
MIXED math WITH year
/CRITERIA=CIN(95) MXITER(100) MXSTEP(10) SCORING(1)
SINGULAR(0.000000000001) HCONVERGE(0, ABSOLUTE)
LCONVERGE(0, ABSOLUTE) PCONVERGE(0.000001, ABSOLUTE)
/FIXED=year | SSTYPE(3)
/METHOD=REML
/PRINT=SOLUTION TESTCOV
/RANDOM=INTERCEPT | SUBJECT(tchrid) COVTYPE(VC).
```

The −2 REML log-likelihood value for this reduced two-level model is 2170.3, and the corresponding value for Model 8.1 was 2123.6 (a difference of 46.7). We compute the p-value for the likelihood ratio test using the following syntax:

```
* Hypothesis 8.1 .
COMPUTE hyp81pvalue = 0.5 * (1 - CDF.CHISQ(46.7,1)).
EXECUTE.
```

This syntax will compute a new variable in the SPSS data set, containing the p-value associated with this test as a constant value for all cases in the data set. We have very strong evidence ($p < 0.001$) against the null hypothesis in this case, and would choose to retain the random student effects in the model; there is clear evidence of substantial between-student variance in performance on the math test, as was apparent in the initial data summary.

We test Hypothesis 8.2 using a similar approach. We first fit a reduced model without the random teacher effects, and then compute the likelihood ratio test statistic and p-value:

```
* Hypothesis 8.2 .
MIXED math WITH year
```

```
/CRITERIA=CIN(95) MXITER(100) MXSTEP(10) SCORING(1)
SINGULAR(0.000000000001) HCONVERGE(0, ABSOLUTE)
LCONVERGE(0, ABSOLUTE) PCONVERGE(0.000001, ABSOLUTE)
/FIXED=year | SSTYPE(3)
/METHOD=REML
/PRINT=SOLUTION TESTCOV
/RANDOM=INTERCEPT | SUBJECT(studid) COVTYPE(VC).

COMPUTE hyp82pvalue = 0.5 * (1 - CDF.CHISQ(79.5,1)).
EXECUTE.
```

We have even stronger evidence against the null hypothesis in this case, and would also choose to retain the random teacher effects in this model. Collectively, we can conclude that there is substantial variance among both students and teachers in performance on the math test. Unfortunately, there are no easy ways to extract the predicted random effects (EBLUPs) associated with teachers and students when using the MIXED command to fit models with crossed random effects in the current version of SPSS.

Finally, we test Hypothesis 8.3 with regard to the fixed effect of YEAR by examining the *t*-test for the fixed year effect included in the SPSS output for Model 8.1. The resulting *p*-value ($p = 0.016$, based on a Satterthwaite approximation of the denominator degrees of freedom) suggests that the fixed effect of YEAR is significantly different from zero, and the positive estimated coefficient suggests that the average performance of the students is increasing significantly over time. At this point, the model diagnostics examined in other chapters could (and should) also be examined for Model 8.1.

8.4.3   R

We assume that the **sat** data frame object created in the initial data summary (Subsection 8.2.2) has been attached to R's working memory. Importantly, the lme() function in the R package **nlme** cannot fit models with crossed random effects, so we only fit Model 8.1 and test the three hypotheses using the lmer() function in the R package **lme4** in this section.

Assuming that it has already been installed from a Comprehensive R Archive Network (CRAN) mirror, the lme4 package first needs to be loaded, so that the lmer() function can be used in the analysis:

```
> library(lme4)
```

We then fit Model 8.1 to the SAT score data using the lmer() function:

```
> # Model 8.1.
> model8.1.fit <- lmer(math ~ year + (1|studid) + (1|tchrid), REML = T)
```

We describe each part of this specification of the lmer() function:

- model8.1.fit is the name of the object that contains the results of the fitted linear mixed model with crossed random effects.

- The first argument of the function, math ~ year + (1|studid) + (1|tchrid), is the model formula, which defines the response variable (math), and the terms with associated fixed effects in the model (year).

- The (1|studid) and (1|tchrid) terms in the model formula indicate that a random effect associated with the intercept should be included for each level of the categorical random factor `studid`, and each level of the random factor `tchrid`. The `lmer()` function will automatically recognize whether the levels of these random factors are crossed or nested when estimating the variances of the random effects.

- The final argument of the function, `REML = T`, tells R that REML estimation should be used for the desired covariance parameters in the model. This is the default estimation method for the `lmer()` function.

After the function is executed, estimates from the model fit can be obtained using the `summary()` function:

```
> summary(model8.1.fit)
```

Additional results of interest for this LMM fit can be obtained by using other functions in conjunction with the `model8.1.fit` object. For example, predicted values (EBLUPs) of the random effects for each student and each teacher can be displayed using the `ranef()` function:

```
> ranef(model8.1.fit)
```

> **Software Note:** As noted in earlier chapters, the `lmer()` function only produces *t*-statistics for the fixed effects, with no corresponding *p*-values. This is primarily due to the lack of agreement in the literature over appropriate degrees of freedom for these test statistics. In general, we recommend use of the `lmerTest` package in R for users interested in testing hypotheses about parameters estimated using the `lmer()` function, and we illustrate the use of this package later in this analysis.

We now test Hypothesis 8.1 by fitting a reduced form of Model 8.1 without the random student effects:

```
> # Hypothesis 8.1.
> model8.1.fit.nostud <- lmer(math ~ year + (1|tchrid), REML = T)
> summary(model8.1.fit.nostud)
```

The likelihood test statistic is calculated by subtracting the -2 REML log-likelihood value for Model 8.1 (the reference model) from that for this reduced model (the value of the test statistic, based on the output provided by the `summary()` function, is $2170 - 2124 = 46$). The test statistic has a null distribution that is a mixture of χ_1^2 and χ_2^2 distributions with equal weights of 0.5, so the `anova()` function cannot be used for the *p*-value. Instead, we calculate a *p*-value for the test statistic as follows:

```
> 0.5*(1 - pchisq(46,0)) + 0.5*(1 - pchisq(46,1))
```

See Subsection 8.5.1 for details. The test statistic is significant ($p < 0.001$), so we decide to reject the null hypothesis and retain the random student effects in the model.

We use a similar approach for testing Hypothesis 8.2, extracting the -2 REML log-likelihood (or `REMLdev`) from the `summary()` function output for this reduced model excluding the random teacher effects:

```
> # Hypothesis 8.2.
> model8.1.fit.notchr <- lmer(math ~ year + (1|studid), REML = T)
> summary(model8.1.fit.notchr)
```

The value of this test statistic, based on the output provided by the `summary()` function, is $2203 - 2124 = 79$. We compute the asymptotic p-value for this test statistic:

```
> 0.5*(1 - pchisq(79,0)) + 0.5*(1 - pchisq(79,1))
```

The resulting p-value ($p < 0.0001$) provides strong evidence for retaining the random teacher effects in the model. There is clear evidence of substantial variation among both students and teachers.

Finally, we load the `lmerTest` package, which enables approximate t-tests of fixed-effect parameters in models fitted using the `lmer()` function.

```
> library(lmerTest)
```

We once again fit Model 8.1:

```
> # Model 8.1.
> model8.1.fit <- lmer(math ~ year + (1|studid) + (1|tchrid), REML = T)
```

We then apply the `summary()` function to the new model fit object:

```
> summary(model8.1.fit)
```

In the resulting output, the p-value computed for the approximate t-statistic ($p = 0.016$, based on a Satterthwaite approximation of the denominator degrees of freedom) suggests that the fixed effect of YEAR is significantly different from zero, and the positive estimated coefficient suggests that the average performance of the students is increasing significantly over time. At this point, the model diagnostics examined in other chapters could (and should) also be examined for Model 8.1.

8.4.4 Stata

We begin the analysis by importing the comma-separated values file containing the SAT score data (`school_data_final.csv`) into Stata:

```
. insheet using "C:\temp\school_data_final.csv", comma
```

We now proceed with fitting Model 8.1. We use the following `mixed` command to fit Model 8.1 to the SAT score data:

```
. * Model 8.1.
. mixed math year || _all: R.studid || _all: R.tchrid, variance reml
. estat ic
```

In this `mixed` command, MATH is listed as the dependent variable, followed by the predictor variable that has an associated fixed effect in Model 8.1 (YEAR). Note that YEAR is being treated as a continuous covariate, given that it is not preceded by `i.` (Stata's factor notation). A fixed intercept term is included in the model by default.

We next indicate the crossed random effects using two successive `||` symbols in the command. The first random factor, STUDID, follows the first `||` symbol, and is specified in a unique fashion. The `_all:` notation indicates that the entire data set is treated as one "cluster," and `R.studid` indicates that STUDID is a categorical random factor, where the effects of the levels of that factor randomly vary for the one "cluster." We then indicate the second random factor crossed with STUDID following a second `||` symbol, using the same general concept with the categorical TCHRID factor. We note that this command does not specify an explicit nesting structure, given that the entire data set is treated as one "subject."

Finally, after a comma, we specify the `variance` option, which requests estimates of the variances of the random effects in the output (rather than estimates of standard deviations), and the `reml` option, requesting REML estimation of the parameters in this model (rather than the default ML estimation). The subsequent `estat ic` command requests that Stata compute selected information criteria (e.g., AIC, BIC) for this model and display them in the output.

After fitting Model 8.1 and generating the parameter estimates and model information criteria in the Stata results window, we now test Hypothesis 8.1 by removing the random student effects from Model 8.1, and performing a likelihood ratio test. We first refit Model 8.1 without the random student effects:

```
. * Model 8.1, no random student effects.
. mixed math year || _all: R.tchrid, variance reml
. estat ic
```

This reduced model could also be fitted using this command, shown in earlier chapters for simple two-level models:

```
. * Model 8.1, no random student effects.
. mixed math year || tchrid:, variance reml
. estat ic
```

The –2 REML log-likelihood value for this reduced two-level model (found by multiplying the REML log-likelihood displayed by the `estat ic` command by –2) is 2170.3, and the corresponding value for Model 8.1 was 2123.6 (a difference of 46.7). We compute the p-value for the likelihood ratio test using the following syntax:

```
. * Hypothesis 8.1 .
. di 0.5*chi2tail(1,46.7)
```

Based on the resulting p-value, we have very strong evidence ($p < 0.001$) against the null hypothesis in this case, and would choose to retain the random student effects in the model; there is clear evidence of substantial between-student variance in performance on the math test, as was apparent in the initial data summary.

We test Hypothesis 8.2 using a similar approach. We first fit a reduced model without the random teacher effects, and then compute the likelihood ratio test statistic and p-value:

```
. * Model 8.1, no random teacher effects.
. mixed math year || _all: R.studid, variance reml
. estat ic
```

```
. * Hypothesis 8.2 .
. di 0.5*chi2tail(1,79.5)
```

We have even stronger evidence against the null hypothesis in this case, and would also choose to retain the random teacher effects in this model. Collectively, we can conclude that there is substantial variance among both students and teachers in performance on the math test.

Once a model with crossed random effects has been fitted using the `mixed` command, the `predict` post-estimation command can be used to generate new variables in the Stata data set containing predicted values of each random effect. For example, we save the EBLUPs of the random school and teacher effects, respectively, in two new variables (**b1** and **b2**) using the following **predict** command (after fitting Model 8.1):

```
. * Model 8.1.
. mixed math year || _all: R.studid || _all: R.tchrid, variance reml
. predict b*, reffects
```

Finally, we test Hypothesis 8.3 with regard to the fixed effect of YEAR by examining the z-statistic for the fixed year effect included in the Stata output for Model 8.1. The resulting p-value ($p = 0.003$) suggests that the fixed effect of YEAR is significantly different from zero, and the positive estimated coefficient suggests that the average performance of the students is increasing significantly over time. At this point, the model diagnostics examined in other chapters could (and should) also be examined for Model 8.1.

8.4.5   HLM

8.4.5.1   Data Set Preparation

When using the HLM software to fit linear mixed models with crossed random effects to data sets with two crossed random factors like the SAT score data, three separate data sets need to be prepared:

1. The **Level-1 data set**: Each row in this data set corresponds to an observation on a unit of analysis (including a unique measurement on the dependent variable) for a given combination of the two crossed random factors. In the context of the present SAT score study, each row of the Level-1 data set represents a measurement on a student–teacher combination in a given year. This data set is similar in structure to the data set displayed in Table 8.1. The Level-1 data set for this example includes STUDID, TCHRID, the dependent variable (MATH), and the predictor variable of interest (YEAR).

2. The **Row-Level data set**: When thinking about the crossed random factors in a data set like the SAT score data, one can think of a matrix like that shown in Section 8.2.1, where the unique levels of one random factor define the rows, and the unique levels of the second random factor define the columns. This distinction between the row factor and the column factor is essentially arbitrary in HLM, but the row-level data set contains one row per unique level of one of the two random factors. In this example, we consider STUDID as the random factor defining the "rows" of this cross-tabulation, and this data set therefore has one row per student. We include the unique student ID (STUDID), in addition to a variable recording the number of measures collected on each student (MEASURES), and we sort the data set is ascending order by STUDID. Although we don't actually analyze the MEASURES variable in this example, HLM requires at least one non-ID variable to be included in this data set. In general, additional student-level covariates of interest in the analysis would be included in the row-level data set.

3. The **Column-Level data set**: This data set has one row per unique level of the second random factor. In this example, this data set has one row per unique level of TCHRID, or per teacher. We include the TCHRID variable in this data set, along with a variable recording the number of measures collected on each teacher (MEASURES_T), and sort the data set in ascending order by TCHRID. Similar to the row-level data set, the column-level data set needs to have at least

one non-ID variable for the teachers, and additional teacher-level covariates could also be included in this data set.

After these three data sets have been created, we can proceed to create the multivariate data matrix (MDM), and fit Model 8.1.

8.4.5.2 Preparing the MDM File

In the main HLM menu, click **File, Make new MDM file** and then **Stat package input**. In the window that opens, select **HCM2** to fit a hierarchical linear model with crossed random effects associated with two (2) random factors, and click **OK**. Select the **Input File Type** as **SPSS/Windows**.

To prepare the MDM file for Model 8.1, locate the **Level-1 Specification** area, and **Browse** to the location of the Level-1 data set. Click **Open** after selecting the Level 1 SPSS file, click the **Choose Variables** button, and select the following variables: STUDID (click "rowid" for the STUDID variable, because this random factor identifies the "rows" in the cross-classification), TCHRID (click "colid" for the TCHRID variable, because this random factor identifies the "columns" in the cross-classification), and both MATH and YEAR (click "in MDM" for each of these variables). Click **OK** when finished.

Next, locate the **Row-Level Specification** area, and **Browse** to the location of the row-level (student-level) SPSS data set. Click **Open** after selecting the file, and click the **Choose Variables** button to include STUDID (click "rowid") and the variable indicating the number of measures on each student, MEASURES (click "in MDM"). Again, although we don't analyze the MEASURES variable in this example, HLM requires that at least one numeric variable be included in the MDM file from both the row and column data sets. Click **OK** when finished.

Next, locate the **Column-Level Specification** area, and **Browse** to the location of the column-level (teacher-level) SPSS data set. Click **Open** after selecting the file, and click the **Choose Variables** button to include TCHRID (click "colid") and the variable indicating the number of measures on each teacher, MEASURES_T (click "in MDM"). Again, although we don't analyze the MEASURES_T variable in this example, HLM requires that at least one numeric variable be included in the MDM file from both the row and column data sets. Click **OK** when finished.

After making these choices, select **No** for **Missing Data?** in the Level-1 data set, because we do not have any missing data in this analysis. In the upper-right corner of the MDM window, enter a name with a .mdm extension for the MDM file (e.g., SAT.mdm). Save the .mdmt template file under a new name (click **Save mdmt file**), and click **Make MDM**.

After HLM has processed the MDM file, click the **Check Stats** button to display descriptive statistics for the variables in the Level-1, row, and column data files (this is not optional). Click **Done** to begin building Model 8.1.

8.4.5.3 Model Fitting

In the model building window, identify MATH as the **Outcome variable**. To add more informative subscripts to the models (if they do not already appear), click **File** and **Preferences**, and then choose **Use level subscripts**.

To complete the specification of Model 8.1, we first add the fixed effect of the uncentered YEAR variable to the model. We choose uncentered because this variable has already been centered at the data management stage. The Level 1 model is displayed in HLM as follows (where i is an index for the individual observations, j is an index for students, and k is an index for teachers):

Model 8.1: Level 1 Model

$$\text{MATH}_{ijk} = \pi_{0jk} + \pi_{1jk}(\text{YEAR}_{ijk}) + e_{ijk}$$

The Level 2 equation for the random intercept specific to a given observation (π_{0jk}) includes a constant fixed effect, θ_0, a random effect associated with the student (b_{00j}), which allows the intercept to vary randomly across students, and a random effect associated with the teacher (c_{00k}), which allows the intercept to vary randomly across teachers. The Level 2 equation for the random coefficient for YEAR specific to a student–teacher combination, π_{1jk}, is simply defined by a constant fixed effect (θ_1), although we could allow the effect of year to randomly vary across both students and teachers:

Model 8.1: Level 2 Model

$$\pi_{0jk} = \theta_0 + b_{00j} + c_{00k}$$
$$\pi_{1jk} = \theta_1$$

We display the overall LMM by clicking the **Mixed** button in the model building window:

Model 8.1: Overall Mixed Model

$$\text{MATH}_{ijk} = \theta_0 + \theta_1 * \text{YEAR}_{ijk} + b_{00j} + c_{00k} + e_{ijk}$$

This model is the same as the general specification of Model 8.1 introduced in Subsection 8.3.1.1, although the notation is somewhat different.

After specifying Model 8.1, click **Basic Settings** to enter a title for this analysis (such as "SAT Score Data: Model 8.1") and a name for the output (.html) file that HLM generates when fitting this model. Note that the default outcome variable distribution is Normal (Continuous).

Click **OK** to return to the model-building window. We note that there is not an option to choose either REML or ML estimation under **Other Settings** and **Estimation Settings**; HLM only provides ML estimation as an option for models with crossed random effects.

Click **File** and **Save As** to save this model specification in a new .hlm file. Finally, click **Run Analysis** to fit the model. After the estimation of the parameters in Model 8.1 has finished (one may need to enter "y" for estimation to proceed past the predetermined maximum number of iterations), click **File** and **View Output** to see the resulting estimates if they do not automatically appear in a new browser window.

In the other software procedures, we tested Hypotheses 8.1 and 8.2 by fitting nested models that excluded either the random student effects or the random teacher effects, and then performed likelihood ratio tests. HLM automatically produces tests of the null hypothesis that the variance of the random effects associated with a given random factor is zero, and in the case of a model with crossed random effects, two of these hypothesis tests are displayed in the output [see Raudenbush & Bryk (2002) for details]. These two tests strongly suggest that the null hypothesis of zero variance in the random effects should be rejected for both students and teachers, indicating that we should retain both crossed random effects in the model.

Hypothesis 8.3 can be tested by examining the p-value for the fixed effect of YEAR (θ_1 in the HLM notation). We have strong evidence against the null hypothesis ($p = 0.001$), and conclude that there is a significant positive increase in the expected math score as a function of YEAR.

At this point, fixed effects of additional covariates could be added to the model in an effort to explain variance in the two sets of random effects, and residual diagnostics could be examined. As illustrated in earlier chapters, HLM enables users to save external residual files in the format of a statistical package of their choosing (e.g., SPSS, Stata) in the **Basic Settings** window.

8.5   Results of Hypothesis Tests

The test results reported in this section were calculated based on the analysis in R.

8.5.1   Likelihood Ratio Tests for Random Effects

Hypothesis 8.1.   The random effects (u_i) associated with the intercept for each student can be omitted from Model 8.1.

The likelihood ratio test statistic for Hypothesis 8.1 is calculated by subtracting the –2 REML log-likelihood for Model 8.1 from that for a reduced version of Model 8.1 excluding the random student effects. This difference is calculated as $2170 - 2124 = 46$. Because the null hypothesis value for the variance of the random student effects is on the boundary of the parameter space (i.e., zero), the asymptotic null distribution of this test statistic is a mixture of χ_0^2 and χ_1^2 distributions, each with equal weights of 0.5 (Verbeke & Molenberghs, 2000). To evaluate the significance of the test, we calculate the p-value as follows:

$$p\text{-value} = 0.5 \times P(\chi_0^2 > 46) + 0.5 \times P(\chi_1^2 > 46) < 0.001$$

We reject the null hypothesis and retain the random effects associated with students in Model 8.1. We perform a similar test of **Hypothesis 8.2** for the random teacher effects, and arrive at a similar conclusion.

We noted earlier that the HLM software provides slightly different chi-square tests of these hypotheses, each of which led to the same conclusions; interested readers can consult Raudenbush & Bryk (2002) for more details.

8.5.2   Testing the Fixed Year Effect

Hypothesis 8.3.   The fixed effect of YEAR can be omitted from Model 8.1.

In this chapter, we tested this hypothesis using t- or z-statistics produced in the output by the various software procedures, which are computed using the ratio of the estimated fixed effect to its estimated standard error. Regardless of the slight variance in these statistics (and their degrees of freedom, if applicable) across the procedures in SAS, SPSS, R, Stata, and HLM, we arrived at the same conclusion, rejecting the null hypothesis and concluding that the fixed effect of YEAR is positive and significant.

8.6   Comparing Results across the Software Procedures

Table 8.2 shows that the results for Model 8.1 generally agree across the software procedures, in terms of the fixed-effect parameter estimates and their estimated standard errors. The

TABLE 8.2: Comparison of Results for Model 8.1

Estimation Method	SAS: proc mixed	SPSS: MIXED	R: lmer() function	Stata: mixed	HLM: HCM2
	REML	**REML**	**REML**	**REML**	**ML**
Fixed-Effect Parameter	*Estimate (SE)*	*Estimate (SE)*	*Estimate (SE)*	*Estimate (SE)*	*Estimate (SE)*
β_0 (Intercept)	597.38(8.38)	597.38(8.38)	597.38(8.38)	597.38(8.38)	597.71(7.53)
β_1 (YEAR)	29.05(9.63)	29.05(9.63)	29.05(9.63)	29.05(9.63)	28.56(8.64)
Covariance Parameter	*Estimate (SE)*	*Estimate (SE)*	*Estimate (n.c.)*	*Estimate (SE)*	*Estimate (n.c.)*
σ_i^2 (Students)	338.41(67.84)	338.41(67.84)	338.41	338.41(67.84)	340.79
σ_j^2 (Teachers)	762.90(396.09)	762.94(396.12)	762.94	762.94(396.12)	604.90
σ^2 (Residuals)	238.30	238.30	238.30	238.30	237.91
Model Information Criteria					
−2 REML log-likelihood	2123.6	2123.6	2124.0	2123.6	2135.9
AIC	2129.6	2129.6	2134.0	2133.6	*n.c.*
BIC	2123.6	2140.0	2151.0	2150.9	*n.c.*

Note: (n.c.) = not computed

Note: 234 Measures at Level 1; 122 students and 12 teachers

five software procedures also generally agree in terms of the values of the estimated variance components for the two sets of crossed random effects, σ_i^2 and σ_j^2, and their standard errors, when reported.

We note that the value of the –2 REML log-likelihood is the same across the four software procedures using REML estimation (with the difference in R just due to rounding). The Hierarchical Cross-Classified Model (HCM2) procedure in HLM uses ML estimation only, leading to the different estimates in Table 8.2 and the different value of the –2 ML log-likelihood. We also once again note that there is some disagreement in the values of the information criteria (AIC and BIC), because of different calculation formulas that are used in the different procedures.

We also note that the `lmer()` function in R and the HLM procedure do not compute standard errors for the estimated variance components. The various methods discussed in this chapter for testing hypotheses about the variance components can be used to make inferences about the variance components in these cases.

8.7 Interpreting Parameter Estimates in the Final Model

The results that we present in this section were obtained by fitting Model 8.1 to the SAT score data, using REML estimation in Stata.

8.7.1 Fixed-Effect Parameter Estimates

The fixed-effect parameter estimates, standard errors, significance tests, and 95% confidence intervals obtained by fitting Model 8.1 to the SAT score data in Stata are reported in the following output:

```
                                             Wald chi2(1) = 9.11
    Log restricted-likelihood = -1061.8139   Prob > chi2  = 0.0025
   -----------------------------------------------------------------
    gcf      Coef.    Std. Err.    z     P > |z|   [95 pct Conf. Interval]
   -----------------------------------------------------------------
    year    29.04963  9.626692   3.02    0.003     10.18166   47.9176
    _cons  597.3811   8.378181  71.30    0.000    580.96020  613.8021
```

Based on these estimates, we would conclude that the expected value of the SAT math score at Grade 4 for the individuals in this school is 597.38, with a 95% confidence interval for this mean of (580.96, 613.80). We would also conclude that the fixed effect of YEAR on the SAT score is significant ($p = 0.003$), with a one-year (or one-grade) increase resulting in an expected change of 29.05 in the SAT score.

We can use the `margins` and `marginsplot` post-estimation commands in Stata to visualize this relationship. Note that these commands need to be submitted immediately after fitting Model 8.1.

```
. * Model 8.1, with plots of marginal predicted values.
. mixed math year || _all: R.studid || _all: R.tchrid, variance reml
. margins, at(year = (-1,0,1))
. marginsplot
```

We note that we are using the `margins` command to plot expected values of the dependent variable (SAT math score) at values −1, 0, and 1 on the predictor variable YEAR.

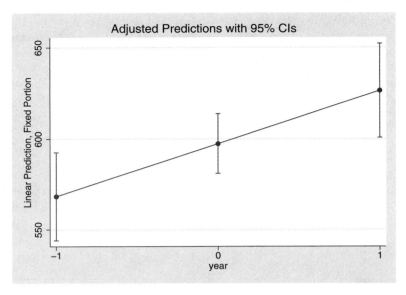

FIGURE 8.4: Predicted SAT math score values by YEAR based on Model 8.1.

We then immediately follow the `margins` command, which displays the marginal predicted values in the output, with the `marginsplot` command to plot the predicted values (in addition to 95% confidence intervals for the predicted values). The resulting plot is shown in Figure 8.4.

8.7.2   Covariance Parameter Estimates

The estimated covariance parameters obtained by fitting Model 8.1 to the SAT score data using the `mixed` command in Stata with REML estimation are reported in the following output:

```
Random-effects Parameters    Estimate    Std. Err.    [95 pct Conf. Interval]
---------------------------------------------------------------------------

_all: Identity
var(R.studid)                338.4091    67.8428      228.453    501.2879

_all: Identity
var(R.tchrid)                762.9415    396.1218     275.7685  2110.7560

var(Residual)                238.2957    32.8589      181.8624   312.2407

LR test vs. linear regression: chi2(4) = 162.50    Prob > chi2 = 0.0000

Note: LR test is conservative and provided only for reference.
```

We first see that the variance of the random student effects, `var(R.studid)`, is estimated to be 338.4091. The 95% confidence interval suggests that this parameter is in fact greater than zero; we illustrated the use of a likelihood ratio test for a more formal (and asymptotic) test of this hypothesis earlier in the chapter. We next see that the estimated variance of the random teacher effects, `var(R.tchrid)`, is 762.9415. The 95% confidence interval for this variance component also suggests that the variance is greater than zero. Finally, we see the estimated residual variance (238.2957). Based on these results, it is fairly clear that between-student and between-teacher variance represent substantial portions of the overall variance in the SAT math scores.

8.8 The Implied Marginal Variance-Covariance Matrix for the Final Model

In this section, we capitalize on the ability of `proc mixed` in SAS to compute the implied marginal variance-covariance matrix for observations in the SAT score data set based on Model 8.1, and examine the unique structure of this matrix. The matrix of implied marginal variances and covariances for the SAT score observations (and the corresponding correlation matrix) can be obtained in SAS by including the `v` and `vcorr` options in either of the two `random` statements for Model 8.1:

```
random int / subject = studid v vcorr;
```

In models with crossed random effects, the default in SAS is no longer to display the implied marginal variance-covariance matrix for the first "subject" in the data set (e.g., Chapter 5), given that there are now multiple "subject" variables crossed with each other. When fitting these models, `proc mixed` will display this matrix for all observations in the data set, which can be prohibitive when working with larger data sets (in this case, we would see a 234×234 matrix). In this illustration, examination of this matrix is instructive for understanding how the estimated variance components in a model with crossed random effects determine the marginal variances and covariances of observations in the data set with crossed random factors.

In the table below, we consider the 6×6 submatrix in the upper-left corner of the full matrix. This submatrix represents the first six observations in the SAT score data set, shown earlier in Table 8.1, where we have three unique students (with 1, 3, and 2 observations, respectively), and four unique teachers. We first see that the estimated marginal variance of a given observation (on the diagonal of the submatrix below) is simply defined by the sum of the three estimated variance components ($338.41 + 762.90 + 238.30 = 1339.61$). The implied marginal covariance of observations on the same *teacher* (e.g., observations 1 and 2) is defined by the estimated variance of the random teacher effects (762.90). The estimated covariance of observations on the same *student* (e.g., observations 2 and 3) is defined by the estimated variance of the random student effects (338.41). The matrix then has empty cells where there were no covariances, due to the fact that a given pair of observations was collected on unique students and teachers (e.g., observations 1 and 4).

Row	Col1	Col2	Col3	Col4	Col5	Col6
1	1339.61	762.90			762.90	
2	762.90	1339.61	338.41	338.41	762.90	
3		338.41	1339.61	338.41		
4		338.41	338.41	1339.61		
5	762.90	762.90			1339.61	338.41
6					338.41	1339.61

The "blank" cells in this matrix indicate the need to use estimation algorithms for **sparse matrices** that was discussed in Chapter 2 when fitting models with crossed random effects.

For example, in order to compute estimates of standard errors for the estimated fixed effects in this model, we need to invert the "sparse" implied marginal variance-covariance shown above.

The implied marginal correlation matrix is simply determined by dividing all cells in the implied marginal variance-covariance matrix by the estimated total variance on the diagonal. The implied intraclass correlation of observations on the same student is simply the proportion of the total variance due to students, while the implied intraclass correlation of observations on the same teacher is simply the proportion of the total variance due to teachers. The implied marginal correlation matrix for the SAT score data set is shown in the output below.

Row	Col1	Col2	Col3	Col4	Col5	Col6
1	1.0000	0.5695			0.5695	
2	0.5695	1.0000	0.2526	0.2526	0.5695	
3		0.2526	1.0000	0.2526		
4		0.2526	0.2526	1.0000		
5	0.5695	0.5695			1.0000	0.2526
6					0.2526	1.0000

We see that there is a stronger correlation of observations on the same teacher compared to observations from the same student, suggesting that the teacher tends to be a stronger determinant of performance on the test.

8.9   Recommended Diagnostics for the Final Model

Using the procedures outlined in earlier chapters in the various software packages, we recommend performing the following set of diagnostic analyses when fitting models with crossed random effects:

- Generate predicted values (EBLUPs) of the two (or more) crossed random effects, and examine the distributions of the random effects using normal Q–Q plots. For example, when adding the solution option to each of the two random statements using to fit Model 8.1 in SAS proc mixed, we identify several students and teachers with extreme values, and the data for these "subjects" could be examined in more detail to make sure that there are not any extreme outliers or data entry errors.

- Generate residuals and fitted values for the dependent variable based on the final model, and use normal Q–Q plots and scatter plots to examine assumptions of normality and constant variance for the residuals. Consider whether important covariates (or functions of covariates) have been omitted if these assumptions seem violated, and consider transforming the dependent variable if needed. Alternative generalized linear mixed models for non-normal outcomes (outside of the scope of this book) may be needed if these suggested remedies do not fix apparent violations of these assumptions.

- Plot the residuals as a function of each of the continuous predictor variables included in the model, to make sure that there are no systematic patterns in the residuals as a function of the predictors (indicating possible model misspecification).

The sensitivity of model estimates (and the corresponding inferences) to removing extreme subjects or extreme observations should also be examined.

8.10 Software Notes and Additional Recommendations

We noted some important differences between the five software procedures in this chapter when fitting models with crossed random effects:

1. There are no straightforward ways to extract the predicted random effects (EBLUPs) associated with teachers and students when using the `MIXED` command to fit models with crossed random effects in the current version of SPSS. The other four procedures discussed in this chapter provide options for computing these EBLUPs.

2. To fit models with crossed random effects in R, we need to use the `lmer()` function in the `lme4` package.

3. The `lme4` package version of the `lmer()` function in R does not provide p-values for computed test statistics based on fitted models. As a result, we recommend using the `lmerTest` package in R to test hypotheses about parameters estimated using the `lmer()` function.

4. When using the HLM package to fit models with crossed random effects, only ML estimation is allowed. This will lead to slight differences in parameter estimates relative to other procedures that may be using REML by default.

Because fitting models with crossed random effects can be more intensive computationally than fitting models with nested random effects, we recommend that analysts only include random intercepts associated with the various levels of crossed random factors. This will capture essential features of the dependence of observations based on the multiple random factors, and enable estimation of components of variance due to the random factors. Additional (crossed) random effects (e.g., random coefficients associated with a given predictor variable) can be included in models and tested if there is explicit research interest in between-subject (or between-cluster) variance in these coefficients, but this can lead to computational difficulties, especially when attempting to fit models to larger data sets than the one considered in this chapter.

A

Statistical Software Resources

A.1 Descriptions/Availability of Software Packages

A.1.1 SAS

SAS is a comprehensive software package produced by the SAS Institute, Inc., which has its headquarters in Cary, North Carolina. SAS is used for business intelligence, scientific applications, and medical research. SAS provides tools for data management, reporting, and analysis. `proc mixed` is a procedure located within the SAS/STAT software package, a collection of procedures that implement statistical analyses. The current version of the SAS/STAT software package at the time of this publication is SAS Release 9.3, which is available for many different computing platforms, including Windows and UNIX. Additional information on ordering and availability can be obtained by calling 1-800-727-0025 (United States), or visiting the following web site: `http://www.sas.com/nextsteps/index.html`.

A.1.2 IBM SPSS Statistics

IBM SPSS Statistics (referred to simply as "SPSS" in this book) is a comprehensive statistical software package produced by International Business Machines (IBM) Corporation, headquartered in Armonk, New York. SPSS's statistical software, or the collection of procedures available in the Base version of SPSS and several add-on modules, is used primarily for data mining, data management and database analysis, market and survey research, and research of all types in general. The Linear Mixed Models (`MIXED`) procedure in SPSS is part of the Advanced Statistics module that can be used in conjunction with the Base SPSS software. The current version of the SPSS software package at the time of this publication (Version 22) is available for Windows, MacOS, and Linux desktop platforms. Additional information on ordering and availability can be obtained by calling 1-800-543-2185, or visiting the following web site: `http://www-01.ibm.com/software/analytics/spss/products/statistics/`.

A.1.3 R

R is a free software environment for statistical computing and graphics, which is available for Windows, UNIX, and MacOS platforms. R is an open source software package, meaning that the code written to implement the various functions can be freely examined and modified. The `lme()` function for fitting linear mixed models can be found in the `nlme` package, which automatically comes with the R software, and the newer `lmer()` function for fitting linear mixed models can be found in the `lme4` package, which needs to be downloaded by users. The newest version of R at the time of this publication is 3.1.0 (April 2014), and all analyses in this book were performed using at least Version 2.15.1. To download the base R software or any contributed packages (such as the `lme4` package) free of charge, readers can visit any of the Comprehensive R Archive Network (CRAN) mirrors listed at the following web site: `http://www.r-project.org/`.

This web site provides a variety of additional information about the R software environment.

A.1.4 Stata

Stata is a statistical software package for research professionals of all disciplines, offering a completely integrated set of commands and procedures for data analysis, data management, and graphics. Stata is produced by StataCorp LP, which is headquartered in College Station, Texas. The `mixed` procedure for fitting linear mixed models was first available in Stata Release 13, which is currently available for Windows, Macintosh, and UNIX platforms. For more information on sales or availability, call 1-800-782-8272, or visit: `http://www.stata.com/order/`.

A.1.5 HLM

The HLM software program is produced by Scientific Software International, Inc. (SSI), headquartered in Lincolnwood, Illinois, and is designed primarily for the purpose of fitting hierarchical linear models. HLM is not a general-purpose statistical software package similar to SAS, SPSS, R, or Stata, but offers several tools for description, graphing and analysis of hierarchical (clustered and/or longitudinal) data. The current version of HLM (HLM 7) can fit a wide variety of hierarchical linear models, including generalized HLMs for non-normal response variables (not covered in this book). A free student edition of HLM 7 is available at the following web site: `http://www.ssicentral.com/hlm/student.html`.

More information on ordering the full commercial version of HLM 7, which is currently available for Windows, UNIX systems, and Linux servers, can be found at the following web site: `http://www.ssicentral.com/ordering/index.html`.

A.2 Useful Internet Links

The web site for this book, which contains links to electronic versions of the data sets, output, and syntax discussed in each chapter, in addition to syntax in the various software packages for performing the descriptive analyses and model diagnostics discussed in the example chapters, can be found at the following link: `http://www.umich.edu/~bwest/almmussp.html`.

A very helpful web site introducing matrix algebra operations that are useful for understanding the calculations presented in Chapter 2 and Appendix B can be found at the following link: `http://www.sosmath.com/matrix/matrix.html`.

In this book, we have focused on procedures capable of fitting linear mixed models in the HLM software package and four general-purpose statistical software packages. To the best of our knowledge, these five software tools are in widespread use today, but these by no means are the only statistical software tools available for the analysis of linear mixed models. The following web site provides an excellent survey of the procedures available in these and other popular statistical software packages, including MLwiN: `http://www.bristol.ac.uk/cmm/learning/mmsoftware/`.

B

Calculation of the Marginal Variance-Covariance Matrix

In this appendix, we present the detailed calculation of the marginal variance-covariance matrix, V_i, implied by Model 5.1 in Chapter 5 (the analysis of the Rat Brain data). This calculation assumes knowledge of simple matrix algebra.

$$V_i = Z_i D Z_i' + R_i$$

$$= \begin{pmatrix} 1 \\ 1 \\ 1 \\ 1 \\ 1 \\ 1 \end{pmatrix} (\sigma_{int}^2)(1 \quad 1 \quad 1 \quad 1 \quad 1 \quad 1) + \begin{pmatrix} \sigma^2 & 0 & 0 & 0 & 0 & 0 \\ 0 & \sigma^2 & 0 & 0 & 0 & 0 \\ 0 & 0 & \sigma^2 & 0 & 0 & 0 \\ 0 & 0 & 0 & \sigma^2 & 0 & 0 \\ 0 & 0 & 0 & 0 & \sigma^2 & 0 \\ 0 & 0 & 0 & 0 & 0 & \sigma^2 \end{pmatrix}$$

Note that the Z_i design matrix has a single column of 1s (for the random intercept for each animal in Model 5.1). Multiplying the Z_i matrix by the D matrix, we have the following:

$$Z_i D = \begin{pmatrix} \sigma_{int}^2 \\ \sigma_{int}^2 \\ \sigma_{int}^2 \\ \sigma_{int}^2 \\ \sigma_{int}^2 \\ \sigma_{int}^2 \end{pmatrix}$$

Then, multiplying the above result by the transpose of the Z_i matrix, we have

$$Z_i D Z_i' = \begin{pmatrix} \sigma_{int}^2 \\ \sigma_{int}^2 \\ \sigma_{int}^2 \\ \sigma_{int}^2 \\ \sigma_{int}^2 \\ \sigma_{int}^2 \end{pmatrix} (1 \quad 1 \quad 1 \quad 1 \quad 1 \quad 1) = \begin{pmatrix} \sigma_{int}^2 & \sigma_{int}^2 & \sigma_{int}^2 & \sigma_{int}^2 & \sigma_{int}^2 & \sigma_{int}^2 \\ \sigma_{int}^2 & \sigma_{int}^2 & \sigma_{int}^2 & \sigma_{int}^2 & \sigma_{int}^2 & \sigma_{int}^2 \\ \sigma_{int}^2 & \sigma_{int}^2 & \sigma_{int}^2 & \sigma_{int}^2 & \sigma_{int}^2 & \sigma_{int}^2 \\ \sigma_{int}^2 & \sigma_{int}^2 & \sigma_{int}^2 & \sigma_{int}^2 & \sigma_{int}^2 & \sigma_{int}^2 \\ \sigma_{int}^2 & \sigma_{int}^2 & \sigma_{int}^2 & \sigma_{int}^2 & \sigma_{int}^2 & \sigma_{int}^2 \\ \sigma_{int}^2 & \sigma_{int}^2 & \sigma_{int}^2 & \sigma_{int}^2 & \sigma_{int}^2 & \sigma_{int}^2 \end{pmatrix}$$

For the final step, we add the 6×6 R_i matrix to the above result to obtain the V_i matrix:

$$V_i = Z_i D Z_i' + R_i$$

$$= \begin{pmatrix} \sigma_{int}^2 + \sigma^{2\iota} & \sigma_{int}^2 & \sigma_{int}^2 & \sigma_{int}^2 & \sigma_{int}^2 & \sigma_{int}^2 \\ \sigma_{int}^2 & \sigma_{int}^2 + \sigma^2 & \sigma_{int}^2 & \sigma_{int}^2 & \sigma_{int}^2 & \sigma_{int}^2 \\ \sigma_{int}^2 & \sigma_{int}^2 & \sigma_{int}^2 + \sigma^2 & \sigma_{int}^2 & \sigma_{int}^2 & \sigma_{int}^2 \\ \sigma_{int}^2 & \sigma_{int}^2 & \sigma_{int}^2 & \sigma_{int}^2 + \sigma^2 & \sigma_{int}^2 & \sigma_{int}^2 \\ \sigma_{int}^2 & \sigma_{int}^2 & \sigma_{int}^2 & \sigma_{int}^2 & \sigma_{int}^2 + \sigma^2 & \sigma_{int}^2 \\ \sigma_{int}^2 & \sigma_{int}^2 & \sigma_{int}^2 & \sigma_{int}^2 & \sigma_{int}^2 & \sigma_{int}^2 + \sigma^2 \end{pmatrix}$$

We see how the small sets of covariance parameters defining the $\boldsymbol{D}$ and $\boldsymbol{R}_i$ matrices (σ^2_{int} and σ^2, respectively) are used to obtain the implied marginal variances (on the diagonal of the $\boldsymbol{V}_i$ matrix) and covariances (off the diagonal) for the six observations on an animal i.

Note that this marginal $\boldsymbol{V}_i$ matrix implied by Model 5.1 has a compound symmetry covariance structure (see Subsection 2.2.2.2), where the marginal covariances are restricted to be positive due to the constraints on the $\boldsymbol{D}$ matrix in the LMM ($\sigma^2_{int} > 0$). We could fit a marginal model without random animal effects and with a compound symmetry variance-covariance structure for the marginal residuals to allow the possibility of negative marginal covariances.

C

Acronyms/Abbreviations

Definitions for selected acronyms and abbreviations used in the book:

AIC	=	Akaike Information Criterion
ANOVA	=	Analysis of Variance
AR(1)	=	First-order Autoregressive (covariance structure)
BIC	=	Bayes Information Criterion
CS	=	Compound Symmetry (covariance structure)
DIAG	=	Diagonal (covariance structure)
det	=	Determinant
df	=	Degrees of freedom
(E)BLUE	=	(Empirical) Best Linear Unbiased Estimator
(E)BLUP	=	(Empirical) Best Linear Unbiased Predictor (for random effects)
EM	=	Expectation-Maximization (algorithm)
EMMEANS	=	Estimated Marginal MEANS (from SPSS)
GLS	=	Generalized Least Squares
HET	=	Heterogeneous Variance Structure
HLM	=	Hierarchical Linear Model
ICC	=	Intraclass Correlation Coefficient
LL	=	Log-likelihood
LMM	=	Linear Mixed Model
LRT	=	Likelihood Ratio Test
LSMEANS	=	Least Squares MEANS (from SAS)
MAR	=	Missing at Random
ML	=	Maximum Likelihood
MLM	=	Multilevel Model
N–R	=	Newton–Raphson (algorithm)
ODS	=	Output Delivery System (in SAS)
OLS	=	Ordinary Least Squares
REML	=	Restricted Maximum Likelihood
UN	=	Unstructured (covariance structure)
VC	=	Variance Components (covariance structure)

Bibliography

Akaike, H. (1973). Information theory and an extension of the maximum likelihood principle. In E. Petrov, & F. Csaki (eds.), *Second International Symposium on Information Theory and Control*, pp. 267–281. Akademiai Kiado.

Allison, P. (2001). *Missing Data: Quantitative Applications in the Social Sciences*. Newbury Park, CA: Sage Publications.

Anderson, D., Oti, R., Lord, C., & Welch, K. (2009). Patterns of growth in adaptive social abilities among children with autism spectrum disorders. *Journal of Abnormal Child Psychology*, *37*(7), 1019–1034.

Asparouhov, T. (2006). General multi-level modeling with sampling weights. *Communications in Statistics, Theory and Methods*, *35*(3), 439–460.

Asparouhov, T. (2008). Scaling of sampling weights for two-level models in Mplus 4.2. Available at `http://www.statmodel.com/download/Scaling3.pdf`.

Bottai, M., & Orsini, N. (2004). A new stata command for estimating confidence intervals for the variance components of random-effects linear models. Presented at the United Kingdom Stata Users' Group Meetings, London, United Kingdom, June 28–29.

Brown, H., & Prescott, R. (2006). *Applied Mixed Models in Medicine*, Second Edition. New York, NY: John Wiley and Sons.

Carle, A. (2009). Fitting multilevel models in complex survey data with design weights: Recommendations. *BMC Medical Research Methodology*, *9*(49), 1–13.

Carlin, B. P., & Louis, T. A. (2009). *Bayesian Methods for Data Analysis*, Third Edition. London: Chapman & Hall / CRC Press.

Casella, G., & Berger, R. (2002). *Statistical Inference*. North Scituate, MA: Duxbury Press.

Claeskens, G. (2013). Lack of fit, graphics, and multilevel model diagnostics. In M. Scott, J. Simonoff, & B. Marx (eds.), *The Sage Handbook of Multilevel Modeling*, pp. 425–443. London: Sage Publications.

Cooper, D., & Thompson, R. (1977). A note on the estimation of the parameters of the autoregressive-moving average process. *Biometrika*, *64*(3), 625–628.

Crainiceanu, C., & Ruppert, D. (2004). Likelihood ratio tests in linear mixed models with one variance component. *Journal of the Royal Statistical Society, Series B*, *66*, 165–185.

Davidian, M., & Giltinan, D. (1995). *Nonlinear Models for Repeated Measurement Data*. London: Chapman & Hall.

Dempster, A., Laird, N., & Rubin, D. (1977). Maximum likelihood from incomplete data via the EM algorithm. *Journal of the Royal Statistical Society, Series B*, *39*(1), 1–38. With discussion.

Diggle, P. J., Heagerty, P. J., Liang, K.-Y., & Zeger, S. L. (2002). *Analysis of Longitudinal Data*, Second Edition, vol. 25 of Oxford Statistical Science Series. Oxford: Oxford University Press.

Douglas, C., Demarco, G., Baghdoyan, H., & Lydic, R. (2004). Pontine and basal forebrain cholinergic interaction: Implications for sheep and breathing. *Respiratory Physiology and Neurobiology*, *143*(2-3), 251–262.

Enders, C. (2013). Centering predictors and contextual effects. In M. Scott, J. Simonoff, & B. Marx (eds.), *The Sage Handbook of Multilevel Modeling*, pp. 89–107. London: Sage Publications.

Faraway, J. (2005). *Linear Models with R*. London: Chapman & Hall / CRC Press.

Fellner, W. (1987). Sparse matrices, and the estimation of variance components by likelihood equations. *Communications in Statistics-Simulation*, *16*(2), 439–463.

Galecki, A. (1994). General class of covariance structures for two or more repeated factors in longitudinal data analysis. *Communications in Statistics-Theory and Methods*, *23*(11), 3105–3119.

Galecki, A., & Burzykowski, T. (2013). *Linear Mixed-Effects Models using R: A Step-by-Step Approach*. New York, NY: Springer.

Geisser, S., & Greenhouse, S. (1958). An extension of box's results on the use of the f distribution in multivariate analysis. *The Annals of Mathematical Statistics*, *29*(3), 885–891.

Gelman, A. (2007). Struggles with survey weighting and regression modeling. *Statistical Science*, *22*(2), 153–164.

Gelman, A., Carlin, J., Stern, H., & Rubin, D. (2004). *Bayesian Data Analysis*. London: Chapman and Hall / CRC Press.

Gelman, A., & Hill, J. (2006). *Data Analysis Using Regression and Multilevel / Hierarchical Models*. New York, NY: Cambridge University Press.

Gregoire, T., Brillinger, D., Diggle, P., Russek-Cohen, E., Warren, W., & Wolfinger, R. (1997). *Modeling Longitudinal and Spatially Correlated Data: Methods, Applications and Future Directions*. New York, NY: Springer-Verlag.

Gurka, M. (2006). Selecting the best linear mixed model under reml. *The American Statistician*, *60*(1), 19–26.

Harville, D. A. (1977). Maximum likelihood approaches to variance component estimation and to related problems. *Journal of the American Statistical Association*, *72*(358), 320–340. With a comment by J. N. K. Rao and a reply by the author.

Heeringa, S., West, B., & Berglund, P. (2010). *Applied Survey Data Analysis*. New York, NY: Chapman & Hall / CRC Press.

Helms, R. (1992). Intentionally incomplete longitudinal designs: 1. Methodology and comparison of some full span designs. *Statistics in Medicine*, *11*(14–15), 1889–1913.

Huynh, H., & Feldt, L. (1976). Estimation of the box correction for degrees of freedom from sample data in the randomized block and split plot designs. *Journal of Educational Statistics*, *1*(1), 69–82.

Jackman, S. (2009). *Bayesian Analysis for the Social Sciences*. New York, NY: Wiley.

Jennrich, R. I., & Schluchter, M. D. (1986). Unbalanced repeated-measures models with structured covariance matrices. *Biometrics*, *42*(4), 805–820.

Kenward, M., & Roger, J. (1997). Small sample inference for fixed effects from restricted maximum likelihood. *Biometrics*, *53*(3), 983–997.

Korn, E., & Graubard, B. (1999). *Analysis of Health Surveys*. New York, NY: Wiley.

Laird, N., Lange, N., & Stram, D. (1987). Maximum likelihood computations with repeated measures: Application of the EM algorithm. *Journal of the American Statistical Association*, *82*(397), 97–105.

Laird, N., & Ware, J. (1982). Random-effects models for longitudinal data. *Biometrics*, *38*(4), 963–974.

Lindstrom, M. J., & Bates, D. M. (1988). Newton-Raphson and EM algorithms for linear mixed-effects models for repeated-measures data. *Journal of the American Statistical Association*, *83*(404), 1014–1022.

Little, R. J. A., & Rubin, D. B. (2002). *Statistical Analysis with Missing Data*, Second Edition. Wiley Series in Probability and Statistics. Hoboken, NJ: Wiley-Interscience [John Wiley & Sons].

Liu, C., & Rubin, D. (1994). The ECME algorithm: A simple extension of EM and ECM with faster monotone convergence. *Biometrika*, *81*(4), 633–648.

Loy, A., & Hofmann, H. (2014). Hlmdiag: A suite of diagnostics for hierarchical linear models in r. *Journal of Statistical Software*, *56*(5), 1–28.

McCulloch, C. E., Searle, S. R., & Neuhaus, J. M. (2008). *Generalized, Linear, and Mixed Models*, Second Edition. Wiley.

Molenberghs, G., & Verbeke, G. (2005). *Models for Discrete Longitudinal Data*. Berlin: Springer-Verlag.

Morrell, C. (1998). Likelihood ratio testing of variance components in the linear mixed-effects model using restricted maximum likelihood. *Biometrics*, *54*(4), 1560–1568.

Morrell, C., Pearson, J., & Brant, L. (1997). Linear transformations of linear mixed-effects models. *The American Statistician*, *51*(4), 338–343.

Nelder, J. (1977). A reformulation of linear models. *Journal of the Royal Statistical Society, Series A*, *140*(1), 48–77.

Ocampo, J. (2005). Effect of porcelain laminate contour on gingival inflammation. Master's thesis, University of Michigan School of Dentistry.

Patterson, H. D., & Thompson, R. (1971). Recovery of inter-block information when block sizes are unequal. *Biometrika*, *58*, 545–554.

Pfeffermann, D., Skinner, C., Holmes, D., Goldstein, H., & Rasbash, J. (1998). Weighting for unequal selection probabilities in multilevel models. *Journal of the Royal Statistical Society, Series B*, *60*(1), 23–40.

Pinheiro, J., & Bates, D. (1996). Unconstrained parameterizations for variance-covariance matrices. *Statistics and Computing*, *6*, 289–296.

Pinheiro, J., & Bates, D. (2000). *Mixed-Effects Models in S and S-PLUS*. Berlin: Springer-Verlag.

Rabe-Hesketh, S., & Skrondal, A. (2006). Multilevel modeling of complex survey data. *Journal of the Royal Statistical Society, Series A, 169*, 805–827.

Rao, C. (1972). Estimation of variance of covariance components in linear models. *Journal of the American Statistical Association, 67*(337), 112–115.

Raudenbush, S., & Bryk, A. (2002). *Hierarchical Linear Models: Applications and Data Analysis Methods*. Newbury Park, CA: Sage Publications.

Raudenbush, S., Bryk, A., & Congdon, R. (2005). *HLM 6: Hierarchical Linear and Nonlinear Modeling*. Lincolnwood, IL: Scientific Software International.

Robinson, G. (1991). That blup is a good thing: The estimation of random effects. *Statistical Science, 6*(1), 15–32. Discussion: pp. 32-51.

Schabenberger, O. (2004). Mixed model influence diagnostics. Presented at the Twenty-Ninth Annual SAS Users Group International Conference, Montreal, Canada; May 9–12. Paper 189-29.

Searle, S., Casella, G., & McCulloch, C. (1992). *Variance Components*. New York, NY: John Wiley.

Self, S. G., & Liang, K.-Y. (1987). Asymptotic properties of maximum likelihood estimators and likelihood ratio tests under nonstandard conditions. *Journal of the American Statistical Association, 82*(398), 605–610.

Snijders, T., & Bosker, R. (1999). *Multilevel Analysis: An Introduction to Basic and Advanced Multilevel Modeling*. Newbury Park, CA: Sage Publications.

Spybrook, J., Bloom, H., Congdon, R., Hill, C., Martinez, A., & Raudenbush, S. (2011). Optimal design plus empirical evidence: Documentation for the "Optimal Design" software version 3.0. Available from `www.wtgrantfoundation.org`.

Steele, R. (2013). Model selection for multilevel models. In M. Scott, J. Simonoff, & B. Marx (eds.), *The Sage Handbook of Multilevel Modeling*, pp. 109–125. London: Sage Publications.

Stram, D., & Lee, J. (1994). Variance components testing in the longitudinal mixed effects model. *Biometrics, 50*(4), 1171–1177.

Valliant, R., Dever, J., & Kreuter, F. (2013). *Practical Tools for Designing and Weighting Survey Samples*. New York, NY: Springer.

van Breukelen, G., & Moerbeek, M. (2013). Design considerations in multilevel studies. In M. Scott, J. Simonoff, & B. Marx (eds.), *The Sage Handbook of Multilevel Modeling*, pp. 183–199. London: Sage Publications.

Veiga, A., Smith, P., & Brown, J. (2014). The use of sample weights in multivariate multilevel models with an application to income data collected using a rotating panel survey. *Journal of the Royal Statistical Society, Series C (Applied Statistics), 63*(1), 65–84.

Verbeke, G., & Molenberghs, G. (2000). *Linear Mixed Models for Longitudinal Data*. New York, NY: Springer-Verlag.

Verbyla, A. (1990). A conditional derivation of residual maximum likelihood. *The Australian Journal of Statistics, 32*(2), 227–230.

West, B., & Elliott, M. (Forthcoming in 2014). Frequentist and bayesian approaches for comparing interviewer variance components in two groups of survey interviewers. *Survey Methodology*.

Winer, B., Brown, D., & Michels, K. (1991). *Statistical Principles in Experimental Design*. New York, NY: McGraw-Hill.

Index